Algebraic Theory of Processes

MIT Press Series in the Foundations of Computing
Michael Garey and Albert Meyer, editors

Complexity Issues in VLSI: Optimal Layouts for the Shuffle-Exchange Graph and Other Networks, by Frank Thomson Leighton, 1983.

Equational Logic as a Programming Language, by Michael J. O'Donnell, 1985

General Theory of Deductive Systems and Its Applications, by S. Yu. Maslov, 1987

Resource Allocation Problems: Algorithmic Approaches, by Toshihide Ibaraki and Naoki Katoh, 1988.

Algebraic Theory of Processes, by Matthew Hennessy, 1988.

Algebraic Theory of Processes

Matthew Hennessy

The MIT Press
Cambridge, Massachusetts
London, England

This book was set in Times New Roman by Asco Trade Typesetting Ltd., Hong Kong, and printed and bound by Halliday Lithograph in the United States of America.

Library of Congress Cataloging-in-Publication Data

Hennessy, Matthew.
Algebraic theory of processes.
(MIT Press series in the foundations of computing)
Bibliography: p.
Includes index.
1. Electronic data processing—Mathematics. 2. Programming languages (Electronic computers)—Semantics. 3. Algebra, Abstract. I. Title. II. Series.
QA76.9.M35H46 1988 004'.01'51 87-29699
ISBN 0-262-08171-7

Contents

Series Foreword

Theoretical computer science has now undergone several decades of development. The "classical" topics of automata theory, formal languages, and computational complexity have become firmly established, and their importance to other theoretical work and to practice is widely recognized. Stimulated by technological advances, theoreticians have been rapidly expanding the areas under study, and the time delay between theoretical progress and its practical impact has been decreasing dramatically. Much publicity has been given recently to breakthroughs in cryptography and linear programming, and steady progress is being made on programming language semantics, computational geometry, and efficient data structures. Newer, more speculative, areas of study include relational databases, VLSI theory, and parallel and distributed computation. As this list of topics continues expanding, it is becoming more and more difficult to stay abreast of the progress that is being made and increasingly important that the most significant work be distilled and communicated in a manner that will facilitate further research and application of this work. By publishing comprehensive books and specialized monographs on the theoretical aspects of computer science, the series on Foundations of Computing provides a forum in which important research topics can be presented in their entirety and placed in perspective for researchers, students, and practitioners alike.

Michael R. Garey
Albert R. Meyer

Preface

This book is the result of a course given at Aarhus University in the Spring term of 1985. It presents a semantic theory of communicating processes and a logical proof system for reasoning about them. The approach relies heavily on abstract algebra but the book starts more or less from first principles. It is designed to be self-contained; the problems addressed are motivated from the standpoint of computer science, and all the required algebraic concepts are covered, assuming only some basic mathematics.

Most of the research reported here was originally carried out with partial funding by the SERC. Part I of the book was written while visiting the University of Aarhus and I should like to acknowledge the hospitality of the Aarhus faculty and staff. I also thank Marek Bednarezyk and Mogens Nielsen for their detailed comments on a draft of the ms.

Various versions of parts of the manuscript have been typed by K. Møller in Aarhus and E. Kerse and A. Fleming in Edinburgh, all of whom I should like to thank. I would especially like to thank F. Williams at Sussex, who has been efficient and tireless in her retyping of the endless versions of the manuscript.

Algebraic Theory of Processes

Introduction

To describe processes, we naturally need a language. A large number of languages already exist for this purpose and we concern ourselves with only that small variety, often loosely referred to as algebraic languages. We first informally give one example of such a language. This leads to a discussion of the problems encountered when dealing with these languages. Finally, we outline the contents of the book.

A Language for Describing Processes

We consider a process to be a machine for performing actions in some prescribed manner. We prefer to interpret these actions as being communications, but much of the mathematical development in this book can be explained and understood relative to a predefined uninterpreted set of actions, which we call Act. A simple recursive definition of such a process is

$$A \Longleftarrow aA. \tag{$*$}$$

In this introduction we will use uppercase letters for names of processes and lowercase letters for names of actions. The definition (∗) may be read as: A is a process which can perform the action a and then become the process A once more. In short it can perform an arbitrary sequence of a actions. This is represented diagramatically in figure I.1. These diagrams are called *flow graphs* and are designed to give a static view of the potential of processes.

The slightly more complicated process

$$R \Longleftarrow abR$$

is also represented in figure I.1. This is a process which can perform the sequence of actions

$$a, b, a, b, \ldots .$$

The arrows in flow graphs emanating from processes represent potential actions or communications. We use unbroken arrows to represent immediate potential actions, whereas broken ones represent potential actions which may become possible later in the life of the process. Both A and R can immediately perform an a action, whereas R can only perform a b action when it has already performed an a action; the action b is not immediately possible. The process AB defined by

$$AB \Longleftarrow aAB + bAB$$

$$A \Longleftarrow aA$$

$$R \Longleftarrow abR$$

$$AB \Longleftarrow aAB + bAB$$

$$A_c \Longleftarrow \bar{a}A_c$$

$$R_c \Longleftarrow \bar{a}\bar{b}R_c$$

$$Z \Longleftarrow a\bar{b}Z + \bar{c}dZ$$

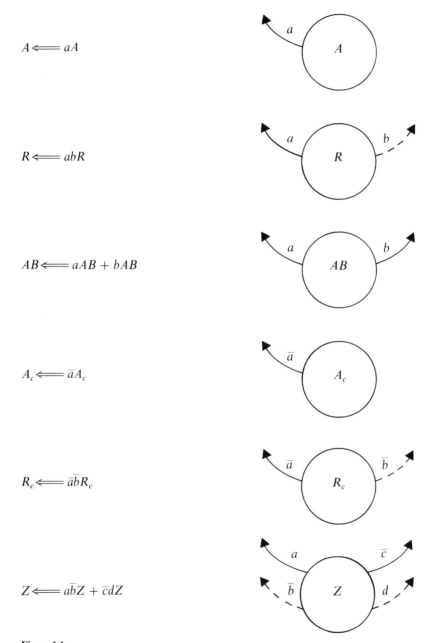

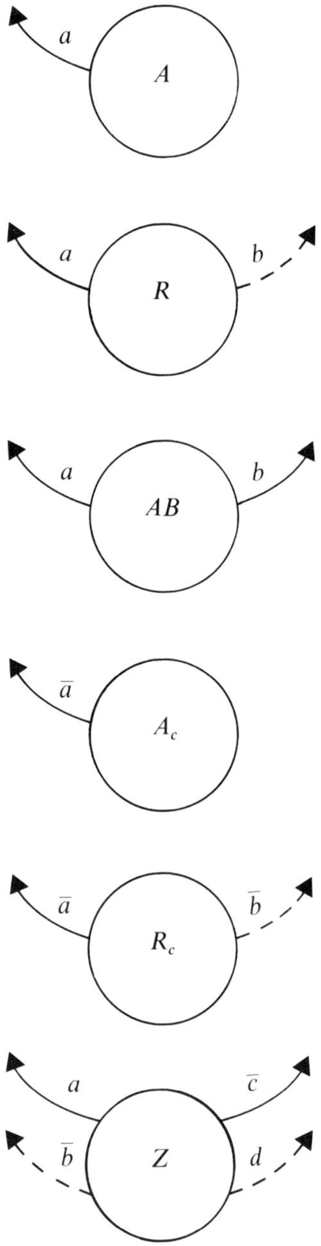

Figure 1.1
Flow graphs.

is also represented in figure I.1. Here "+" denotes some form of choice. At every point in time AB can perform either an a or b action (but not both) and thereby be transformed once more into AB. More specifically, if asked to perform an a action it will perform it and be transformed once more into AB; if asked to perform b it will also do it with the same effect; if asked to perform both it will perform only one, chosen at random. So it can perform any sequence of actions from $\{a, b\}$.

It is important to realize that the flow graphs do not convey all the information inherent in the corresponding definition. For example both X and Y have the same flow graph:

$$X \Longleftarrow acX + bdX$$

$$Y \Longleftarrow adY + bcY.$$

Flow graphs give a static picture of the potential of a process at a particular point in time. This potential varies dynamically as the process evolves. For example the flow graph of both X and Y is depicted in figure I.2 i). If X performs a, the flow graph is given in ii) whereas if it performs b the flow graph is that of iii). The situation is reversed for process Y; ii) gives the proper picture after b iii) after a.

Communication

The flow-graph representation suggests a natural interpretation of processes. The process A can be viewed as a *communicating agent* which has one *port* through which it may communicate. This port may be called the a-port and the action a interpreted as "emitting an output signal through the port a" or "sending a signal to a communication channel named a." Under this interpretation AB represents an agent with two ports, an a-port and a b-port; at any point in time it is prepared to emit an output signal through either one of these. Similarly, both X and Y represent communicating agents which have four ports labeled a, b, c, d, and the agents have the potentials for outputting signals at these ports at various times during their life cycle. The action of outputting a signal has a natural complement: inputting a signal. Let us use "barred" actions such as $\bar{a}$, $\bar{b}$, $\bar{c}$ to denote such complementary actions. For example

$$A_c \Longleftarrow \bar{a}A_c$$

represents a cyclic process which can repeatedly perform the action $\bar{a}$, as depicted in figure I.1. Interpreted as a communicating agent it has one communication port $\bar{a}$ through which it can continuously receive a signal from a communication channel named a. Similarly

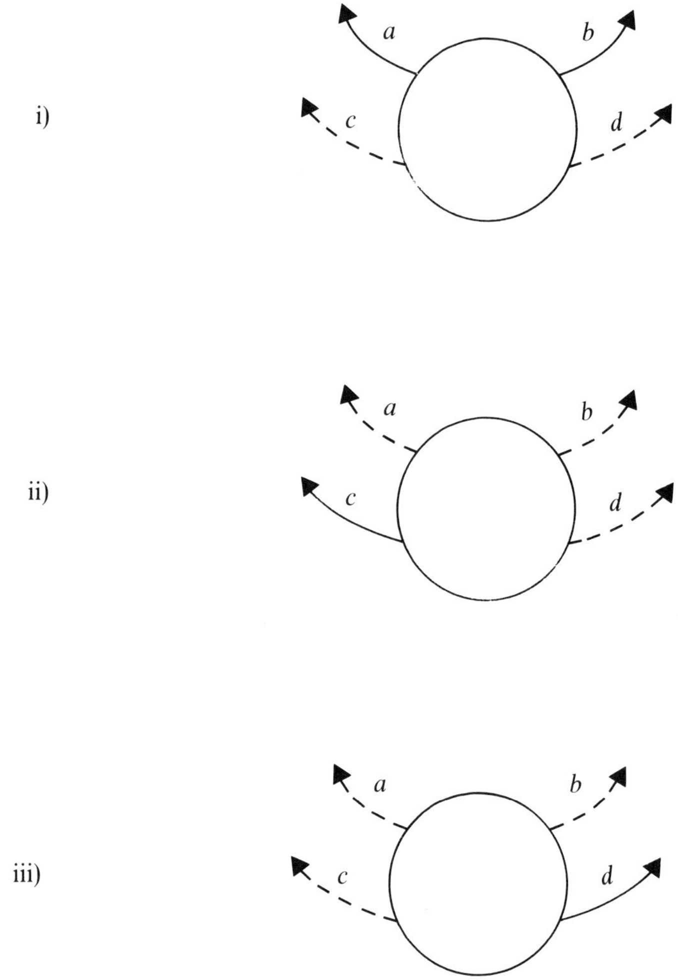

Figure I.2
Flow graphs.

$R_c \Longleftarrow \bar{a}\bar{b}R_c$

represents a process which has two communication ports labeled $\bar{a}$, $\bar{b}$ and it can cyclically receive signals at port $\bar{a}$ followed by port $\bar{b}$ from channels a, b respectively. We emphasize the implicit conventions which have just been introduced:

—every port is either an input or output port
—input can only be received via input ports which are labeled with $^{-}$
—output can only be emitted via output ports which are named without using $^{-}$.

Of course, a process may be able to receive and send signals, albeit using different ports. One such example is

$Z \Longleftarrow a\bar{b}Z + \bar{c}dZ$

depicted in figure I.1. It has two output ports a, d and two input ports $\bar{b}$, $\bar{c}$. Initially it can either receive a signal at port $\bar{c}$ or emit a signal at port a. If it does the former, it may then output a signal at port d whereas if it does the latter it may subsequently receive an input at port $\bar{b}$. Conceptually we may think of each pair of complementary ports as being linked via a communication channel. If process P has an output port c and Q has an input port $\bar{c}$, then there is a possibility of communication between P and Q. If P emits a signal at port c this signal may pass through the conceptual channel linking P and Q, to be input by Q at port c. Communication, or synchronization, is viewed as the simultaneous occurrence of two complementary actions, one sending a signal to a channel and the other receiving a signal from the same channel. For example consider the cyclic processes

$C1 \Longleftarrow acC1$

$C2 \Longleftarrow b\bar{c}C2.$

We may construct a new process by placing these two in parallel. At a descriptive level we can write

$P \Longleftarrow C1 \mid C2,$

i.e. P is a new process which consists of two subprocesses $C1$ and $C2$ running in parallel. The corresponding flow graphs are given in figure I.3. Note that in the flow graph for P the potential communication between $C1$ and $C2$ is included. A broken arrow is used because this communication via a conceptual channel linking port c in $C1$ and port $\bar{c}$ in $C2$ is not immediately possible.

However if the subprocess $C1$ performs action a and $C2$ performs action b, the agent evolves to that depicted in iii). Here this internal communication is immediately possible. If it is performed the original state depicted in ii) is regained. Of course $C1$ is not compelled to use the conceptual channel linking it with $C2$. The arrow labeled c indicates that $C1$ may instead choose to emit the signal through port c to be picked up by some process other than $C1$ in the environment. Presumably this other process also has an input port labeled $\bar{c}$.

Restriction

In the subprocess $C1$ of P, in the last example, the port c is used both to communicate with the other subprocess $C2$ and with the environment. To describe processes in which ports are used only for internal communication we introduce some further notation, called hiding or restriction.

If X is a process and C is any set of channels then $X\backslash C$ describes a process whose behavior is similar to that of X except that the channels in C may not be used to communicate with the environment. Alternatively we can say that the ports which X could use in connection with these channels are no longer available for external use. For example

$$RP \Longleftarrow P\backslash\{c\}$$

is a process which has two subprocesses and only two ports a, b or at least only two ports visible from the environment. The corresponding flow graph is given in figure I.3 iv). Note that the link between the subprocesses $C1$ and $C2$ is retained because it does not conflict with the restriction on the external use of ports. This applies to the overall process P and not the constituent subprocesses $C1$ and $C2$. Consequently when RP has performed the actions a and b, thereby evolving to a situation similar to that depicted in of figure I.3 iii), the *only* possible next move is the internal communication between the subprocesses $C1$ and $C2$.

The use of hiding may introduce a sophisticated type of nondeterminism. For example consider the process NAB given by

$$X1 \Longleftarrow cadX1$$

$$X2 \Longleftarrow \bar{c}\bar{d}X2$$

$$X3 \Longleftarrow cbdX3$$

$$NAB \Longleftarrow (X1 \mid X2 \mid X3)\backslash\{c, d\}.$$

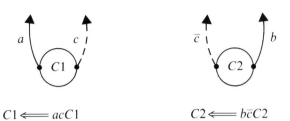

i) $C1 \Leftarrow acC1$ $\quad$ $C2 \Leftarrow b\bar{c}C2$

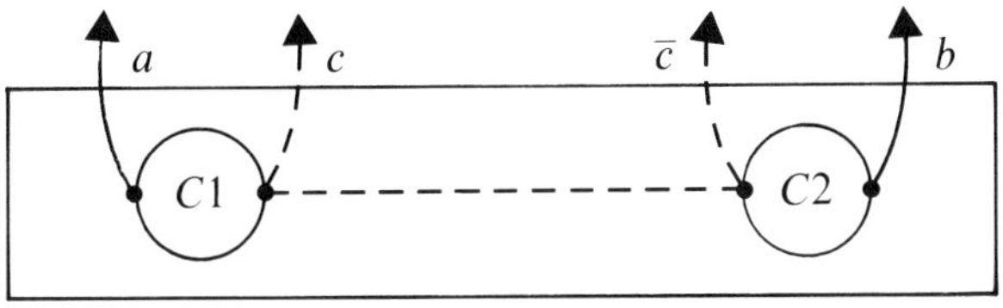

ii) $P \Leftarrow C1 \mid C2$

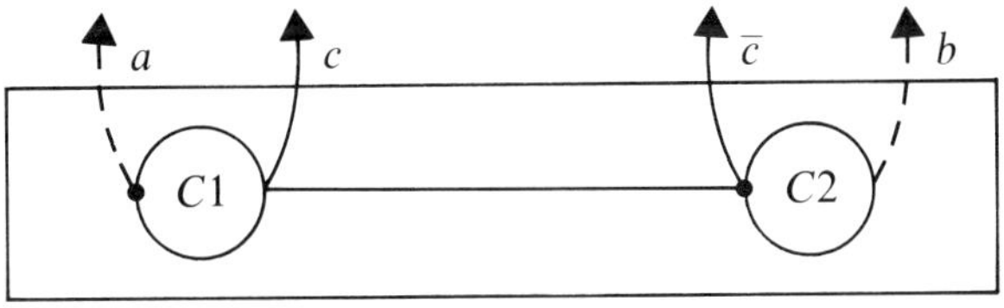

iii) a possible evolution of P

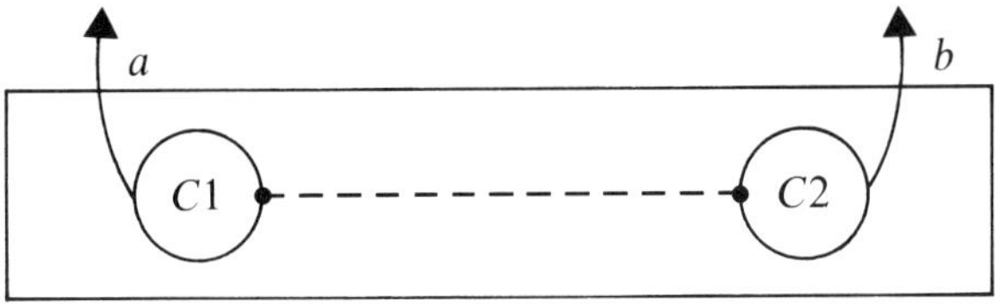

iv) $RP \Leftarrow P\backslash\{c\}$

Figure I.3
Flow graphs with communication.

The flow graphs are given in figure I.4 a). In this particular process the only immediately possible actions are the two internal communications, one between $X1$ and $X2$, the other between $X2$ and $X3$. Moreover only one of these communications can be performed but the choice is completely autonomous to the device. It is not possible to influence it from the outside. Consequently whether port a or port b becomes available for communication is not controllable by any user. This machine is inherently nondeterministic.

Various forms of deadlock can also be represented. For example the process $A\backslash\{a\}$, where A is the cyclic process

$$A \Longleftarrow aA,$$

can perform no action: A can only perform the action a which is forbidden by the restriction. It is convenient to have a direct description in the language for this process which can perform no actions; we use the symbol NIL. For example

$$MD \Longleftarrow \bar{a}bMD + \bar{c}NIL$$

represents a process which can receive an input at port $\bar{a}$ or port $\bar{c}$; if it receives at the latter it can perform no more communications whereas if it receives at the former it can subsequently emit a signal at port b and then reconstitute itself. A more subtle form of "doing nothing" is called "infinite chatter," where a process may indefinitely communicate internally. An extreme example is

$$IC \Longleftarrow (B \mid B_c)\backslash\{b\}$$

where

$$B \Longleftarrow bB$$

$$B_c \Longleftarrow \bar{b}B_c.$$

Here IC can compute indefinitely; messages are sent indefinitely from subprocess B to subprocess B_c but at no point can the process communicate with the outside world. The flawed process MD, given above can be modified to

$$MIC \Longleftarrow \bar{a}bMIC + \bar{c}IC,$$

which has a slightly different effect.

Even more subtle processes can be defined. Consider

$$ABB \Longleftarrow (AB \mid B_c)\backslash\{b\},$$

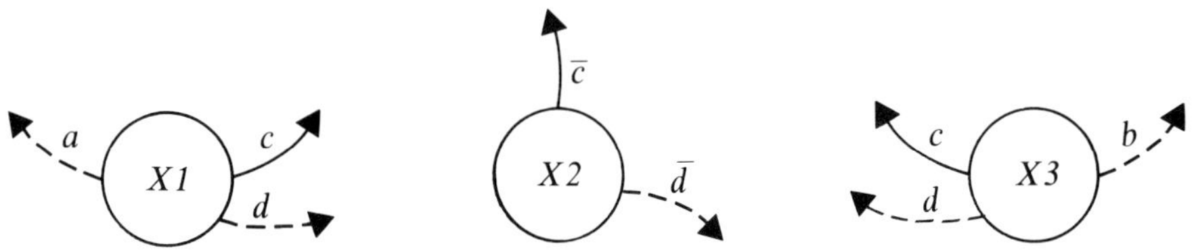

$X1 \Leftarrow cadX1$ $\quad$ $X2 \Leftarrow \bar{c}\bar{d}$ $\quad$ $X3 \Leftarrow cbdX3$

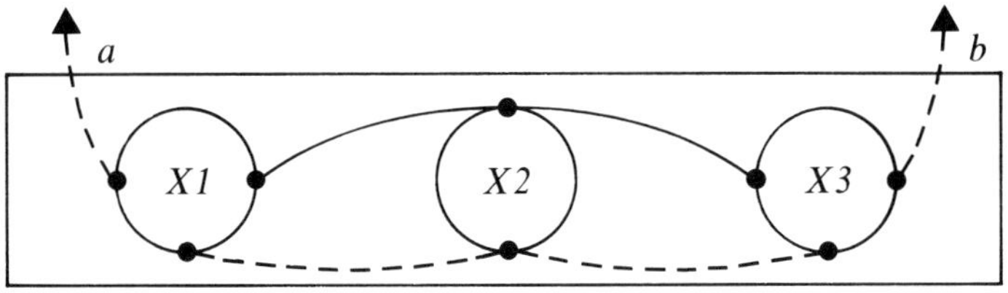

a) $NAB \Leftarrow (X1 \mid X2 \mid X3)\backslash\{c, d\}$

$AB \Leftarrow aA + bB$ $\quad$ $B_c \Leftarrow \bar{b}B_c$

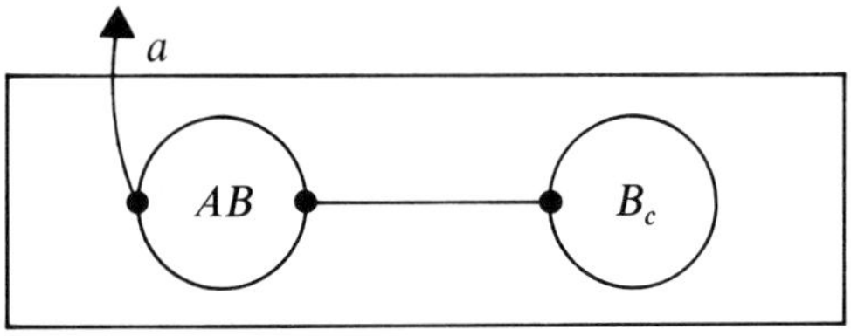

b) $AAB \Leftarrow (AB \mid B_c)\backslash\{b\}$

Figure I.4

where we recall

$AB \Longleftarrow aAB + bAB$

$B_c \Longleftarrow \bar{b}B_c.$

This process is described in figure I.4 b). At each point in time there is contention between a possible external action, emitting a signal at the output port a, and an internal action, communication between the two subprocesses AB and B_c.

The last few examples are pathological but they do serve a purpose. Hiding or restricting ports is a very powerful concept. It enables us to abstract away from details of the inner workings of processes. However any theory of processes based on a language which allows such hiding will have to address such pathological descriptions.

This completes our informal description of a language. We could of course give more methods of constructing processes, but those we have already introduced provide a sufficient basis for the following discussion.

Discussion

The language described in the previous section is a typical example of the so-called algebraic languages for processes. Let us refer to it as **EPL**, an Example Process Language. It consists of a facility for recursive definitions together with a small number of methods for defining new processes from existing ones, called combinators. **EPL** is based on the pure **CCS** of Milner 1980; other variations on this theme can be seen in Hoare 1985, Bergstra and Klop 1984, de Bakker and Zucker 1982, and Austry and Boudol 1984. These are variations in that they choose alternative methods for combining processes, the essential differences being different views of nondeterminism and parallelism. These are represented by different combinators in the various languages.

EPL can be used to represent the behavior of processes (be they software systems, hardware systems, or a combination of both) at some level of abstraction. It can be used either in a prescriptive manner or in a purely descriptive manner. For example RP in figure I.3 may be viewed as a detailed description of an actual system. It gives details of the number of subprocesses involved, *C1* and *C2*, their behavior, as given by their respective definitions, and how the subprocesses interact. **EPL** is used here in a simple descriptive manner.

Used prescriptively, descriptions in **EPL** represent desired or projected behaviors of a system. The actual system may not exist or indeed it may never exist. After an examination of the projected behaviors, as represented by the prescriptive description, it may be found wanting and modified. The description S, defined by

$$S \Longleftarrow abS + baS,$$

is a trivial example of **EPL** used in this manner. It prescribes a system which can cyclically offer output signals at ports a and b. It gives no clue as to how this behavior is to be attained. It is a description of *what* the behavior should be rather than *how* the behavior is to be attained. Such descriptions are often called specifications.

To sum up, **EPL** can be used for two different functions for which two different languages are often used. It can be used to describe specifications of processes, such as S, and processes themselves, such as RP. We will call the latter type of description an *implementation*.

Immediate questions arise: When are two descriptions essentially the same? When are they simply two different descriptions of essentially the same system? These questions are of particular relevance when one of the descriptions, D_1 say, is viewed as a specification and the other, D_2, is viewed as an implementation. Saying D_1 and D_2 are essentially the same or behaviorally equivalent, which we write as $D_1 \sim D_2$, can be interpreted as saying the process D_2 is a correct implementation of the specification D_1. For example is the process RP a correct implementation of the specification S? Are they descriptions at different conceptual levels of the same underlying process? Even without this partition of descriptions into specifications and implementations the relation $\sim$ is of interest. If two processes P_1, P_2 are essentially the same, $P_1 \sim P_2$, then they offer the same functionality. So one may be replaced with impunity by the other without affecting the overall behavior of a larger system. For example the larger system could be a travel agency and P_1, P_2 two different airline reservation systems. If these are behaviorally equivalent, then one can be replaced by the other without affecting the functioning of the travel agency.

To be of value, a descriptive language such as **EPL** should come equipped with a well-understood notion of behavioral equivalence, or "is a correct implementation of," between processes.

When the language in question is some standard sequential language for describing algorithms, such as PASCAL, the nature of behavioral equivalence is obvious: two such programs are equivalent if they compute the same function. For more general descriptive languages this does not even make sense. By and large, descriptions in these language do not compute functions. The essence of the underlying processes lies in their ability or inability to offer certain communications at given points in time. One common method of defining a behavioral equivalence is using a denotational model. See e.g. Hoare 1985 and deBakker and Zucker 1982. Here every description is mapped to a unique object in a domain of meanings. Two descriptions are then considered to be equivalent if they are mapped to the same object, i.e. have the same meaning. This approach has many advantages, particularly if the so-called Scott-Strachey scheme (Scott and Strachey 1971) is followed. Here the models have very elegant mathematical properties and the meaning function from descriptions to objects is structure-oriented: the meaning of a description is determined in some well-defined way by the meaning of its constituent subparts. Under this scheme the mathematical properties of the model are transferred to the descriptive language, giving rise to proof systems which operate directly on expressions in the language.

To a large extent the presentation of such a model deflects or obscures the central question: which descriptions *should* be identified? Certain identifications present no problem. For example the pair of descriptions

$A \mid B \qquad B \mid A$

should be identified since the theory is of behavior rather than form. Also the descriptions S and RP probably should be identified. RP is an implementation of the specification embodied in S. Both descriptions have the same intuitive external behavior and if we are willing to abstract from internal behavior they should be identified. The pair AB and NAB is more problematic. To identify them would be to ignore much of the subtlety of the behavior of processes. Although they offer the same potential behavior, AB represents a much more robust version. However to *justify* distinguishing them one needs a coherent view or understanding of nondeterminism.

The pairs NIL, IC or the derivative pair MD, MIC present similar problems. Do we distinguish them or identify them? And what justification can we offer? Equally important is the ramification of this justification for pairs of descriptions which offer different but related behavior.

A final example is the pair A, ABB. To justify a decision on this pair we need a firm understanding of the extent to which internal computations can influence external behavior.

Of course any particular model decides between identifying and distinguishing these pairs. However the justification lies hidden in the definition of the model or in the mind of the designer of the model. We argue that this justification which underlies a model is its most important feature and should always be elucidated. Rather than starting with a model of processes, we develop a rationale for distinguishing or identifying them. This involves first of all formalizing how these processes actually work. This gives an intentional view of processes using an *operational semantics*. We then formalize how they can be used, in terms of how they work, from which is derived a behavioral equivalence: two processes are equivalent if no hypothetical user can ever distinguish between them. We subsequently build a denotational model and justify its relevance by relating it to this notion of behavioral equivalence. The relation is that of full abstraction; two processes are identified in the model if and only if they have the same behavior. In other words the denotational model and the behavioral equivalence induce exactly the same identifications between processes.

As we have noted, denotational models facilitate the development of proof systems for the equivalence of descriptions. In our case we obtain a set of laws or equations between processes, or more formally their descriptions. These are semantics-preserving, in the sense that they are true both in the denotational model and of the behavioral equivalence. They give rise to a transformational proof system; if one description, e.g. that of an actual implementation, can be transformed into another description, perhaps of a specification, using the laws, then both descriptions are semantically the same. Moreover we will show that the semantics is completely determined by these laws, and we will indirectly obtain a complete proof system.

Outline of the Book

In this book we offer a particular semantic theory of processes and develop its consequences. We are not primarily concerned with the particular language **EPL** and its use in the formal description of systems and their specification. Instead we concentrate on the development of the mathematical concepts and notation necessary to present and explain the formal semantics of a wide range of languages, of which **EPL** is a typical example. One theme of this formal semantics is that parallelism or concurrency is explained in terms of nondeterminism. A parallel process is equivalent to a nondeterministic one obtained by interleaving the actions of its constituent subprocesses. This is also true of the semantic theories presented in Milner 1980, Hoare 1985, and related work. Of course it is not true of Reisig 1984. Consequently the semantic theory we present is essentially a theory of nondeterminism. Parallelism or concurrency is a derived notion explained in terms of this theory of nondeterminism. It is therefore not surprising that most of the book is concerned with nondeterministic but sequential processes.

Part I Here we treat finite processes only. In terms of **EPL** this corresponds to banning the use of recursion and forgetting about parallelism and hiding. What remains is a trivial language for describing finite nondeterministic but sequential processes. But it provides a sufficient basis to develop the concepts we require. Stripped in this manner, **EPL** is nothing more than a word algebra over a set of combinators

Σ; each combinator corresponds to a different method of combining existing processes to form new ones. A denotational model in this setting is simply a particular Σ-algebra. Chapter 1 consists of a short course in abstract algebras, which provides the necessary mathematical background. We develop the idea that the syntax of a language is the free algebra (or word algebra) over a set of combinators, and the freeness condition provides a method of assigning to every program a unique meaning in any particular model. Moreover by choosing the model carefully, ensuring that it is initial with respect to a set of equations, we automatically obtain a sound and complete proof system for the language relative to this model. This proof system is essentially the equational theory generated by the set of equations.

In chapter 2 we first develop our behavioral view of processes in the general setting of labeled transition systems. This takes the form of a relation $\sqsubseteq$ (a preorder rather than an equivalence) between processes. Intuitively $p \sqsubseteq q$ means q passes every test or experiment that p passes. A particular experiment is supposed to represent one way of using a process. An equivalence, called **Testing Equivalence**, can be obtained by defining $p \approx q$ if and only if $p \sqsubseteq q$ and $q \sqsubseteq p$. That is p and q pass exactly the same tests. However the more primitive relation $\sqsubseteq$ is of independent interest. Because of the way we define tests $p \sqsubseteq q$ implies that q is a more deterministic version of p. So $\sqsubseteq$ could be taken as a formalization of "is a correct implementation of."

We then proceed to apply the general theory to our particular language, a stripped-down version of **EPL**. We develop a denotational model called **Finite Acceptance Trees** which properly reflects Testing Equivalence. Using the theory developed in chapter 1 we also give a sound and complete proof system for this model. This involves showing that the model is initial relative to a set of equations. One corollary of this result is that Testing Equivalence is determined by the corresponding equational theory.

Part II This is essentially a rerun of part I except that we allow recursive definitions. The mathematical background required here is that of the Σ-continuous algebras developed in chapter 3. The denotational models we use, Acceptance Trees, are particular examples of these mathematical structures. They are the infinite extensions of the finite trees of part I. Chapter 4 contains a fairly standard exposition of denotational semantics in an algebraic setting. We then extend the behavioral view of processes, Testing Equivalence, to recursive processes and relate it to the denotational model.

Part III We extend the results of part II to the full language **EPL**. The behavioral view is extended to arbitrary expressions in **EPL** and we show that this extended behavioral equivalence is also determined by an equational theory. This equational theory is derivable, in a technical sense, from that developed in part II. As a result we can use the same denotational models, **Acceptance Trees**, for the full language **EPL**.

We end with a discussion of how similar languages can be treated in an analogous fashion.

Mathematical Notation

We use the usual notation of set theory. Sets are often defined by a property and we write $\{x, x \text{ has property } P\}$ to denote the set consisting of all those elements which enjoy the property P, and $a \in X$ to denote that a is an element of the set X. Union and intersection of sets are denoted as usual by $\cup$ and $\cap$, respectively, and $X \setminus Y$ denotes the set of all elements of X which are not in Y.

The *Cartesian product* of two sets A, B is denoted $A \times B : A \times B = \{\langle a, b \rangle$, and $a \in A$ and $b \in B\}$. We often use A^n to denote $A \times A \times \cdots \times A$ (n times), and an element of A^n is written $\langle a_1, \ldots, a_n \rangle$.

A *function* f from A to B, written $f: A \rightarrow B$, is a mapping of the elements of A to the elements of B, i.e., it associates with each $a \in A$ a unique element $f(a) \in B$. It can be viewed as a subset of $A \times B$ which satisfies i) for every $a \in A$ there is a $b \in B$ such that $\langle a, b \rangle \in f$ and ii) $\langle a, b \rangle \in f$ and $\langle a, b' \rangle \in f$ implies $b = b'$.

A function f is *surjective* if for every $b \in B$ there exists some $a \in A$ such that $f(a) = b$; it is *injective* if $f(a) = f(a')$ implies $a = a'$. We will introduce various notations for functions as the need arises. We will most often be dealing with functions from A^n to A for some $n \geq 0$. These will be referred to as functions over A of arity n. If $n = 0$ these are simply constants, i.e. elements of A.

We use the usual notion of *composition* of functions. If $f: A \rightarrow B$ and $g: B \rightarrow C$, then $g \circ f: A \rightarrow C$ is the function defined by $(g \circ f)(a) = g(f(a))$ for every $a \in A$. We also use id_A to denote the identity function over the set A, defined by $id_A(a) = a$ for every $a \in A$.

A *relation* R over a set A is simply a subset of $A \times A$. It is

reflexive	if $\langle x, x \rangle \in R$ for every $x \in A$
symmetric	if $\langle x, y \rangle \in R$ implies $\langle y, x \rangle \in R$
transitive	if $\langle x, y \rangle \in R$, $\langle y, z \rangle \in R$ implies $\langle x, z \rangle \in R$
antisymmetric	if $\langle x, y \rangle \in R$, $\langle y, x \rangle \in R$ implies $x = y$.

An *equivalence* relation over A is a relation R which is reflexive, symmetric, and transitive. It partitions A into equivalence classes and we use $[x]_R$ or simply $[x]$ to denote the *equivalence class* $\{y, \langle x, y\rangle \in R\}$ of x.

A *partial order* over A is a relation which is reflexive, antisymmetric, and transitive. We often use symbols such as $\leq$, $\sqsubseteq$, etc. to denote partial orders and for emphasis we frequently include mention of A in the definition of a particular partial order: thus we say $\langle A, \leq\rangle$ is a partial order meaning $\leq$ is a partial order over the set A.

We make frequent use of a slightly weaker type of relation. A *preorder* over A is a relation which is reflexive and transitive. There is a standard method of generating an equivalence relation from an arbitrary preorder. If $\leq$ is a preorder, let $a \sim b$ if $a \leq b$ and $b \leq a$. Then $\sim$ is an equivalence relation, called the kernel of $\leq$.

Let A be an arbitrary set of symbols. We use A^* to denote the set of sequences of symbols from A and A^+ for the set of nonempty sequences. The empty sequence is written as ε, so $A^* = A^+ \cup \{\varepsilon\}$. Concatenation is represented by juxtaposition: ss' is the sequence which results from concatenating the sequences s and s'. Finally s is a prefix of s' if there is some sequence s'' such that $ss'' = s'$, i.e. s can be extended to s'.

Throughout the notes we rely heavily on the notion of *inductive definitions* which we now explain. Let N denote the set of natural numbers. The set of even natural numbers, EV, satisfies

i) $0 \in EV$
ii) $x \in EV$ implies $x + 2 \in EV$.

One can prove that EV is the *smallest* subset of N which satisfies these conditions: a subset X of N which satisfies

a) $0 \in X$
b) $x \in X$ implies $x + 2 \in X$.

In fact this can serve as a definition of the set EV: EV is the least set X which satisfies (a) and (b). This is an inductive definition. Clause (a) is called the base clause and b) the induction clause, and in general there may be many of each. Inductive definitions are very convenient as they automatically provide a powerful proof method for deriving properties of the elements of the inductive set or relation. We can characterize the elements of EV as those numbers y for which we can prove the statement

"$y \in EV$"

using the rules

i) $0 \in EV$
ii) $x \in EV$ implies $x + 2 \in EV$.

To prove a property of elements of EV we may now use induction on the number of applications of these rules necessary to prove $y \in EV$. For example to show the property $NODD$—"is not an odd number"—of an arbitrary y in EV there are two cases:

i) The proof of "$y \in EV$" uses only one application. Then y must be 0 and obviously 0 is not an odd number.
ii) The proof of "$y \in EV$" uses $k + 1$ applications. Then the last rule used must be ii) and $y = x + 2$ where $x \in EV$. Moreover the proof of "$x \in EV$" takes only k applications. By induction x is not odd and therefore $x + 2$, i.e. y, is not odd.

In the particular case of the inductive set EV this form of induction amounts to:

to show every element of EV has the property P it is sufficient to prove

a) base case : 0 has property P

b) inductive case: $x + 2$ has property P on the assumption that x has property P.

Similarly to define a function $f: EV \to X$ it is sufficient to

a) base case : define $f(0)$

b) inductive case: define $f(x + 2)$ assuming $f(x)$ is already defined.

As another example of an inductive definition consider any relation $R \subseteq A \times A$. Let R^* be the least relation which satisfies

i) $a \in A$ implies $\langle a, a \rangle \in R^*$
ii) $\langle x, y \rangle \in R$ implies $\langle x, y \rangle \in R^*$
iii) $\langle x, y \rangle \in R$, $\langle y, z \rangle \in R^*$ implies $\langle x, z \rangle \in R^*$.

R^* is called the *reflexive transitive closure of* R. Its definition contains two base clauses, i) and ii), and one inductive clause, iii). Once more these clauses form a proof system for proving statements of the form

"$\langle x, y \rangle \in R^*$"

on which a form of induction can be based. In this case it amounts to:

to show every $\langle x, y \rangle$ in R^* has property P it is sufficient to prove

a) base case: i) $\langle a, a \rangle$ has property P for every a in A

ii) $\langle x, y \rangle$ has property P for every pair $\langle x, y \rangle$ in R

b) inductive case: if $\langle x, y \rangle \in R$ and $\langle y, z \rangle$ has property P then $\langle x, z \rangle$ has property P.

Note that the second base clause, $\langle x, y \rangle \in R$, implies that $\langle x, y \rangle \in R^*$ can actually be derived from i) and iii) and is therefore redundant. However, it has been included for the sake of clarity.

By and large inductive definitions are thoroughly explained as they arise in the following chapters.

I FINITE PROCESSES

1 Algebras

In this chapter we introduce the basic mathematical concepts used throughout the book. The first section introduces the notions of a *signature*, a collection of formal function symbols or combinators, and a *Σ-algebra* for an arbitrary signature Σ. A Σ-algebra is nothing more than an interpretation for the combinators over a set called the *carrier* of the algebra. One particular interpretation has as a carrier the set of terms or words which can be constructed using the combinators. This *term algebra* has particular significance in subsequent developments and its mathematical properties are explained using the idea of *initiality*.

The *syntax* of a language corresponds to a term algebra for a specific signature and a *semantics* is given to the language by specifying a particular Σ-algebra. An example is given in §1.3, where a simple language for finite nondeterministic processes is introduced and given two different semantics. Each of these is completely determined by two different sets of *equations* over the combinators of the language. Technically this means that they are initial in the class of all algebras which satisfy the equations. *Equational classes* are explained in §1.2, where initiality is linked to *sound* and *complete equational proof systems*.

Both of the semantic interpretations in §1.3 seem reasonable and, in order to assess their relevance, another ingredient is required: a behavioral theory of processes. This is the topic of chapter 2. To use this behavioral theory we have to consider as semantic domains Σ-algebras whose carriers have some extra structure. These are developed in §1.4, where the basic mathematical results of §1.1 and §1.2 are reiterated for *Σ-po algebras*, Σ-algebras whose carriers are partial orders.

We end this chapter with a discussion of *full abstraction*. The behavioral theory of the next chapter induces a semantic relation $\lessapprox$ on processes. We define what it means for an interpretation to be fully abstract with respect to such a relation and relate full abstraction to the existence of sound and complete proof systems.

1.1 Basic Definitions

A *signature* is a (finite or infinite) set of formal *functional symbols*. For example, the four function symbols {*Zero*, *Succ*, *Pred*, *Plus*}, would be an appropriate signature for the natural numbers. We use Σ to denote a typical signature.

Each function symbol has associated with it an *arity* which gives the number of arguments of the function it represents. For example, the arity of *Plus* is 2, of *Succ* and *Pred* is 1, and of *Zero* is 0. *Zero* is in fact a constant, which for the sake of uniformity we call a function symbol of arity 0. Formally the arity of a signature is a mapping, arity : $\Sigma \to N$. With each symbol f in Σ it associates its arity, arity(f), a natural number. We use Σ_n to denote the set of function symbols in Σ of arity n.

If Σ is a signature, a *Σ-algebra* is a pair $\langle A, \Sigma_A \rangle$ where

i) A is a set, called the carrier
ii) Σ_A is a set of functions $\{f_A, f \in \Sigma\}$,

such that if arity$(f) = n$ then f_A is a function from $A^n \to A$.

Thus a Σ-algebra is simply an interpretation of the signature Σ. It consists of a set A and an interpretation over A of every function symbol in Σ. A Σ-algebra can also be viewed as a set with structure. The carrier A has a structure imposed by the functions Σ_A in the sense that the elements of A can only be manipulated or accessed using these functions. Naturally a given signature can have many different interpretations, even different interpretations over the same carrier. However, we will often abbreviate $\langle A, \Sigma_A \rangle$ by A, when the functions Σ_A are apparent from the context.

EXAMPLE 1.1.1 Consider the signature $\Sigma = \{Zero, Succ, Pred, Plus\}$ with the arities as indicated above. Then $\langle N, \Sigma_N \rangle$ is a Σ-algebra where

i) N is the set of natural numbers
ii) Σ_N is given by:
   a) $Zero_N$ is the number 0
   b) $Succ_N : N \to N$ is defined by: $Succ_N(n) = n + 1$
   c) $Pred_N : N \to N$ is defined by: $Pred_N(0) = 0$
      $Pred_N(n + 1) = n$
   d) $Plus_N : N^2 \to N$ is defined by: $Plus_N(n, m) = n + m$,
      where $+$ denotes the usual arithmetic plus.

$\langle N, \Sigma_N \rangle$ represents the "natural" interpretation of the function symbols in Σ over the natural numbers. We could define other interpretations over the natural numbers, for example by assigning the number 1 to *Zero* or the function $\times$ (multiplication) to the symbol *Plus*.

EXAMPLE 1.1.2 Let Σ be as in the previous example. Then $\langle Z, \Sigma_Z \rangle$ is also a Σ-algebra, where

i) Z is the set of positive and negative numbers
ii) Σ_Z is given by:
   a) $Zero_Z$ is the number 0
   b) $Succ_Z : Z \to Z$ is defined by: $Succ_Z(n) = n + 1$
   c) $Plus_Z : Z^2 \to Z$ is defined by: $Plus_Z(n, m) = n + m$
   d) $Pred_Z : Z \to Z$ is defined by: $Pred_Z(n) = n - 1$.

For every signature Σ there is a particularly important Σ-algebra called the *term algebra* for Σ. Term algebras will play a central role in what follows. These algebras are purely formal objects. The carriers consist of sequences of symbols or strings, called *terms*, which are constructed using the function symbols in Σ. The functions of the term algebras merely manipulate these terms.

Let T_Σ, the set of terms over Σ, be the least set of strings which satisfies

i) if $f \in \Sigma$ has arity 0 then the string consisting of the symbol f is in T_Σ
ii) if $f \in \Sigma$ has arity $k > 0$ then the string of the form $f(t_1, \ldots, t_k)$ is in T_Σ, whenever $t_1, \ldots, t_k$ are strings in T_Σ.

Thus the elements in T_Σ are strings consisting of the symbols "(" , ")" and " , ", together with the symbols from Σ, which can be constructed using the rules i) and ii) above. Note that if Σ contains no function symbol of arity 0, i.e. no constants, then T_Σ is empty. The functions of the term algebra merely construct new terms in the obvious way. For f in Σ of arity k let $f_{T_\Sigma}: T_\Sigma^k \to T_\Sigma$ be the function which maps a tuple of terms $\langle t_1, \ldots, t_k \rangle$ onto the terms $f(t_1, \ldots, t_k)$. If f has arity 0 then this means that f_{T_Σ} is simply the constant in T_Σ consisting of the string "f."

EXAMPLE 1.1.3 Let Σ be as in example 1.1.1. Then T_Σ contains the terms:

Zero, *Succ*(*Zero*), *Succ*(*Succ*(*Zero*)), ...

Pred(*Zero*), *Pred*(*Pred*(*Zero*)), ...

Succ(*Pred*(*Zero*)), *Pred*(*Succ*(*Zero*)), ...

Plus(*Zero*, *Zero*), *Plus*(*Pred*(*Zero*), *Zero*), ...

In fact T_Σ contains any term which can be constructed from these function symbols. The function $Succ_{T_\Sigma}$ takes a term t and returns the term *Succ*(t). So, for example, if we apply the function $Succ_{T_\Sigma}$ to the term *Pred*(*Succ*(*Zero*)), which is an element of T_Σ, we obtain the element of T_Σ, *Succ*(*Pred*(*Succ*(*Zero*))).

Term algebras are syntactic in nature. For us they play the role of the syntax of a language. Indeed, many programming languages or formal languages in general can be constructed so that their syntax is the term algebra for some particular signature Σ. Their semantics, on the other hand, can also be viewed as a Σ-algebra. Indeed, defining the (denotational) semantics of a language amounts to constructing a particular Σ-algebra which satisfies some particular constraints.

To explain the mathematical significance of term algebras we need to introduce

Σ-*homomorphisms*. These are simply functions between carriers of Σ-algebras which preserve the structure imposed by viewing them as Σ-algebras.

Let $\langle A, \Sigma_A\rangle$, $\langle B, \Sigma_B\rangle$ be two Σ-algebras and h a function from A to B. Then h is a Σ-homomorphism if for every f in Σ of arity k:

$$h(f_A(a_1, \dots, a_k)) = f_B(h(a_1), \dots, h(a_k)). \qquad (*)$$

EXAMPLE 1.1.4 Let Σ be as in the three previous examples and let $in: N \to Z$ be the injection function: $in(n) = n$ for every n in N. Then in is *NOT* a homomorphism between $\langle N, \Sigma_N\rangle$ and $\langle Z, \Sigma_Z\rangle$. This is because (*) is not satisfied for the function symbol *Pred*. For $Pred_N(0) = 0$ and so $in(Pred_N(0)) = 0$. On the other hand, $Pred_Z(in(0)) = Pred_Z(0) = -1$.

EXAMPLE 1.1.5 Let Σ be as in the previous examples. Let EV denote the set of even natural numbers $\{0, 2, 4, \dots\}$. Let Σ_{EV} be defined by:

a) $Zero_{EV} = 0$
b) $Succ_{EV}: EV \to EV$ is defined by $Succ_{EV}(n) = n + 2$
c) $Pred_{EV}: EV \to EV$ is defined by $Pred_{EV}(0) = 0$
$Pred_{EV}(n) = n - 2$ if $n > 0$
d) $Plus_{EV}: EV^2 \to EV$ is defined by $Plus_{EV}(n, m) = n + m$.

Then $\langle EV, \Sigma_{EV}\rangle$ is a Σ-algebra.

Let $in: N \to EV$ be the injective function: $in(n) = 2n$ for every n in N. Then in is a Σ-homomorphism. To prove this we need to show that (*) holds with in in place of h, for every function symbol f in Σ. In each case the result follows by simple arithmetic. For example:

i) $in(Succ_N(n)) = in(n + 1) = 2n + 2$ and $Succ_{EV}(in(n)) = Succ_{EV}(2n) = 2n + 2$
ii) $in(Plus_N(n, m)) = in(n + m) = 2n + 2m$ and $Plus_{EV}(in(n), in(m))$
$= Plus_{EV}(2n, 2m) = 2n + 2m.$

EXAMPLE 1.1.6 Let *PAIRS* denote N^2, i.e. the set of pairs of natural numbers. For every f in Σ, Σ being the signature of the previous examples, let f_{PAIRS} be defined by

a) $Zero_{PAIRS} = \langle 0, 0\rangle$
b) $Succ_{PAIRS}(\langle n, m\rangle) = \langle n + 1, m + 1\rangle$
c) $Pred_{PAIRS}(\langle 0, m\rangle) = \langle 0, 0\rangle$
$Pred_{PAIRS}(\langle n + 1, m\rangle) = \langle n, 0\rangle$
d) $Plus_{PAIRS}(\langle n, m\rangle, \langle n', m'\rangle) = \langle n + n', m + m'\rangle$.

Then $\langle PAIRS, \Sigma_{PAIRS} \rangle$ is a Σ-algebra. In this interpretation the successor and addition functions act pointwise on their arguments; for example, $Succ_{PAIRS}$ increments both elements of the pair. On the other hand, the predecessor function decrements the first element if possible but always reduces the second to zero.

Let $h : PAIRS \rightarrow N$ be defined by $h(\langle n, m \rangle) = n$. Then h is a Σ-homomorphisms. Once more we must check that (*) holds for every f in Σ. As in the previous example, this follows by simple arithmetic.

This example shows that although homomorphisms preserve the structure of algebras they can collapse the domain algebras in the sense that many elements of the domain may be mapped onto the same element in the range.

In general Σ-homomorphisms behave in much the same way as ordinary functions over sets. For example, if $f : A \rightarrow B$, $g : B \rightarrow C$ are Σ-homomorphisms, so is their composition, $g \circ f : A \rightarrow C$. Moreover, the identity function over A is also a Σ-homomorphism over any Σ-algebra $\langle A, \Sigma_A \rangle$. The most fundamental property of term algebras is expressed in terms of homomorphisms.

THEOREM 1.1.7 For every Σ-algebra $\langle A, \Sigma_A \rangle$ there exists a unique Σ-homomorphism $i_A : T_\Sigma \rightarrow A$.

The definition of the carrier of the term algebra T_Σ given above is inductive. It is the least set of strings which contains the constant symbols and which are closed under the operations of the term algebra, i.e. the operations f_Σ, for each f in Σ. The inductive nature of T_Σ gives a very powerful proof method for deriving properties of terms. To show that the property P holds of all terms in T_Σ it is sufficient to

i) prove P holds of all constant symbols in Σ
ii) assuming P holds of the terms $t_1, \ldots, t_k$, prove P holds of the term $f(t_1, \ldots, t_k)$ for every f in Σ of arity k, $k > 0$.

This is called *structural induction* as the induction is actually on the syntactic structure of the terms. One can also use structural induction to define relations or functions over T_Σ. For example, to define a function g on T_Σ it is sufficient to

i. define the result of applying g to the constant function symbols
ii. define the result of applying g to $f(t_1, \ldots, t_k)$ in terms of $g(t_1), \ldots, g(t_k)$, for every f in Σ of arity k, $k > 0$.

In fact, both of these conditions can be collapsed into one:

Assuming the result of applying g to the terms $t_1, \ldots, t_k$, has been defined, define the result of applying g to $f(t_1, \ldots, t_k)$, for each f in Σ of arity $k \geq 0$.

This is equivalent to both i. and ii. A similar simplification of proof by structural induction can be made. Both these uses of structural induction appear in the proof of theorem 1.1.7.

Proof of Theorem 1.1.7 We must prove two statements: 1) the homomorphism exists and 2) it is unique.

1) We define i_A by structural induction on terms. Every term in T_Σ is of the form $f(t_1, \ldots, t_k)$ for some f in Σ of arity k. We may assume by induction that $i_A(t_1), \ldots, i_A(t_k)$ have been defined. Then define $i_A(f(t_1, \ldots, t_k))$ to be $f_A(i_A(t_1), \ldots, i_A(t_k))$. In this way we have defined i_A for every element in T_Σ. It is trivial to check that i_A is a homomorphism: Let f be in Σ with arity k. Then

$$\begin{aligned} i_A(f_{T_\Sigma}(t_1, \ldots, t_k)) &= i_A(f(t_1, \ldots, t_k)) \text{ by the definition of } f_{T_\Sigma} \\ &= f_A(i_A(t_1), \ldots, i_A(t_k)) \text{ by the definition of } i_A. \end{aligned}$$

2) To show that i_A is unique we prove that it coincides with every Σ-homomorphism from T_Σ to A. So let h be any such Σ-homomorphism. We prove, by structural induction, that $i_A(t) = h(t)$ for every t in T_Σ. A typical element of T_Σ is of the form $f(t_1, \ldots, t_k)$, where f is in Σ with arity k. Then

$$\begin{aligned} i_A(f(t_1, \ldots, t_k)) &= f_A(i_A(t_1), \ldots, i_A(t_k)) && \text{by the definition of } i_A \\ &= f_A(h(t_1), \ldots, h(t_k)) && \text{by structural induction} \\ &= h(f_{T_\Sigma}(t_1, \ldots, t_k)) && \text{since } h \text{ is a } \Sigma\text{-homomorphism} \\ &= h(f(t_1, \ldots, t_k)) && \text{by the definition of } f_{T_\Sigma}. \end{aligned}$$

Since every element of T_Σ is of the form $f(t_1, \ldots, t_k)$ for some function symbol f it follows that i_A and h coincide. □

If we view T_Σ as the syntax of a language and a Σ-algebra $\langle A, \Sigma_A \rangle$ as a semantic domain or interpretation, then this theorem states that every expression or term in the language has a unique meaning in $\langle A, \Sigma_A \rangle$: there is only one way to interpret the language in the semantic domain. There is also a sense in which it can be interpreted as saying that T_Σ is the "least" Σ-algebra. In general a Σ-algebra $\langle A, \Sigma_A \rangle$ makes identifications between terms. $\langle N, \Sigma_N \rangle$ from example 1.1.1 identifies *Pred*(*Succ*(*Zero*)) with *Zero* and $\langle Z, \Sigma_Z \rangle$ from example 1.1.2 identifies *Succ*(*Pred*(*Zero*)) with *Zero*. In general $\langle A, E_A \rangle$ identifies two terms t_1, t_2 if $i_A(t_1) = i_A(t_2)$. Then T_Σ is the Σ-algebra which makes the least number or identifications. In fact it makes none. To see this we apply the theorem with A equal to T_Σ; there exists a unique homomorphism i_{T_Σ} from T_Σ to itself. Since the identity function is a Σ-homomorphism it follows that i_{T_Σ} must be the identity, i.e. $i_{T_\Sigma}(t_1) = i_{T_\Sigma}(t_2)$ if and only if t_1 is syntactically the same as t_2.

DEFINITION 1.1.8 Let $\mathscr{C}$ be a class of Σ-algebras. Then a Σ-algebra I in $\mathscr{C}$ is *initial in* $\mathscr{C}$ if for every Σ-algebra J in $\mathscr{C}$ there exists a unique Σ-homomorphism from I to J.

Theorem 1.1.7 can now be rephrased to read:

T_Σ is initial in the class of all Σ-algebras.

Throughout the book we will see various examples of initial algebras.

A Σ-homomorphism $f: A \rightarrow B$ is called a *Σ-isomorphism* if it is a bijection. In this case A and B are said to be isomorphic as Σ-algebras. If A and B are isomorphic then they are essentially the same in the sense that they cannot be differentiated using Σ. Any statement in terms of the function symbols in Σ which is true of A will also be true of B and vice versa. They may of course be intrinsically different in the sense that the nature of the objects in the two carriers are totally different. But viewed structurally by Σ they are identical.

PROPOSITION 1.1.9 *A and B are isomorphic as Σ-algebras if and only if there exist two Σ-homomorphisms, $h: A \rightarrow B$, $g: B \rightarrow A$ such that*

a) $h \circ g = id_B$
b) $g \circ h = id_A$.

Proof i) Suppose $h: A \rightarrow B$ is a bijective Σ-homomorphism. We define a function g which satisfies the requirements of the proposition. Let $g: B \rightarrow A$ be defined by

$$g(b) = a \quad \text{if} \quad h(a) = b.$$

Then g is in fact a function since for every b in B there exists exactly one a such that $h(a) = b$. Also g is a Σ-homomorphism since for every f in Σ of arity k:

$$\begin{aligned} h(f_A(g(b_1), \ldots, g(b_k))) &= f_B(h \circ g(b_1), \ldots, h \circ g(b_k)) \quad \text{since } h \text{ is a } \Sigma\text{-homomorphism} \\ &= f_B(b_1, \ldots, b_k) \quad \text{by definition of } q. \end{aligned}$$

Therefore, again by the definition of g,

$g(f_B(b_1, \ldots, b_k)) = f_A(g(b_1), \ldots, g(b_k))$.

Finally, it is trivial to show $h \circ g = id_B$, $g \circ h = id_A$.

ii) Conversely, suppose g, h satisfy the conditions of the proposition. We show that h is bijective from which it will follow that A and B are isomorphic as Σ-algebras.

a) Suppose $h(a_1) = h(a_2)$. Then $g \circ h(a_1) = g \circ h(a_2)$. Since $g \circ h$ is the identity it follows that $a_1 = a_2$ and so h is injective.

b) Let $b \in B$. Define a to be $g(b)$. Then $h(a) = h \circ g(b) = b$ since $h \circ g$ is the identity. Therefore h is surjective. □

As an immediate corollary to this characterization theorem we have that initial algebras are unique up to isomorphism.

COROLLARY 1.1.10 If I_1, I_2 are initial in a class $\mathscr{C}$ of Σ-algebras then they are isomorphic.

Proof Since I_1 is initial in $\mathscr{C}$ and I_2 is a particular Σ-algebra in $\mathscr{C}$ there exists a (unique) Σ-homomorphism $i_1 : I_1 \to I_2$. Similarly, since I_2 is initial there exists a unique Σ-homomorphism $i_2 : I_2 \to I_1$. It follows that $i_2 \circ i_1 : I_1 \to I_1$ is a Σ-homomorphism. Since $id_{I_1} : I_1 \to I_1$ is also a Σ-homomorphism we can conclude from the initiality of I_1 that $i_2 \circ i_1 = id_{I_1}$. Reasoning in a similar way about I_2 we can conclude that $i_1 \circ i_2 = id_{I_2}$. So i_1, i_2 satisfy the conditions given in proposition 1.1.9 and it follows that I_1, I_2 are isomorphic. □

We will encounter various types of Σ-algebras. In every one of these variations initial algebras will always be unique (up to isomorphism). Moreover in these variations we take as our notion of isomorphism the existence of two homomorphisms which satisfy the requirements in proposition 1.1.9.

We end this section by introducing notation which will be convenient in subsequent developments. We use $\underline{x}$ for the sequence $x_1, \ldots, x_k$. The actual length of the sequence, k, will be apparent from the context in which it is used. Sometimes the expression $f(\underline{t})$ will be used as shorthand for $f(t_1, \ldots, t_k)$, where k is the arity of the function symbol f. It may also be used to represent the sequence $f(x_1), \ldots, f(x_k)$; in each case what we mean will be clear from the context. For example, we can reformulate the definition of a Σ-homomorphism as:

h is a Σ-homomorphism from A to B if for every function symbol f in Σ, $h(f_A(\underline{a})) = f_B(h(\underline{a}))$.

In this case $f_A(\underline{a})$ means $f_A(a_1, \ldots, a_k)$, where k is the arity of f, while $h(\underline{a})$ means $h(a_1), \ldots, h(a_k)$. Similarly we can define T_Σ to be the least set of strings satisfying

i) if f is in Σ with arity 0, then f is in T_Σ
ii) if f is in Σ with arity k, $k > 0$ and $\underline{t}$ is in T_Σ, then $f(\underline{t})$ is in T_Σ.

We can be even more terse and say that T_Σ is the least set of strings which satisfies

i′) if f is in Σ and $\underline{t}$ is in T_Σ, then $f(\underline{t})$ is in T_Σ.

In this case the length of the sequence $\underline{t}$ is assumed to be the arity of f. If this is 0 then $\underline{t}$ is the empty sequence and $f(\underline{t})$ is simply f.

1.2 Equational Classes

Particular collections of Σ-algebras can be singled out using equations. For example both $\langle N, \Sigma_N \rangle$ and $\langle Z, \Sigma_Z \rangle$ of examples 1.1.1 and 1.1.2 satisfy

$$Pred(Succ(x)) = x, \qquad (*)$$

but $\langle PAIRS, \Sigma_{PAIRS} \rangle$ of example 1.1.6 does not because with x instantiated by the pair $\langle 1, 1 \rangle$, $Pred(Succ(\langle 1, 1 \rangle))$ is $\langle 1, 0 \rangle$, which is different than $\langle 1, 1 \rangle$. On the other hand $\langle Z, \Sigma_Z \rangle$ satisfies

$$Succ(Pred(x)) = x, \qquad (**)$$

while this is not true of $\langle N, \Sigma_N \rangle$: with x equal to 0 we have (in $\langle N, \Sigma_N \rangle$) $Succ(Pred(0)) = 1$, which is different than 0. In this section we investigate classes of algebras which are determined by equations such as (∗) and (∗∗).

An equation is determined by two terms which may contain variables. The evaluation of such terms in a Σ-algebra is in respect to an assignment of values to these variables. A Σ-algebra then satisfies an equation if this evaluation of the two terms coincides with respect to every possible assignment of values to variables. We first examine structure-preserving equivalences and how they can be used to factor Σ-algebras. We will see later how equations generate such equivalences and the relation between these equivalences and the class of Σ-algebra which satisfy the equations.

A *Σ-congruence* is the natural extension of the notion of equivalence relation to Σ-algebras: it is simply an equivalence relation which preserves the structure induced by Σ. Let $\langle A, \Sigma_A \rangle$ be a Σ-algebra. A relation C over A is a Σ-congruence if

i) C is an equivalence relation
ii) for every f in Σ, if $\langle \underline{a}, \underline{a}' \rangle \in C$ then $\langle f_A(\underline{a}), f_A(\underline{a}') \rangle \in C$.

Note that ii) is shorthand for

ii) for every f in Σ of arity k, if $\langle a_i, a_i' \rangle \in C$ for every i, $0 \leq i \leq k$, then $\langle f_A(a_1, \ldots, a_k), f_A(a_1', \ldots, a_k') \rangle \in C$.

Let A/C be the set of equivalence classes induced by C. To be more precise, for every a in A let $[a]_C$ denote the set of all elements of A related to a by C, $\{a', \langle a, a' \rangle \in C\}$. Then $A/C = \{[a]_C, a \in A\}$. For each f in Σ we can define a mapping over A/C by

$$f_{A/C}([a_1]_C, \ldots, [a_k]_C) = [f_A(a_1, \ldots, a_k)]_C.$$

LEMMA 1.2.1 a) $\langle A/C, \Sigma_{A/C} \rangle$ is a Σ-algebra.
b) The natural injection mapping in : $A \to A/C$, defined by $\text{in}(a) = [a]_C$ is a Σ-homomorphism.

Proof a) It is sufficient to prove that the definition of $f_{A/C}$, given above, does indeed define a function. We must show that the result does not depend on the particular representatives of $[a_i]_C$ chosen. Let a_i' be in $[a_i]_C$, $0 \leq i \leq k$. So $\langle a_i, a_i' \rangle \in C$ for each i in the range. Since C is a Σ-congruence it follows that $\langle f(\underline{a}), f(\underline{a}') \rangle \in C$, i.e. $f(\underline{a}') \in [f(\underline{a})]_C$.

b) The proof is left to the reader; it follows immediately from the definition of $f_{A/C}$.

□

EXAMPLE 1.2.2 Consider the Σ-algebra of example 1.1.6, $\langle PAIRS, \Sigma_{PAIRS} \rangle$. Let B be the equivalence relation defined by

$$\langle\langle n, m \rangle, \langle n', m' \rangle\rangle \in B \quad \text{if} \quad n = n'.$$

Then B is a Σ-congruence. To prove this it is necessary to show that B is preserved by all of the functions $Succ_{PAIRS}$, $Pred_{PAIRS}$, $Plus_{PAIRS}$. For each $n \in N$ there is a distinct congruence class $\{\langle n, m \rangle, m \in N\}$ and the factored Σ-algebra is isomorphic to $\langle N, \Sigma_N \rangle$.

EXAMPLE 1.2.3 Let A be any Σ-algebra. For t, t' in T_Σ let $t =_A t'$ if $i_A(t) = i_A(t')$. This relation $=_A$ is obviously an equivalence relation over T_Σ. Moreover it is a Σ-congruence. For if $f \in \Sigma$ then $i_A(f_{T_\Sigma}(\underline{t})) = f_A(i_A(\underline{t}))$. So if $i_A(\underline{t}) = i_A(\underline{t}')$ then $i_A(f_{T_\Sigma}(\underline{t})) = i_A(f_{T_\Sigma}(\underline{t}'))$, i.e. $\underline{t} =_A \underline{t}'$ implies $f_{T_\Sigma}(\underline{t}) =_A f_{T_\Sigma}(\underline{t}')$.

We are primarily interested in Σ-congruences over the term algebra T_Σ. If C is such a Σ-congruence we say that the *Σ-algebra A satisfies C* if $i_A(t) = i_A(t')$ whenever

$\langle t, t' \rangle \in C$. In other words, A satisfies C if all pairs of terms which are equivalent with respect to C are identified when interpreted in A. Let $\mathscr{C}(C)$ be the class of all Σ-algebras which satisfy C.

THEOREM 1.2.4 (Initiality for Congruences) The Σ-algebra T_Σ/C is initial in the class $\mathscr{C}(C)$.

Proof a) We must first prove that T_Σ/C is in fact in $\mathscr{C}(C)$. The natural injection mapping $in : T_\Sigma \to T_\Sigma/C$ is a Σ-homomorphism and since T_Σ is initial in the class of all Σ-algebras it must coincide with $i_{T_\Sigma/C}$, because the latter is unique. So let $\langle t, t' \rangle$ be in C. Then

$$\begin{aligned} i_{T_\Sigma/C}(t) &= [t]_C \\ &= [t']_C \quad \text{since } \langle t, t' \rangle \in C \\ &= i_{T_\Sigma/C}(t'). \end{aligned}$$

b) Let A be in $\mathscr{C}(C)$. Define $h : T_\Sigma/C \to A$ by $h([t]_C) = i_A(t)$. This is a well-defined function since if t' is in $[t]_C$ then $\langle t, t' \rangle \in C$ and $i_A(t) = i_A(t')$ because A satisfies C. Moreover h is a Σ-homomorphism since

$$h(f_{T_\Sigma/C}([\underline{t}]_C)) = h([f(\underline{t})]_C) = i_A(f(\underline{t})) = f_A(i_A(\underline{t})) = f_A(h([\underline{t}]_C)).$$

c) We prove that this h is the unique homomorphism from T_Σ/C to A. Let $h' : T_\Sigma/C \to A$ be any Σ-homomorphism. Define $i'_A : T_\Sigma \to A$ by $i'_A(t) = h'[t]_C$. Then $i'_A(f_{T_\Sigma}(\underline{t})) = h'([f(\underline{t})]_C) = h'(f_{T_\Sigma/C}([\underline{t}]_C)) = f_A(h'[\underline{t}]_C) = f_A(i'_A(\underline{t}))$. So i'_A is a Σ-homomorphism. However T_Σ is initial in the class of all Σ-algebras so that i'_A must coincide with i_A. It follows that $h'([t]_C) = i'_A(t) = i_A(t) = h([t]_C)$, i.e., h' coincides with h. □

Remark This theorem is in fact a generalization of the Initiality result, theorem 1.1.7; the latter can be obtained by taking the trivial Σ-congruence Id which equates no terms, i.e. $\langle t, t' \rangle \in Id$ if and only if t is syntactically the same as t'.

We will be interested in this theorem for a particular kind of Σ-congruence, those generated by sets of equations. To define formally how equations generate Σ-congruences we need some new concepts. Let X be a set of variables. We use $x, x_1, x_2, \ldots$ to range over the variables in X. We can extend any signature Σ to a new signature $\Sigma(X)$, which has all the function symbols of Σ as before and in addition each x in X is a function symbol in $\Sigma(X)$ of arity 0. The nonstandard notation $\Sigma(X)$ serves to emphasize that the new constants will play a special role in what follows. We also use the nonstandard notation $T_\Sigma(X)$ to denote the term algebra for the signature $\Sigma(X)$. The terms in T_Σ are also in $T_\Sigma(X)$. In the sequel when we refer to terms (or more correctly Σ-terms) we usually mean elements of T_Σ, which are often called *closed terms* or ground terms in the literature since they contain no variables. Elements of $T_\Sigma(X)$ will be called *open terms* if we wish to emphasize that they may contain variables.

Equations are now simply pairs of elements from the term algebra $T_\Sigma(X)$. For example if Σ is $\{Zero, Succ, Pred, Plus\}$ then $Pred(Succ(x))$, $Succ(Pred(x))$, x, etc., are terms in $T_\Sigma(X)$, provided x is in X.

The term algebra $T_\Sigma(X)$ has a special significance, quite similar to that of T_Σ. Let A be any Σ-algebra. An *A-assignment for X* is a mapping $\rho_A : X \to A$. So ρ_A associates with every variable x in X an element $\rho_A(x)$ in A.

THEOREM 1.2.5 (Freeness) If A is a Σ-algebra and ρ_A an A-assignment for X then there exists a unique Σ-homomorphism h_A from $T_\Sigma(X)$ to A such that $h_A(x) = \rho_A(x)$ for every x in X.

Proof Similar to that of the initiality result, theorem 1.1.7.

1) Define h_A by structural induction on terms in $T_\Sigma(X)$.

a) If t is a variable in X let $h_A(t) = \rho_A(t)$.

b) Otherwise t has the form $f(\underline{t})$ for some f in Σ. In this case let $h_A(f(\underline{t})) = f_A(h_A(\underline{t}))$. Then h_A is defined on all of $T_\Sigma(X)$. The proof that it is a $\Sigma(X)$-homomorphism is similar to the proof that i_A is a Σ-homomorphism in theorem 1.1.7.

2) Let $h' : T_\Sigma(X) \to A$ be another $\Sigma(X)$-homomorphism which coincides with ρ_A on X. We show by structural induction on t that $h(t) = h'(t)$ for every t in $T_\Sigma(X)$.

a) t is a variable in X.

Then both $h(t)$ and $h'(t)$ coincide with $\rho_A(t)$.

b) Otherwise t has the form $f(\underline{t}')$ for some f in Σ. So

$h(f(\underline{t}')) = f_A(h(\underline{t}'))$ since h is a Σ-homomorphism

$= f_A(h'(\underline{t}'))$ by structural induction

$= h'(f(\underline{t}'))$ since h' is a Σ-homomorphism. □

Note that once more this theorem can be viewed as a generalization of theorem 1.1.7; it can be obtained by taking X to be the empty set. The import of the theorem is that every Σ-term with variables can be interpreted or evaluated uniquely in a Σ-algebra A, provided the variables have been bound to elements of A. As we will see this gives us a convenient method of saying formally when a Σ-algebra satisfies a set of equations. In the sequel it will be convenient to denote the Σ-homomorphism given in theorem 1.2.5 as ρ_A. This slight abuse of notation should not cause problems; notice that when ρ_A is applied to elements of T_Σ it actually coincides with i_A. This can be seen from the definition in the proof of the theorem 1.2.5.

The notation of A-assignments allows us to define substitution into terms in a natural way. A *substitution* is a $T_\Sigma(X)$-assignment, i.e. it associates with every variable in X a term in $T_\Sigma(X)$. The application of the substitution ρ to the term t is written as $t\rho$ and is called an *instantiation* of t. Strictly speaking, a substitution should be written as $\rho_{T_{\Sigma(X)}}$ but for convenience we omit the subscript. Moreover, to emphasise the syntactic nature of substitutions we write $t\rho$ instead of $\rho(t)$. Also, to conform with the more usual notation for substitution, we will often write this as $t[\underline{t}'/\underline{x}]$ where $\underline{x}$ is the finite sequence of variables which occur in t and t_i' is $\rho(x_i)$ for every x_i in the sequence $\underline{x}$. If each $\rho(x_i)$ is a closed term, i.e. an element of T_Σ, then ρ is called a *closed substitution* and $t\rho$ a *closed instantiation* of t. Note that closed instantiations are always closed terms, i.e. they are in T_Σ. The following lemma states that substitutions behave as one would expect with respect to assignments in general. If ρ_A is an A-assignment and ρ a substitution, then $\rho_A \circ \rho$ can also be viewed as an A-assignment: it associates with each x the result of evaluating the term $\rho(x)$ according to the A-assignment ρ_A as in theorem 1.2.5. (In the expression $\rho_A \circ \rho$ we are using ρ_A as a $\Sigma(X)$-homomorphism rather than as a simple A-assignment.) Using the Freeness theorem this A-assignment, $\rho_A \circ \rho$ can be extended uniquely to arbitrary terms in $T_\Sigma(X)$.

LEMMA 1.2.6 (Substitution Lemma) For every A-assignment ρ_A and every substitution ρ, the unique extension of the A-assignment $\rho_A \circ \rho$ to $T_\Sigma(X)$ is given by the function $h(t) = \rho_A(t\rho)$.

Proof This lemma states that one can either make a substitution into t first and then evaluate according to an A-assignment ρ_A or first evaluate in A the terms to be substituted and then evaluate t using this modified assignment. The function h is obviously a Σ-homomorphism and it also agrees with the A-assignment $\rho_A \circ \rho$ on variables. According to the Freeness theorem there is a unique such Σ-homomorphism and therefore h and $\rho_A \circ \rho$ coincide as Σ-homomorphisms. □

A particular instance of this lemma is when the A-assignment is another substitution. We then have that $t(\rho \circ \rho') = (t\rho')\rho$.

We now define what it means for an algebra to satisfy a set of equations. For t, t' in $T_\Sigma(X)$ let $t =_A t'$ if for every A-assignment ρ_A, $\rho_A(t) = \rho_A(t')$. When applied to elements of T_Σ this coincides with the definition given in example 1.2.3.

A *Σ-equation* is a pair of terms $\langle t, t' \rangle$ each of which is in $T_\Sigma(X)$. They will be often written in the form

$$t = t'.$$

A relation R over $T_\Sigma(X)$ *satisfies* a set of equations E if $E \subseteq R$. A Σ-algebra A *satisfies* a set of equations E if $E \subseteq =_A$, i.e. for every Σ-equation $\langle t, t' \rangle$ in E and every A-assignment ρ_A, $\rho_A(t) = \rho_A(t')$. If we apply this definition to the equations (*), (**) given at the beginning of this section, we see that the statements made about N, Z and $PAIRS$ are indeed true.

Let $\mathscr{C}(E)$ be the class of Σ-algebras which satisfy the equations E.

THEOREM 1.2.7 (Initiality for Equations) For every set of Σ-equations E, $\mathscr{C}(E)$ has an initial Σ-algebra.

The remainder of this section is devoted to proving this theorem and examining its consequences. The proof is essentially an application of the corresponding theorem about congruences; the initial algebra of $\mathscr{C}(E)$ can be exhibited in the form T_Σ/C for some particular congruence C. There are a number of different ways of defining C. The method we use has the advantage of connecting initial algebras with certain kinds of proof systems. We define a system of equational deductions whereby the equations in E may be used to derive statements of the form

$$t = t',$$

where t, t' are terms in $T_\Sigma(X)$.

The system, which we denote by **DED**(E) is a formalization of the intuitive principles:

—everything is equal to itself

—things equal to the same thing are equal to each other

—equals may be substituted for equals

—the equations in E generate equals.

It consists of a set of six rules. Each rule is presented in the form

$$\frac{\text{PREMISE}}{\text{CONCLUSION}}.$$

1. *Reflexivity* $\dfrac{}{t = t}$

2. *Symmetry* $\dfrac{t = t'}{t' = t}$

3. *Transitivity* $\dfrac{t = t', t' = t''}{t = t''}$

4. *Substitution* $\dfrac{t_1 = t'_1, \ldots, t_k = t'_k}{f(t_1, \ldots, t_k) = f(t'_1, \ldots, t'_k)}$ for every f in Σ of arity k

5. *Instantiation* $\dfrac{t = t'}{t\rho = t'\rho}$ for every substitution ρ.

6. *Equations* $\dfrac{}{t = t'}$ for every equation $\langle t, t' \rangle$ in E

Figure 1.1
The proof system $\mathbf{DED}(E)$.

To apply this rule one must already have derived each statement in the premise. The application of the rule then derives the statement in the conclusion. The rules are given in figure 1.1.

A *proof* is a sequence of statements

$$t_1 = t'_1, t_2 = t'_2, \ldots, t_k = t'_k, \ldots$$

such that each statement can be derived by applying any of the rules to statements earlier in the sequence. This implies that the first statement of the proof must be an instance of rule 1 or rule 6 since these are the only rules with empty premises. We write $\vdash_E t = t'$ if there is a proof whose last statement is $t = t'$; we say that $t = t'$ is a *theorem* of $\mathbf{DED}(E)$. It will be more convenient sometimes to recast this notion of theorem as a relation between terms. Let $=_E$ be the relation on $T_\Sigma(X)$ defined by

$$t =_E t' \quad \text{if and only if} \quad \vdash_E t = t'.$$

Because of rules 1–3, $=_E$ is an equivalence relation and because of rule 4 it is a Σ-congruence.

LEMMA 1.2.8 For t, t' in $T_\Sigma(X)$, if $\vdash_E t = t'$ and A satisfies E then $t =_A t'$. This lemma may be stated more succincty as: $E \subseteq =_A$ implies $=_E \subseteq =_A$.

Proof We use induction on the length of the proof of $t = t'$ in the equational deduction system.

a) The proof is of length 1.

Then $t = t'$ is an instance of rule 1 or rule 6. If it is the former, then t is t' and trivially $\rho_A(t) = \rho_A(t')$ for every A-assignment ρ_A. Otherwise it is an instance of an equation in E. Therefore since A satisfies E, $\rho_A(t) = \rho_A(t')$ for every A-assignment ρ_A.

b) The proof is of length $k + 1$.

Consider the last rule to be applied in this proof of $t = t'$. We examine each possibility in turn.

i) Rules 1, 6. The proof is as in case a).

ii) Rules 2, 3. The proof in these cases is trivial. For example consider rule 3. Then there is a term t'' such that both $t = t''$ and $t'' = t'$ appear earlier in the proof of $t = t'$. So each have a proof of length at most k. By induction we can conclude $\rho_A(t) = \rho_A(t'')$ and $\rho_A(t'') = \rho_A(t')$, i.e. $\rho_A(t) = \rho_A(t')$.

iii) Rule 4. In this case t, t' have the forms $f(\underline{t})$, $f(\underline{t}')$ respectively and each $t_i = t_i'$ has a proof of length at most k. By induction $\rho_A(t_i) = \rho_A(t_i')$ and since ρ_A is a $\Sigma(X)$-homomorphism it follows that $\rho_A(f(\underline{t})) = f_A(\rho_A(\underline{t})) = f_A(\rho_A(\underline{t}')) = \rho_A(f(\underline{t}'))$.

iv) Rule 5. In this case t, t' have the forms $t''\rho$, $t'''\rho$ respectively and $t'' = t'''$ has a proof of length at most k. We must prove $\rho_A(t''\rho) = \rho_A(t'''\rho)$ for an arbitrary A-assignment ρ_A. Consider the A-assignment $\rho_A \circ \rho$. By induction $\rho_A \circ \rho(t'') = \rho_A \circ \rho(t''')$, i.e. $\rho_A(t''\rho) = \rho_A(t'''\rho)$, because of the Substitution lemma 1.2.6. □

As an immediate corollary we have that the class $\mathscr{C}(E)$ is contained in the class satisfying the congruence $=_E$, $\mathscr{C}(=_E)$. Unfortunately the converse is not true; in general $\mathscr{C}(E) \subseteq \mathscr{C}(=_E)$, $\mathscr{C}(E) \neq \mathscr{C}(=_E)$. For A to be in $\mathscr{C}(=_E)$ it is sufficient to guarantee that the equations E are satisfied by all denotable elements whereas they must be satisfied by all elements for A to be in $\mathscr{C}(E)$. This point is addressed in Q17 at the end of the chapter. However the containment is sufficient to prove the initiality theorem for equations.

Corollary 1.2.9 (Initiality for Equations) $T_\Sigma/{=_E}$ is initial in $\mathscr{C}(E)$.

Proof $\mathscr{C}(E) \subseteq \mathscr{C}(=_E)$ and $T_\Sigma/{=_E}$ is initial in $\mathscr{C}(=_E)$ because of theorem 1.2.4. Therefore for every A in $\mathscr{C}(E)$ there exists a unique Σ-homomorphism from $T_\Sigma/{=_E}$ to A. So it remains to show that $T_\Sigma/{=_E}$ is actually in $\mathscr{C}(E)$. Let $\langle t, t' \rangle \in E$ and let ρ be a

$T_\Sigma/{=_E}$-assignment. We must show that $\rho(t) = \rho(t')$. Let ρ' be *any* substitution such that $\rho(x) = [\rho'(x)]$ for every variable x. Then one can prove by structural induction on t that $\rho(t) = [t\rho']$. Now, by applying rule 5 to $t = t'$, with the substitution ρ', we obtain

$t\rho' =_E t'\rho'$ and therefore $\rho(t) = [t\rho'] = [t'\rho'] = \rho(t')$. □

From corollary 1.1.10 we know that $\mathscr{C}(E)$ has a unique initial Σ-algebra (up to Σ-isomorphism) and we denote it by I_E. So $T_\Sigma/{=_E}$ is a particular representation of I_E. This representation does not give much information about the internal structure of I_E. For particular sets of equations we may, however, be fortunate in obtaining intuitively appealing representations of I_E in terms of mathematical objects with which we are familiar. For example, if Σ is $\{Zero, Succ, Pred, Plus\}$ and E_N the set of equations

$$Pred(Succ(x)) = x$$

$$Pred(Zero) = Zero$$

$$Plus(x, Zero) = x$$

$$Plus(x, Succ(y)) = Succ(Plus(x, y)),$$

then one can prove that $\langle N, \Sigma_N \rangle$ is initial in $\mathscr{C}(E_N)$, i.e. $\langle N, \Sigma_N \rangle$ is isomorphic to $T_\Sigma/{=_{E_N}}$. Now this particular representation of I_{E_N}, $\langle N, E_N \rangle$, is rather simple and easier to understand. It gives much more insight into the nature of the initial algebra I_E than the representation $T_\Sigma/{=_{E_N}}$. But, as we will see, the latter representation is also useful.

Let R be a relation on $T_\Sigma(X)$. The proof system **DED**(E) is *sound* with respect to R if

$\vdash_E t = t'$ implies $\langle t, t' \rangle \in R$.

It is *complete* with respect to R if

$\langle t, t' \rangle \in R$ implies $\vdash_E t = t'$.

Soundness means that if we derive $t = t'$ as a theorem in the proof system then $\langle t, t' \rangle \in R$, i.e., we cannot derive false statements about the relation R. *Completeness*, on the other hand, says that any true statement of R can be proven in the logic, i.e. if $\langle t, t' \rangle \in R$ then $t = t'$ is a theorem in the proof system.

THEOREM 1.2.10 (Equational Logic Theorem) If E is a set of equations then

a) **DED**(E) is sound with respect to $=_{I_E}$.
b) **DED**(E) is complete with respect to $=_{I_E}$, restricted to T_Σ.

Proof i) Soundness. I_E satisfies E, as was seen in corollary 1.2.9. So soundness follows immediately from lemma 1.2.8.

ii) Completeness. Suppose $t =_{I_E} t'$, where $t, t' \in T_\Sigma$. Because I_E is isomorphic to $T_\Sigma/=_E$, the unique homomorphism from T_Σ to I_E is simply the natural injection map. So $[t] = [t']$ in $T_\Sigma/=_E$, i.e. $t =_E t'$. □

1.3 Finite Nondeterministic Processes

In this section we give a simple language for describing finite nondeterministic processes. The language will in fact be T_Σ for a particular signature Σ. Each term in T_Σ will represent a machine. A semantics or interpretation for the language is given by defining a particular Σ-algebra **PS**. The meaning of a machine t is thus uniquely defined: it is $i_{\mathbf{PS}}(t)$, where $i_{\mathbf{PS}}$ is the unique Σ-homomorphism from the term algebra T_Σ to the Σ-algebra **PS**. We then show that **PS** is in fact initial in the class E for a particular set of Σ-equations E. As an immediate corollary we have, from theorem 1.2.10, a sound and complete proof system for proving when two machines are semantically equivalent.

A simple view of processes or machines is that they are devices which perform actions. Different kinds of machines perform different types of actions. We will not be concerned, for the moment, with the nature of these actions. We simply presuppose a set *Act* of actions, with a, b, etc., ranging over *Act*. The simplest possible machine is the one which can perform no actions. We introduce a constant, or function symbol of arity 0, to denote this process, NIL. NIL represents a machine which is fully specified in every way. It will never under any circumstances perform any action.

Every action a gives rise to a function symbol of arity 1, which we also denote by the symbol a. If t is a term then the term $a(t)$ represents the machine which can perform the action a and then act like the machine t. This is called *prefixing* since the machine t is prefixed by the action a. Finally, we have a function symbol $+$ of arity 2, which introduces nondeterminism. If t_1, t_2 are terms, then $+(t_1, t_2)$ represents a machine which can act either like the machine t_1 or like t_2.

This completes the description of the signature. So Σ^1, the signature in question, is completely determined by

$\Sigma_0^1 = \{NIL\}$

$\Sigma_1^1 = Act$, where *Act* is the predefined set of actions

$\Sigma_2^1 = \{+\}$

$\Sigma_n^1 = \emptyset$, for $n > 2$.

A *finite nondeterministic machine* is simply an element of the term algebra T_{Σ^1}. To emphasize that we view these terms as processes or machines we also use $\mathbf{M_1}(Act)$ or simply $\mathbf{M_1}$ in place of T_{Σ^1}: the set of machines which perform actions from *Act*. The subscript 1 indicates that this is the first type of machine we consider. In the sequel we will also use letters such as p, q, etc., to range over $\mathbf{M_1}$. It should be emphasized that elements of $\mathbf{M_1}$ are closed terms; they contain no free variables. We introduce the following conventions, which will make these machines more readable:

—the symbol $+$ is treated as an infix symbol so that $+(p, q)$ is written as $(p + q)$

—$a(p)$ is rendered as ap

—brackets are omitted whenever possible, with prefixing having a higher precedence than $+$

—$aNIL$ is often rendered as a.

EXAMPLE 1.3.1

1. $a(bc + a)$ represents a machine which can perform the action a and then either perform the action b followed by the action c or perform the action a again. Formally it is written as the term $a(+(b(c(NIL)), a(NIL)))$.

2. $ab + ac$ represents a machine which can either perform the sequence of actions a followed by b or a followed by c.

3. $a(b + NIL)$ represents a machine which can perform the action a and then either perform the action b or do nothing.

An interpretation for these machines is given by choosing a particular Σ^1-algebra. The choice of interpretation depends on an intuitive notion of what one deems to be important about machines. Different Σ^1-algebras identify different machines and so, to a large extent, the choice is governed by the justification one can offer for these identifications. For example, one could say that one is only interested in the sequences of actions that a machine can perform. So one interprets each machine as a set of these sequences, i.e. as subsets of Act^*. For our very simple kind of machines we require only a very restricted class of subsets. A subset S of Act^* is *prefix closed* if whenever s is in S and s' is a prefix of s, it follows that s' is also in S. Let **PS** be the set of nonempty finite prefix-closed subsets of Act^*. Note that ε is in S for every S in **PS** since ε is a prefix of every string.

Let $NIL_{\mathbf{PS}}$ be the set $\{\varepsilon\}$ which is in **PS**.

Let $a_{\mathbf{PS}} : \mathbf{PS} \to \mathbf{PS}$ be defined by $a_{\mathbf{PS}}(S) = \{\varepsilon\} \cup \{as, s \in S\}$.

Let $+_{\mathbf{PS}} : \mathbf{PS} \to \mathbf{PS}$ be defined by $+_{\mathbf{PS}}(S, S') = S \cup S'$.

Then $\langle \Sigma^1_{\mathbf{PS}}, \mathbf{PS} \rangle$ is a Σ^1-algebra and with every machine p in $\mathbf{M_1}$ we can associate a unique meaning in **PS**, namely $i_{\mathbf{PS}}(p)$. To emphasise that we are using **PS** as a semantic domain or interpretation for the syntactic objects in $\mathbf{M_1}$, we use the notation $\mathbf{PS}[\![p]\!]$ for $i_{\mathbf{PS}}(p)$. In general we will use the brackets $[\![\ ]\!]$ to enclose objects considered to be syntax. It is nothing more than a reminder of intentions. This notation will also be applied to open terms over the language. To every $t \in T_{\Sigma^1}(X)$ and ρ a **PS**-assignment, i.e. a mapping from every variable x to an element of **PS**, the Freeness theorem 1.2.5 associates a unique element of **PS** which we denote $\mathbf{PS}[\![t]\!]\rho$. In the particular case when t is closed, this element is always $i_{\mathbf{PS}}(t)$ regardless of the assignment ρ. So, strictly speaking, $\mathbf{PS}[\![\ ]\!]$ associates with a term a mapping from **PS**-assignments to elements of **PS** but for closed terms we will always identify this constant mapping which always returns $i_{\mathbf{PS}}(t)$ with $i_{\mathbf{PS}}(t)$ itself; i.e. we look on $\mathbf{PS}[\![p]\!]$ as an element of **PS**, for closed terms p. For all of part I of the book we will in fact only use the notation $[\![\ ]\!]$ for such terms.

EXAMPLE 1.3.2

1. $\mathbf{PS}[\![a(NIL + b)]\!]$ is the set of strings $\{\varepsilon, a, ab\}$.
2. $\mathbf{PS}[\![a(b + c)]\!] = \mathbf{PS}[\![ab + ac]\!] = \mathbf{PS}[\![ab + a(b + c)]\!] = \{\varepsilon, a, ab, ac\}$.
3. $\mathbf{PS}[\![a + abc]\!] = \mathbf{PS}[\![a + ab + abc]\!] = \mathbf{PS}[\![abc]\!] = \{\varepsilon, a, ab, abc\}$.

We now turn our attention to an equational characterization of this interpretation. Let E_1 be the set of equations

A1 $\quad x + x = x$

A2 $\quad x + y = y + x$

A3 $\quad x + (y + z) = (x + y) + z$

A4 $\quad x + NIL = x$

PS1 $\quad a(x + y) = ax + ay$

Note that **PS1** is in fact an axiom schema. It represents a set of equations, one for each action a.

LEMMA 1.3.3 **PS** satisfies the equations E_1.

Proof By examining each of the equations in turn. All are trivial. **A1**–**A3** follow from simple properties of set-theoretic union $\cup$. For example, **PS** satisfies **A2** because for any two sets S, S', $S \cup S' = S' \cup S$. **A4** is satisfied by **PS** because for any term p,

$$\begin{aligned}\mathbf{PS}[\![p + NIL]\!] &= \mathbf{PS}[\![p]\!] \cup \mathbf{PS}[\![NIL]\!] \\ &= \mathbf{PS}[\![p]\!] \cup \{\varepsilon\} \\ &= \mathbf{PS}[\![p]\!] \text{ since } \varepsilon \text{ is in every set in } \mathbf{PS}.\end{aligned}$$

Finally, **PS** satisfies **PS1** for every action a because

$$\begin{aligned}\mathbf{PS}[\![a(p+q)]\!] &= a_{\mathbf{PS}}(\mathbf{PS}[\![p+q]\!]) \\ &= a_{\mathbf{PS}}(\mathbf{PS}[\![p]\!] \cup \mathbf{PS}[\![q]\!]) \\ &= \{as, s \in \mathbf{PS}[\![p]\!] \cup \mathbf{PS}[\![q]\!]\} \cup \{\varepsilon\} \\ &= \{as, s \in \mathbf{PS}[\![p]\!]\} \cup \{as, s \in \mathbf{PS}[\![q]\!] \cup \{\varepsilon\} \\ &= a_{\mathbf{PS}}([\![p]\!]) \cup a_{\mathbf{PS}}(\mathbf{PS}[\![q]\!]) \\ &= \mathbf{PS}[\![ap]\!] \cup \mathbf{PS}[\![aq]\!] \\ &= \mathbf{PS}[\![ap + aq]\!].\end{aligned}$$

□

THEOREM 1.3.4 **PS** is initial in $\mathscr{C}(E_1)$.

Proof From the previous lemma we know that **PS** is in $\mathscr{C}(E_1)$. Let A be any other Σ^1-algebra which satisfies the equations E_1. We must show

i) there is a Σ^1-homomorphism h from **PS** to A
ii) any other such Σ^1-homomorphism coincides with h.

i) We first define h on sequences of actions. If s is the sequence $a^1a^2\ldots a^n$, $n \geq 0$, then $h(s)$ is $a_A^1(a_A^2(\ldots a_A^n(NIL_A)\ldots))$. Note that $h(\varepsilon)$ is NIL_A. Now if S is the set $\{s_1, \ldots, s_n\}$, $n > 0$, define $h(S)$ to be $h(s_1) +_A h(s_2) +_A \cdots +_A h(s_n)$. Since A satisfies the axioms **A2**, **A3**, this is in fact well-defined in the sense that $h(S)$ does not depend on any particular enumeration of the elements of S. So h is a function from a superset of **PS** to A. We must prove it is a Σ^1-homomorphism, when restricted to **PS**.

a) Obviously $h(NIL_{\mathbf{PS}}) = NIL_A$, since $NIL_{\mathbf{PS}}$ is $\{\varepsilon\}$.
b) Consider $h(a_{\mathbf{PS}}S)$, where S is $\{s_1, \ldots, s_n\}$, $n > 0$. Because A satisfies **A2**, **A3**, the particular enumeration does not matter and we can write

$$\begin{aligned}
h(a_{\mathbf{PS}}S) &= h(\{as_1, \ldots, as_n\} \cup \{\varepsilon\}) \\
&= h(as_1) +_A \cdots +_A h(as_n) +_A h(\varepsilon) \\
&= a_A h(s_1) +_A \cdots +_A a_A h(s_n) +_A NIL_A \quad \text{by the definition of } h \text{ on sequences} \\
&= a(h(s_1) +_A h(s_2) +_A \cdots +_A h(s_n)) \quad \text{because } A \text{ satisfies the equations } \mathbf{PS1}, \mathbf{A4} \\
&= a_A h(S).
\end{aligned}$$

c) Consider $h(S_1 +_{\mathbf{PS}} S_2)$ where S_1, S_2 are $\{s_1, \ldots, s_n\}$, $\{s'_1, \ldots, s'_m\}$, $n, m \geq 0$, respectively. Then

$$\begin{aligned}
h(S_1) +_A h(S_2) &= h(s_1) +_A \cdots +_A h(s_n) +_A h(s'_1) +_A \cdots +_A h(s'_m) \\
&= h(s''_1) +_A \cdots +_A h(s''_k) \quad \text{where} \quad S_1 \cup S_2 = \{s''_1, \ldots, s''_k\}, \\
&\qquad \text{because } A \text{ satisfies } \mathbf{A1} \text{ and } \mathbf{A2}, \mathbf{A3} \\
&= h(S_1 +_{\mathbf{PS}} S_2).
\end{aligned}$$

We have proven that h is a Σ^1-homomorphism.

ii) Let $h' : \mathbf{PS} \to A$ be any Σ^1-homomorphism. We must prove that $h(S) = h'(S)$ for every set S. The proof is by induction on the size of S.

a) S is the singleton set. Then S must be the set $\{\varepsilon\}$. So

$$\begin{aligned}
h(\varepsilon) &= NIL_A \text{ by definition.} \\
&= h'(NIL_{\mathbf{PS}}) \text{ since } h' \text{ is a } \Sigma\text{-homomorphism} \\
&= h'(\varepsilon).
\end{aligned}$$

b) The size of S is greater than 1.

For any a let $S/a = \{s, as \in S\}$. Each S/a is prefix-closed because S is prefix-closed. Furthermore there is at least one a such that S/a is not empty. For such an a, $S = a_{\mathbf{PS}}(S/a) \cup S'$, where S' is the set of elements in S not starting with a. Then

$$\begin{aligned}
h(S) &= h(a_{\mathbf{PS}}(S/a) \cup S') \\
&= h(a_{\mathbf{PS}}(S/a)) +_A h(S') \quad \text{by definition of } h
\end{aligned}$$

$= a_A(h(S/a)) +_A h(S')$ since h is a Σ^1-homomorphism

$= a_A(h'(S/a)) +_A h'(S')$ by induction since the size of both S/a and S' is less than that of S

$= h'(a_{\mathbf{PS}}(S/a)) +_{\mathbf{PS}} h(S')$ since h' is a Σ^1-homomorphism

$= h'(S)$. □

The import of this theorem is that we have a sound and complete proof system for the semantic equivalence of finite nondeterministic machines. This follows from the Equational Logic theorem 1.2.10. If we use the proof system $\mathbf{DED}(E_1)$, then whenever we can derive $\vdash_{E_1} p = q$ it will be true that $\mathbf{PS}[\![p]\!] = \mathbf{PS}[\![q]\!]$. For example $\vdash_{E_1} ax = ax + a$ and so for every p, $\mathbf{PS}[\![ap]\!] = \mathbf{PS}[\![ap + a]\!]$. Conversely, if p, q are such that $\mathbf{PS}[\![p]\!] = \mathbf{PS}[\![q]\!]$ then there is a proof of $p = q$ in the system $\mathbf{DED}(E_1)$. For example $\mathbf{PS}[\![a(b + c) + acd]\!] = \mathbf{PS}[\![a(b + cd)]\!]$ and so $a(b + c) + acd = a(b + cd)$ can be deduced within the system $\mathbf{DED}(E_1)$.

There are, of course, many other possible semantic interpretations of these machines. For example, we might wish the semantics to reflect the choices possible at each stage in a computation. So that $a(b + c)$ would be considered to be semantically different from $ab + ac$. One interpretation which makes this distinction is *rooted trees*. Let **RT** be the set of finite rooted trees whose branches are labeled by actions from *Act*. **RT** can be viewed as a Σ^1-algebra by defining:

i) $NIL_{\mathbf{RT}}$ to be the trivial tree which has a root and no branches •

ii) $a_{\mathbf{RT}}$ maps the tree

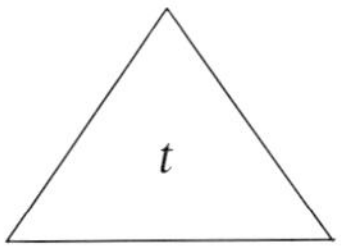

to the tree

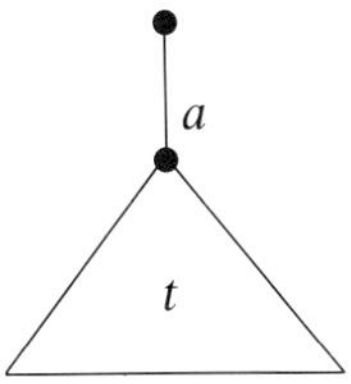

iii) $+_{\mathbf{RT}}$ maps the trees

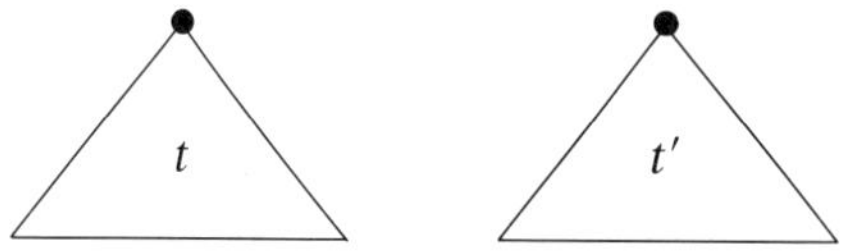

to the tree

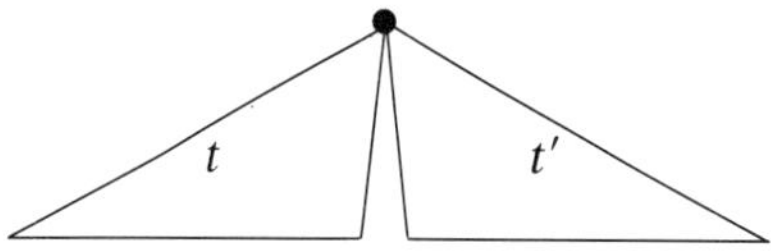

which is obtained by identifying the roots of the two trees t, t'.

EXAMPLE 1.3.5

i) $\mathbf{RT}[\![ab + ac]\!]$ is the tree

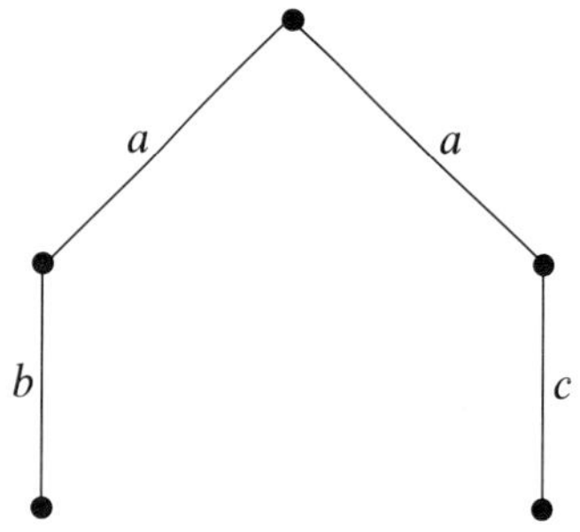

ii) $\mathbf{RT}[\![a(b + c)]\!]$ is the tree

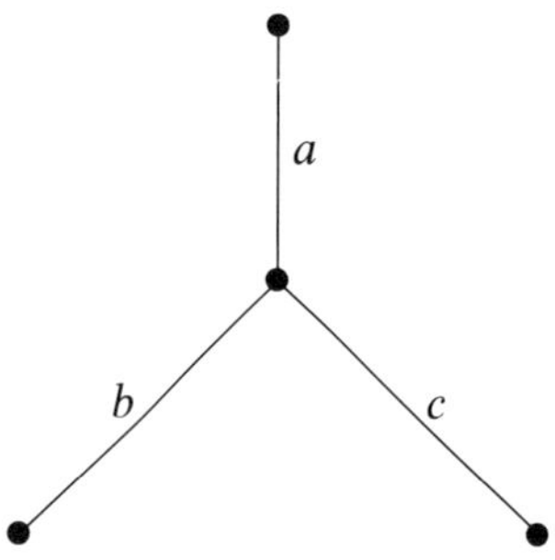

iii) **RT**$[\![a(b + c) + ab]\!]$ is the tree

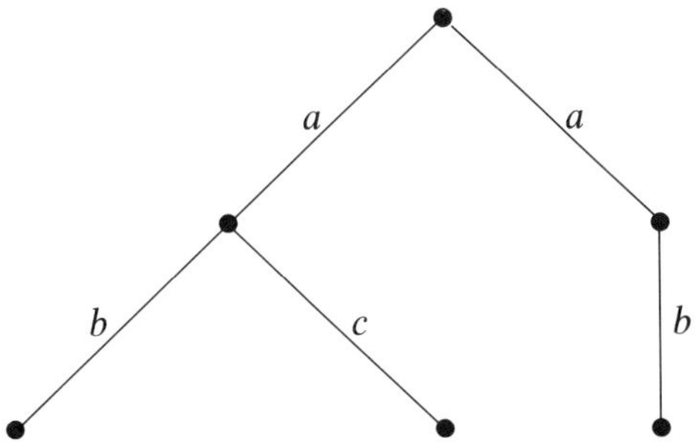

The interpretation **RT** satisfies fewer equations than **PS**. For example, the equation **PS1** is *not* satisfied by **RT**, as the first two examples above show. Also **A1** is not valid for **RT**. For example **RT**$[\![aNIL]\!]$ is the tree

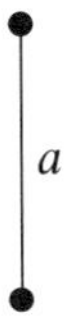

whereas **RT**$[\![aNIL + aNIL]\!]$ is the different tree

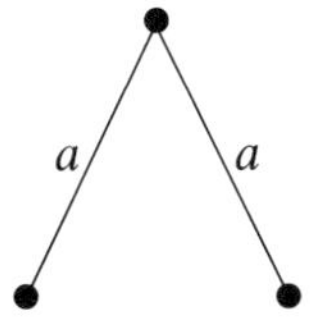

It does, however, satisfy the remaining equations **A2**, **A3**, **A4**. Let E_2 denote this set of equations. We state without proof the following result.

THEOREM 1.3.6 **RT** is initial in the class $\mathscr{C}(E_2)$.

The proof is straightforward and follows the lines of theorem 1.3.4. It gives as a corollary, again due to the Equational Logic theorem, a sound and complete proof system for the semantic equivalence in **RT**, namely **DED**(E_2).

There is no obvious criterion for preferring either of these two semantic models **RT** and **PS**. Each reflects a different intuition about the behavior of finite nodeterministic processes. To evaluate their true merit we must be more definite as to what we mean by the behavior of a process. Processes (machines, programs, etc.) are supposed to be executed. They are computational or algorithmic entities which, when executed, affect and are affected by their environment. This environment may be users or other computational entities, or indeed a combination of both. The models **RT** and **PS** are therefore only of interest if they inform us of the effect of executing these processes or, more generally, of using them. Chapter 2 discusses in detail a particular operational or behavioral view of processes, via an operational semantics. This may then be used as a criterion for determining the intrinsic merit of any proposed semantic interpretations such as **RT** and **PS**. However, because of the form this operational view will take, the semantic domains we require are not simply Σ-algebras; they require some extra structure. This is the topic of the next section.

We end this section with some remarks on the Equational Logic theorem, which says that the proof system **DED**(E) is sound and complete with respect to the relation $=_{I_E}$ over T_Σ. In general there may be many different sets of equations which are both sound and complete with respect to the same relation. Semantically speaking, there may be different sets of equations, E, E', which give the same initial interpretation, i.e. $I_{E'} = I_E$. For example if we replace **A2** by the schema

$$ax + by = by + ax,$$

then the resulting set of equations, E'_1, are still sound and complete with respect to the semantic interpretation **PS**. It is difficult to state a method of preferring one such set of equations over another. From one point of view, we require the smallest possible set. For a smaller set of equations will in general give more insight into the nature or internal organization of the initial algebra. On the other hand, the larger set will give a more powerful proof system. For recall that the completeness applies only to closed terms. There may be many theorems true of open terms which are not derivable in the complete proof system. For example

A2 $\quad x + y = y + x$

is true of **PS**, but it is not derivable in **DED**(E'_1). As another example, the equation

$$Plus(x, y) = Plus(y, x)$$

is true in I_{E_N}, where E_N is the set of equations given at the end of §1.2, but is not a theorem of **DED**(E_N). The completeness result for the proof system merely states that every *closed* instantiation of **A2** is derivable in **DED**(E'_1). So E_1 is clearly a more powerful set of equations than E'_1, and for the purpose of proving relationships between machines the stronger set is to be preferred.

1.4 Inequational Classes

Because of the operational view of processes which we will take in chapter 2, the natural semantic domains we consider will be partial orders. Therefore, we now modify the results in §1.1 and §1.2 to take this into account; we develop a theory of Σ-algebras which are endowed with partial orders. Let $\langle A, \leq_A \rangle$ be a partial order. A function $f: A \to A$ is *monotonic* if $a \leq_A a'$ implies $f(a) \leq_A f(a')$. For functions with arity greater than 1 we need to turn A^k into a partial order. For $\underline{a}, \underline{a}' \in A^k$ let $\underline{a} \leq_A \underline{a}'$ if for each $1 \leq i \leq k$, $a_i \leq_A a_i'$. It is easy to check that $\leq_A$ defined in this manner is a partial order over A^k. Then we say $f: A^k \to A$ is *monotonic* if $\underline{a} \leq_A \underline{a}'$ implies $f(\underline{a}) \leq_A f(\underline{a}')$. Alternatively, we could require $\underline{a} \leq_A' \underline{a}'$ implies $f(\underline{a}) \leq_A f(\underline{a}')$, where $\leq_A'$ is the slightly weaker relation defined by $\underline{a} \leq_A' \underline{a}'$ if there exists one $1 \leq j \leq k$, such that $a_j \leq_A a_j'$ and for every i different from j, $a_i = a_i'$. We leave the reader to check that both conditions are equivalent. We will tend to use the latter.

A *Σ-po algebra* is a triple $\langle A, \leq_A, \Sigma_A \rangle$ where

—A is a set, called the carrier

—$\leq_A$ is a partial order over A

—for each f in Σ of arity k, there is a monotonic function $f_A: A^k \to A$.

EXAMPLE 1.4.1 Let Σ be the signature of example 1.1.1, {*Zero*, *Succ*, *Pred*, *Plus*}. Let $\leq_N$ be the usual "less than or equal to" relation on the natural numbers and let Σ_N be the functions as defined in example 1.1.1. Then $\langle N, \leq_N, \Sigma_N \rangle$ is a Σ-*po* algebra.

As with Σ-algebras, we refer to the Σ-*po* algebra $\langle A, \leq_A, \Sigma_A \rangle$ simply as A, when $\leq_A$ and Σ_A can be inferred from the context. If p, q are recognizably elements of A we also write $p \leq q$ instead of $p \leq_A q$. We develop the theory of Σ-*po* algebras in exactly the same way as that in §1.2 with the obvious modifications; these merely take into account the extra structure. For example, a Σ-*po* homomorphism is a Σ-homomorphism which preserves the extra structure, the partial order. Let $\langle A, \leq_A, \Sigma_A \rangle$, $\langle B, \leq_B, \Sigma_B \rangle$ be Σ-*po* algebras. The function $h: A \to B$ is a Σ-*po* homomorphism if

i) for each f in Σ, $h(f_A(\underline{a})) = f_B(h(\underline{a}))$

ii) $a \leq_A a'$ implies $h(a) \leq_A h(a')$.

It is a *Σ-po isomorphism* if it has an inverse, i.e. a Σ-*po* homomorphism $g: B \to A$ such that $g \circ h = id_A$ and $h \circ g = id_B$. As with Σ-isomorphism, there is an alternative characterization: h is a Σ-*po* isomorphism if and only if

i) it is surjective

ii) $h(a) \leq_B h(a')$ implies $a \leq_A a'$.

We can view every Σ-algebra $\langle A, \Sigma_A \rangle$ in a straightforward way as the Σ-*po* algebra $\langle A, \equiv_A, \Sigma_A \rangle$ where $\equiv_A$ denotes equality on A. In other words, A is endowed with the trivial partial order which relates two elements if and only if they are identical. The functions in Σ_A are therefore automatically monotonic. In particular $T_\Sigma(X)$ is a Σ-*po* algebra and, moreover, it satisfies the freeness condition for Σ-*po* algebras.

THEOREM 1.4.2 (Freeness) If A is a Σ-*po* algebra and ρ_A an A-assignment for X, then there exists a unique Σ-*po* homomorphism h_A from $T_\Sigma(X)$ to A such that $h_A(x) = \rho_A(x)$ for every x in X.

Proof The proof is virtually identical to that of theorem 1.2.5. Let h_A be as defined in that theorem. There it was shown to be a Σ-homomorphism and because the partial order on $T_\Sigma(X)$ is trivial, it is also monotonic. Let h' be any other Σ-*po* homomorphism from $T_\Sigma(X)$ to A which coincides with ρ_A on X. Then h' is also a Σ-homomorphism from the Σ-algebras T_Σ to the Σ-algebra $\langle A, \Sigma_A \rangle$. Applying theorem 1.2.5, we obtain that h' coincides with h_A. □

The Σ-*po* homomorphism h_A given by this theorem coincides with the one obtained by the theorem 1.2.5, when the Σ-*po* algebra A is considered simply as a Σ-algebra. So we will continue to refer to it as ρ_A. If we let X be the empty set of variables, we obtain a unique Σ-*po* homomorphism from T_Σ to A. Again, this actually coincides with i_A given in the Initiality theorem 1.1.7. Moreover, we continue the syntactic convention of §1.3 by writing, where appropriate, $A[\![t]\!]$ to mean the interpretation of the closed term t in the domain $\langle A, \leq_A, \Sigma_A \rangle$, i.e. $i_A(t)$.

If $\langle A, \leq_A, \Sigma_A \rangle$ is a Σ-*po* algebra, a Σ-*preorder* over A is a relation $\sqsubseteq$ over A which satisfies, for every a, a', a'' in A,

i) $a \sqsubseteq a$ (reflexivity)

ii) $a \sqsubseteq a'$, $a' \sqsubseteq a''$ implies $a \sqsubseteq a''$ (transitivity)

iii) $a \leq_A a'$ implies $a \sqsubseteq a'$ (i.e. $\sqsubseteq$ extends $\leq_A$)

iv) $\underline{a} \sqsubseteq \underline{a}'$ implies $f_A(\underline{a}) \sqsubseteq f_A(\underline{a}')$ for every f in Σ (substitutivity)

It is the natural extension of the definiton of Σ-congruence to Σ-*po* algebras, except that we do *not* require antisymmetry.

EXAMPLE 1.4.3 Let A be any Σ-*po* algebra. For t, t' in the Σ-*po* algebra T_Σ let $t \leq_A t'$ if $i_A(t) \leq_A i_A(t')$ and $t =_A t'$ if $i_A(t) = i_A(t')$. Then $\leq_A$ is a Σ-preorder on T_Σ. It preserves the functions in T_Σ because they are interpreted on A as monotonic functions.

This relation $\leq_A$ (and $=_A$) can be extended to $T_\Sigma(X)$ in the natural way by letting $t \leq_A t'$ if $\rho_A(t) \leq_A \rho_A(t')$ for every A-assignment ρ.

As with Σ-congruences and Σ-algebras, we can factor Σ-*po* algebras by Σ-preorders to obtain new Σ-*po* algebras. If $\sqsubseteq$ is a Σ-preorder over A, let $\sim$ denote the kernel of $\sqsubseteq$, i.e. $a \sim a'$ if $a \sqsubseteq a'$ and $a' \sqsubseteq a$. Then $\sim$ is an equivalence over the carrier A. Let $A/\sqsubseteq$ denote the set of equivalence classes. For $[a], [a']$ in $A/\sqsubseteq$ let $[a] \sqsubseteq [a']$ if $a \sqsubseteq a'$ in A. (For notational convenience we abbreviate $[a]_\sim$ by $[a]$ and $\sqsubseteq_{A/\sqsubseteq}$ by $\sqsubseteq$; the required subscript should be obvious from the context.) Note that this definition of $\sqsubseteq$ does not actually depend on any particular representatives of the equivalence classes. For if $a_1 \in [a]$, $a_1' \in [a']$ and $a \sqsubseteq a'$ then $a_1 \sqsubseteq a_1'$ because

$$\begin{aligned} a_1 &\sqsubseteq a \quad \text{since} \quad a_1 \sim a \\ &\sqsubseteq a' \\ &\sqsubseteq a_1' \quad \text{since} \quad a' \sim a_1'. \end{aligned}$$

Moreover, $\sqsubseteq$ is a partial order on $A/\sqsubseteq$. The only nontrivial property of $\sqsubseteq$ to check is antisymmetry. Suppose $[a] \sqsubseteq [a']$ and $[a'] \sqsubseteq [a]$. To prove $[a] = [a']$ it is sufficient to prove $a \sim a'$, i.e. $a \sqsubseteq a'$ and $a' \sqsubseteq a$. The former follows from the fact that $[a] \sqsubseteq [a']$, the latter from $[a'] \sqsubseteq [a]$. Finally we can endow $A/\sqsubseteq$ with functions in the obvious way:

$$f_{A/\sqsubseteq}([a_1], \ldots, [a_k]) = [f_A(a_1, \ldots, a_k)].$$

LEMMA 1.4.4

a) $\langle A/\sqsubseteq, \sqsubseteq, \Sigma_{A/\sqsubseteq} \rangle$ is a Σ-*po* algebra.

b) The natural injection mapping $in: A \to A/\sqsubseteq$ defined by $in(a) = [a]$, is a Σ-*po* homomorphism.

Proof a) We leave the reader to prove that $f_{A/\sqsubseteq}$ is well-defined. This is similar to the proof of lemma 1.2.1 part a), and uses the fact that $\sqsubseteq$ is function-preserving. We show that $f_{A/\sqsubseteq}$ is monotonic. Let $[\underline{a}] \sqsubseteq [\underline{a}']$. Then $\underline{a} \sqsubseteq \underline{a}'$ in A. Since $\sqsubseteq$ is function-preserving, $f(\underline{a}) \sqsubseteq f(\underline{a}')$, i.e. $[f(\underline{a})] \sqsubseteq [f(\underline{a}')]$, i.e. $f_{A/\sqsubseteq}([\underline{a}]) \sqsubseteq f_{A/\sqsubseteq}([\underline{a}'])$.

b) The proof is similar to the corresponding part of lemma 1.2.1. □

We now apply this technical lemma to the particular case of Σ-preorders over T_Σ to obtain an initiality result. Let $\sqsubseteq$ be an arbitrary Σ-preorder over T_Σ. A Σ-*po* algebra $\langle A, \leq_A, \Sigma_A \rangle$ satisfies $\sqsubseteq$ if $t \sqsubseteq t'$ implies $i_A(t) <_A i_A(t')$. Let $\mathscr{C}(\sqsubseteq)$ be the class of all Σ-*po* algebras which satisfy $\sqsubseteq$.

THEOREM 1.4.6 (Initiality for Preorders) The Σ-*po* algebra $T_\Sigma/\sqsubseteq$ is initial in the class $\mathscr{C}(\sqsubseteq)$.

Proof Similar to the Initiality for Congruences theorem 1.2.4. The proof that $T_\Sigma/\sqsubseteq$ is in fact in $\mathscr{C}(\sqsubseteq)$ is similar to the corresponding fact about T_Σ/C in theorem 1.2.4.

Let A be a Σ-*po* algebra which satisfies $\sqsubseteq$. Define $h: T_\Sigma/\sqsubseteq \to A$ by $h([t]) = i_A(t)$. To show it is a well-defined function we must prove that if $t \sim t'$ then $i_A(t) = i_A(t')$. However if $t \sim t'$ then by definition $t \sqsubseteq t'$ and $t' \sqsubseteq t$. Since A satisfies $\sqsubseteq$ it follows that $i_A(t) \leq i_A(t')$ and $i_A(t') \leq i_A(t)$, i.e. $i_A(t) = i_A(t')$. Moreover h is monotonic; for if $[t] \sqsubseteq [t']$, then $t \sqsubseteq t'$ and since A satisfies $\sqsubseteq$ it follows that $h(t) \leq_A h(t')$. The proof that h is a homomorphism and is unique is similar to the analogous results in theorem 1.2.4. □

As in §1.2, we apply this theorem to the particular case when the Σ-preorder $\sqsubseteq$ is generated syntactically by inequations. A Σ-*inequation* is an expression of the form

$$t \leq t',$$

where t, t' are in $T_\Sigma(X)$. A relation R over $T_\Sigma(X)$ *satisfies* a set of inequations E if $E \subseteq R$. Then, as with Σ-algebras, the Σ-*po* algebra $\langle A, \leq_A, \Sigma_A \rangle$ *satisfies* E if $E \subseteq =_A$, i.e. for every $t \leq t'$ in E and every A-assignment ρ_A, $t\rho_A \leq_A t'\rho_A$. We use $\mathscr{C}(E)$ to denote the class of Σ-*po* algebras which satisfy the set of inequations E.

EXAMPLE 1.4.7 Let $\langle N, \leq_N, \Sigma_N \rangle$ be as defined in example 1.4.1. Then it satisfies the inequations

$$\begin{aligned}
Zero &\leq x\\
Pred(Zero) &\leq Zero\\
Pred(Succ(x)) &\leq x\\
x &\leq Pred(Succ(x))\\
Plus(x, Zero) &\leq x\\
x &\leq Plus(x, Zero)
\end{aligned}$$

$Plus(x, Succ(y)) \leq Succ(Plus(x, y))$

$Succ(Plus(x, y)) \leq Plus(x, Succ(y))$.

As might be expected there is a proof system, corresponding to that of §1.2, for inequations. The rules are given in figure 1.2.

We also refer to this proof system as **DED**(E). It is used as before except that we now derive statements of the form $t \leq t'$, where t, t' are in $T_\Sigma(X)$ and we are *not* allowed to use the rule of symmetry. We write $\vdash_E t \leq t'$ if $t \leq t'$ can be derived in this system, and for convenience we will also often write $t \leq_E t'$ to mean $\vdash_E t \leq t'$. We leave it to the reader to prove, in analogy with lemma 1.2.8, the following lemma:

LEMMA 1.4.8 For every t, t' in $T_\Sigma(X)$ if $\vdash_E t \leq t'$ then $t \leq_A t'$ for every Σ-*po* algebra A which satisfies E.

COROLLARY 1.4.9 (Initiality for Inequations) $T_\Sigma/\leq_E$ is initial in $\mathscr{C}(E)$.

Proof Again the proof is virtually identical to the corresponding result about equations, corollary 1.2.9. We leave the reader to check the details. □

Finally, we have a soundness and completeness result for the proof system **DED**(E). As with equations, we let I_E denote the unique (up to isomorphism) initial Σ-*po* algebra in $\mathscr{C}(E)$ for any set of inequations E.

1. *Reflexivity* $$\frac{}{t \leq t}$$

2. *Transitivity* $$\frac{t \leq t',\ t' \leq t''}{t \leq t''}$$

3. *Substitutivity* $$\frac{t_1 \leq t_1', \ldots, t_k \leq t_k'}{f(t_1, \ldots, t_k) \leq f(t_1', \ldots, t_k')}$$ for every f in Σ of arity k

4. *Instantiation* $$\frac{t \leq t'}{t\rho \leq t'\rho}$$ for every substitution ρ

5. *Inequations* $$\frac{}{t \leq t'}$$ for every inequation $t \leq t'$ in E

Figure 1.2
The proof system **DED**(E).

THEOREM 1.4.10 (Equational Logic Theorem) If E is a set of inequations

a) **DED**(E) is sound with respect to $\leq_{I_E}$ over $T_\Sigma(X)$

b) **DED**(E) is complete with respect to $\leq_{I_E}$ over T_Σ.

Proof a) As in the corresponding theorem 1.2.10 for equations.

b) Suppose t, t' are in T_Σ and $t \leq_{I_E} t'$. Because I_E coincides with $T_\Sigma/\leq_E$ we have that $i_{T_\Sigma/\leq_E}(t) < i_{T_\Sigma/\leq_E}(t')$. But $i_{T_\Sigma/\leq_E}$ coincides with the natural injection function in, so that $in(t) \leq in(t')$, i.e. $[t] \sqsubseteq [t']$. From the definition of $\sqsubseteq$ it follows that $t \leq_E t'$, i.e. $\vdash_E t \leq t'$. □

DED(E) will not in general be complete for $\leq_{I_E}$ over $T_\Sigma(X)$. However, we can strengthen part a) of this theorem. A Σ-preorder R over $T_\Sigma(X)$ is *substitution closed* if for every substitution ρ, $\langle t, t' \rangle \in R$ implies $\langle t\rho, t'\rho \rangle \in R$. One can show, using the substitution lemma 1.2.6, that for any Σ-*po* A, the Σ-*po* $\leq_A$ is substitution closed.

PROPOSITION 1.4.11 If E is a set of inequations, then $\leq_E$ is the least substitution closed Σ-preorder which satisfies E.

Proof Obviously $\leq_E$ satisfies E because of rule 5 of **DED**(E). It is substitution closed because of rule 3. Now let R be any substitution closed Σ-preorder which satisfies E. Using exactly the same technique as in lemma 1.2.8, we can prove $\vdash_E t \leq t'$ implies $\langle t, t' \rangle \in R$. □

To emphasise the intimate connection between initiality in $\mathscr{C}(E)$ and the deduction system **DED**(E), we give a stronger result than the Equational Logic Theorem. A Σ-*po* algebra $\langle A, \leq_A, \Sigma_A \rangle$ is *surjective* if for every a in A there is some t in T_Σ such that $i_A(t) = a$. Intuitively A is surjective if every element is denotable by some syntactic object in T_Σ. By the construction of I_E, as a factored algebra of T_Σ, it is obvious that it is surjective.

COROLLARY 1.4.12 The Σ-*po* algebra $\langle A, \leq_A, \Sigma_A \rangle$ is initial in $\mathscr{C}(E)$ if and only if it is surjective and **DED**(E) is sound and complete with respect to $\leq_A$ over T_Σ.

Proof If $\langle A, \leq_A, \Sigma_A \rangle$ is initial in $\mathscr{C}(E)$ it is isomorphic to I_E. By construction this is surjective and the Equational Logic theorem gives that **DED**(E) is sound and complete with respect to $\leq_A$.

Conversely, suppose $\langle A, \leq_A, \Sigma_A \rangle$ satisfies the requirements of the corollary. We must first show that $A \in \mathscr{C}(E)$, i.e. for every A-assignment ρ_A and every inequation

$t \leq t'$ in E, $\rho_A(t) \leq_A \rho_A(t')$. We only know that **DED**(E) is sound with respect to $\leq_A$ restricted to T_Σ, whereas t and t' are in $T_\Sigma(X)$. So the soundness of **DED**(E) cannot be applied directly. However, because A is surjective there exists a closed substitution ρ such that for every x in X $\rho_A \circ \rho(x) = \rho_A(x)$. By the Substitution lemma this means that $\rho_A(u) = \rho_A(u\rho)$ for every u in $T_\Sigma(X)$. Now $t\rho \leq t'\rho$ is derivable in **DED**(E) using rule 5 and, since both $t\rho$ and $t'\rho$ are closed, we can apply the soundness of **DED**(E) with respect to $\leq_A$ for closed terms to obtain $\rho_A(t\rho) \leq \rho_A(t'\rho)$, i.e. $\rho_A(t) \leq \rho_A(t')$. It follows that $A \in \mathscr{C}(E)$.

Let $h: I_E \to A$ be the unique Σ-*po* homomorphism guaranteed by the initiality of I_E. Note that h simply maps the equivalence class $[t]$ to $i_A(t)$. We now construct an inverse for h. For $a \in A$ let $k(a) = [t]$, where $i_A(t) = a$. Because A is surjective, $k(a)$ is defined for every $a \in A$. Moreover, it is a well-defined function from A to I_E: if $i_A(t) = i_A(t')$, i.e. $t \leq_A t'$ and $t' \leq_A t$, then because of the completeness of **DED**(E) with respect to $\leq_A$ we have that $t \leq_E t'$ and $t' \leq_E t$, i.e. $[t] = [t']$ in I_E. We leave it to the reader to check that it is a Σ-*po* homomorphism and the inverse of h. □

If we examine example 1.4.7 we see that many of the inequations of the form $t \leq t'$ also have their inverse $t' \leq t$ present. This situation will often arise and so, for convenience, we will use $t = t'$ as an abbreviation for the two inequations $t \leq t'$ and $t' \leq t$. This allows a more intuitive presentation of sets of inequations. For instance, that of example 1.4.7 may now be rendered as the set

$$Zero \leq x$$

$$Pred(Zero) \leq Zero$$

$$Pred(Succ(x)) = x$$

$$Plus(x, Zero) = x$$

$$Plus(x, Succ(y)) = Succ(Plus(x, y)).$$

This abbreviation or notation allows us to extend the proof system to the rules:

6. *Equality* $\dfrac{t \leq t', t' \leq t}{t = t'}, \quad \dfrac{t = t'}{t \leq t', t' \leq t}, \quad \dfrac{t = t'}{t' = t}.$

Moreover, we can introduce rules for $t = t'$ corresponding to each of the rules 1 to 4 for $t \leq t$, simply by replacing $\leq$ in each by $=$. The resulting system is still sound and complete and it is this extension which we will use in the sequel. We also use $t =_E t'$ to denote $\vdash_E t = t'$. We have somewhat overloaded subscripted equalities and inequalities. However, the meaning will always be clear from the context: if A is an Σ-algebra $t \leq_A t'$ means that when t and t' are evaluated in A with respect to an arbitrary assignment to variables they are related via $\leq_A$; whereas if E is a set of inequations, $t \leq_E t'$ means $t \leq t'$ is a theorem of the deductive system **DED**(E). In fact, the former notation will not be used very frequently as we will tend to use the more suggestive $A[\![\ ]\!]: t \leq_A t'$ means exactly the same as $A[\![t]\!] \leq A[\![t']\!]$.

In chapter 2 we develop an operational viewpoint of machines (terms in T_Σ for some Σ). This will result in various behaviorally motivated relations over machines which we denote by symbols such as $\sqsubseteq$. These relations have a very intuitive interpretation: for machines p, q, $p \sqsubseteq q$ if "q is a more deterministic version of p." Indeed, the multiplicity of behavioral relations stems from the various ways in which one can formalize the statement "is more deterministic than." Nevertheless we may use any one of these relations to evaluate proposed semantic interpretations. Because they are naturally preorders rather than equivalences, it is more suitable to consider as interpretations Σ-*po* algebras rather than simple Σ-algebras.

Let R be a (behaviorally motivated) relation over terms from T_Σ. Viewing R as a criterion for evaluating the suitability of particular interpretations, we say that a given interpretation A (i.e. a Σ-*po* algebra) is *fully abstract with respect to* R if, for every t, t' in T_Σ,

$$\langle t, t' \rangle \in R \quad \text{if and only if} \quad A[\![t]\!] \leq A[\![t']\!].$$

So A relates, and only relates, terms which are related by R. Not surprisingly, there is a strong relationship between full abstraction and initiality.

PROPOSITION 1.4.13 If A is surjective then it is fully abstract with respect to R if and only if

i) R is a Σ-preorder over T_Σ
ii) A is initial in $\mathscr{C}(R)$.

Proof Suppose A is fully abstract with respect to R.

i) Then $\leq_A$, as a relation over T_Σ, coincides with R. In example 1.4.3 it was seen that $\leq_A$, and therefore R, is a Σ-preorder.
ii) Obviously $A \in \mathscr{C}(R)$. Let $h: T_\Sigma/R \to A$ be the unique Σ-*po* homomorphism which is guaranteed by the initiality of T_Σ/R. We can prove h is a bijection in exactly the same manner as in corollary 1.4.12.

Conversely, suppose i) and ii) are satisfied. Since A is in $\mathscr{C}(R)$ it follows that $\langle t, t' \rangle \in R$ implies $A[\![t]\!] \leq A[\![t']\!]$. Also, if $A[\![t]\!] \leq A[\![t']\!]$ then $[t]_R \leq [t']_R$, i.e. $\langle t, t' \rangle \in R$. □

The relationship between full abstraction and initiality is even more revealing when applied to algebras which are initial with respect to sets of equations.

COROLLARY 1.4.14 I_E is fully abstract with respect to R if and only if **DED**(E) is both sound and complete with respect to R.

Proof By the Equational Logic theorem. □

The behaviorally motivated relations over machines, such as those in $\mathbf{M_1}$ which we develop in the next chapter, are based on an intuitive idea of testing: $p \sqsubseteq q$ if q passes every test which p passes. This also implies that q is at least as deterministic as p since the more tests a process passes the more deterministic is its behavior. The kernel of this relation is $\approx$, defined by $p \approx q$ if both p and q pass the same test. We now show that the interpretations **PS** and **RT** are both inadequate: they are not fully abstract with respect to this behavioral relation $\approx$. Of course, this depends on exactly how we formalize "passing a test" or, indeed, what we mean by a test. We anticipate the formal development in the next chapter and give an intuitive explanation here. A primitive form of test, appropriate to $\mathbf{M_1}$, is

please do action a.

These primitive tests can be composed to form more complicated ones, such as:

please do action a and then do action b.

Let e denote this test. Informally we can argue that the machine $a(bNIL + cNIL)$ will *always* succeed when test e is applied. Having performed the action a demanded by e it is in the state $bNIL + cNIL$ where it can do any one of the actions b, c; in particular, it can always do the subsequent action demanded by e, namely b. On the other hand, the machine $abNIL + acNIL$ may or may not pass the test; it is not always *guaranteed*. If, in response to the demand to do a, it performs the second occurrence it is then in the state $cNIL$ in which it cannot comply with the request to perform b.

It follows that the interpretation **PS** is not fully abstract with respect to this behavioral notion of testing. It identifies machines, such as $a(bNIL + cNIL)$ and $abNIL + acNIL$, which are different behaviorally. On the other hand, the interpretation **RT** errs in the other direction. It is not fully abstract because it distinguishes machines which cannot be distinguished by tests. For example, $\mathbf{RT}[\![aNIL + aNIL]\!] \neq \mathbf{RT}[\![aNIL]\!]$ and it is difficult to formulate a type of test we outlined above to distinguish between $aNIL$ and $aNIL + aNIL$.

Somewhere between **PS** and **RT** lies the proper interpretation relative to our proposed $\lessapprox$. In fact because of variations in the formalization of testing we obtain three different variations on $\lessapprox$. If they are Σ^1-preorders the required interpretations are obtained automatically: $T_\Sigma/\lessapprox$ is fully abstract with respect to $\lessapprox$. However, in this form they are of little value since it is not possible to understand the nature of the objects in the interpretation or the relationship between them without appealing directly to the operational behavior as encoded in $\lessapprox$. Instead we develop particular representations of these interpretations which are intuitive and easy to understand. We then show they are isomorphic to I_E for a specific set of inequations E, thereby providing a sound and complete proof system **DED**(E) for the behavioral relations $\lessapprox$.

We end this section with a discussion of a slight lacuna in our notation. Referring to theorem 1.4.6, what we actually mean is that there exists a unique Σ-*po* homomorphism from $T_\Sigma/\sqsubseteq$, considered as a Σ-*po* algebra, to every Σ-*po* algebra A in $\mathscr{C}(\sqsubseteq)$. When objects are Σ-*po* algebras the definition of initiality (1.1.8) does not state that the required Σ-homomorphisms should be Σ-*po* homomorphisms, i.e. should be in addition monotonic. To be quite precise in our definitions we should work within category theory. However, rather than introduce the concepts of category theory in this introductory book, we simply make the convention that if all the objects under discussion are Σ-*po* algebras, then we expect the homomorphisms also to be Σ-*po* homomorphisms. This convention will also be applied when we introduce more structure into our algebras. In general, homomorphisms will be required to preserve whatever structure exists.

Exercises

Q1 If $f: A \longrightarrow B, g: B \longrightarrow C$ are both Σ-homomorphisms, show that $g \circ f: A \longrightarrow C$ is also a Σ-homomorphism.

Q2 Show that if $g \circ f$ is surjective then g is surjective and if $g \circ f$ is injective then f is injective.

Q3 a) Let $h: A \longrightarrow B$ be a Σ-homomorphism between two Σ-algebras. Define the relation $=_h$ over A by

$a =_h a'$ if $h(a) = h(a')$.

Prove $=_h$ is a Σ-congruence.

b) If $h: A \longrightarrow B$ is a Σ-*po* homomorphism between two Σ-*po* algebras let $a \leq_h a'$ if $h(a) \leq h(a')$. Show $\leq_h$ is a Σ-preorder over A.

Q4 The Σ-algebra $\langle B, \Sigma_B \rangle$ is a Σ-subalgebra of the Σ-algebra $\langle A, \Sigma_A \rangle$ if

i) $B \subseteq A$
ii) $f_B(\underline{b}) = f_A(\underline{b})$ for every $\underline{b}$ in B and f in Σ.

If $h: A \longrightarrow B$ is a Σ-homomorphism between Σ-algebras let $h(A) = \{h(a), a \in A\}$. Show $h(A)$ is a subalgebra of B. If h is surjective prove that $h(A)$ is isomorphic to B.

Q5 [Homomorphism Theorem] Let $h: A \longrightarrow B$ be a Σ-homomorphism between two Σ-algebras.

a) Show that the Σ-algebras $h(A)$ and $A/{=_h}$ are isomorphic.

b) Prove that there exists a factorization of h, $f: A \longrightarrow C$, $g: C \longrightarrow B$ for some Σ-algebra C such that f is surjective, g is injective, and $h(a) = g \circ f(a)$ for every $a \in A$.

c) Prove that B is isomorphic to $A/{=_h}$ if and only if h is surjective.

Q6 Extend the results in Q3, Q4, and Q5 to Σ-*po* algebras and conclude that $T_\Sigma/{\leq_A}$ is isomorphic to A if and only if A is surjective.

Q7 Let A be a Σ-*po* algebra. Show that the following are equivalent.

a) A is surjective

b) The only subalgebra of A is A itself

c) Every subalgebra of A is isomorphic to A.

Q8 Let $\mathscr{C}$ be a class of Σ-algebras and I initial in $\mathscr{C}$. Show that the only subalgebra of I in $\mathscr{C}$ is I itself.

Q9 A class of Σ-algebras $\mathscr{C}$ is closed with respect to subalgebras if whenever $A \in \mathscr{C}$ and B is a subalgebra of A then $B \in \mathscr{C}$ also. Show that if $\mathscr{C}$ is closed with respect to subalgebras and I is initial in $\mathscr{C}$ then I is surjective.

Q10 Generalize the definition of Σ-subalgebra to Σ-sub-*po*-algebra and prove the statements in Q4, Q5, Q7, and Q8 for Σ-*po* algebras in place of Σ-algebras.

Q11 If A and B are Σ-*po* algebras and A is surjective prove that there is at most one Σ-*po* homomorphism from A to B.

Q12 Show that $\langle N, \Sigma_N \rangle$ is initial in $\mathscr{C}(E_N)$ where E_N is the set of equations given at the end of §1.2.

Q13 Show that $\langle Z, \Sigma_Z \rangle$ is initial in $\mathscr{C}(E_Z)$ where E_Z is obtained from E_N by replacing the equation $Pred(Zero) = Zero$ with $Succ(Pred(X)) = X$ and adding $Plus(x, Pred(y)) = Pred(Plus(x, y))$.

Q14 Show that $\langle N, \leq_N, \Sigma_N \rangle$ of example 1.4.7 is initial in $\mathscr{C}(E'_N)$, where E'_N is the set of inequations given at the end of §1.4.

Q15 Let $\langle Z, \leq_Z, \Sigma_Z \rangle$ be the obvious Σ-*po* algebra obtained from the Σ-algebra $\langle Z, \Sigma_Z \rangle$. Show that it is initial in $\mathscr{C}(E'_Z)$ where E'_Z is obtained from E'_N in the same way as E_Z is obtained from E_N in Q12.

Q16 Show $\langle PAIRS, \Sigma_{PAIRS} \rangle$ and $\langle N, \Sigma_N \rangle$ are not isomorphic as Σ-algebras.

Q17 a) Using the Substitution lemma show that if A is surjective and $A \in \mathscr{C}(=_E)$, then $A \in \mathscr{C}(E)$.

b) A Σ-*po* algebra A *weakly satisfies* a set of inequations E if for every $\langle t, t' \rangle \in E$ and every closed substitution ρ, $i_A(t\rho) = i_A(t'\rho)$. Prove that if A is surjective it weakly satisfies E if and only if it satisfies E.

c) Prove that if $A \in \mathscr{C}(=_E)$ then it weakly satisfies E.

d) Give an example of a Σ-*po* algebra in $\mathscr{C}(=_E)$ but not in $\mathscr{C}(E)$.

Q18 Show that if there is a fully abstract interpretation with respect to R there is a surjective fully abstract interpretation with respect to R.

Q19 For any set of equations E let $\rho \leq \rho'$ if for every x in X $\rho(x) \leq_E \rho'(x)$.

a) Prove that $\rho \leq_E \rho'$ implies $t\rho \leq_E t\rho'$ for every term t.

b) Show that $\rho \leq_E \rho'$ and $t \leq_E u$ implies $t\rho \leq_E u\rho'$.

These generalized proof rules may be called *generalized substitution*.

2 Testing Processes

In this chapter we develop a behavioral theory of processes, called *Testing equivalence*, and apply it to our simple language for finite nondeterministic machines. A general framework for testing processes is developed in §2.1. Processes are inherently nondeterministic and therefore their reaction to tests may vary from application to application. Because of this, testing gives rise to three different behavioral relations: one in terms of the tests a process always passes, one in terms of those it sometimes passes; and the third is simply a conjunction of the first two. These three relations and the connections between them are discussed in §2.1.

One general method of generating these testing relations is discussed in §2.2. A natural method of giving a description of the operational nature of processes is in terms of *labeled transition systems*. A labeled transition system consists of a set of nodes, representing states of processes, and parametrized next-state relations over the nodes, representing the effects of performing actions. We show how labeled transition systems induce Testing relations. Then by giving an operational semantics to our process language we automatically obtain for it a behavioral theory based on Testing.

In §2.3 we give a new interpretation for our process language **fAT** (*finite Acceptance Trees*), and show that it is fully abstract with respect to one of the trinity of Testing relations. It lies somewhere between the two models defined in chapter 1, **PS** (prefix-closed strings) and **RT** (rooted trees), in terms of their ability to discriminate between processes. In the next section we prove that this model is characterized by a set of equations **A2**. However, this characterization requires a new nondeterministic operator $\oplus$, called *internal nondeterminism*. In §2.5 we revise the behavioral theory to encompass this new operator. This means that we end part I of the book with three different but equivalent views of an extended language $\mathbf{M_2}$:

—a behavioral view, Testing equivalence

—a denotational view, the model **fAT**

—a proof-theoretic view, the sound and complete proof system **DED**(**A2**).

2.1 The Testing Methodology

Perhaps the simplest use of a program is the following:

i) give input to the program

ii) wait for the program to return output.

This use of programs requires at least one and at most two interactions between the user and the program. The first is the communication of input from the user to the program and the second, which may never take place, is the communication of the result from the program to the user. This view of interaction between a user program may be analyzed further by considering the user as experimenting on the program. An experiment is simply a pair consisting of a relevant input and an expected output, $\langle i, o \rangle$. To perform the experiment on the program P we

i) give the input i to the program
ii) wait for the program to return output o'
iii) if output o' arrives then the experiment is a success if o' is the expected output o.

If the programs under consideration are deterministic, i.e. a specific input can lead to at most one output, the input-output behavior of a program is completely determined by the set of experiments which it passes. Contrapositively, we can say that two programs have different behavior if there is an experiment which one passes and the other does not. In other words, two programs are behaviorally equivalent unless there is an experiment which "shows" that they are different.

There are many details to be elaborated and made precise. For example, it is assumed that the only kind of behavior considered is that of input-output. It does not appear to be relevant that one program may take longer to produce an output than another. The set of possible experiments is not made explicit. As it varies, the corresponding testing equivalence induced on programs also varies. Moreover, this notion of testing for differences in behavior is not effective even though we have used terminology which suggests otherwise. For this reason the "show" above is in quotes. There are two ways in which a program can fail a test: i) it may never produce an output to be examined or ii) the output produced is different than that expected. There is no distinction made between these two different kinds of failure since only the successful experiments are taken into consideration. It is the former kind of failure which leads to the ineffective nature of testing. For consider the two programs:

P: input x;	P′: input x;
output 0	L: goto L.

They have different input-output behavior and consequently there is a test which distinguishes between them. For example, P passes the test $\langle 0, 0 \rangle$ whereas P′ fails it. However, it never becomes apparent that P′ fails this (or any other) test. After a finite amount of time one can conclude nothing vis-à-vis P′ and the test $\langle 0, 0 \rangle$. Consequently, although P and P′ are not equivalent with respect to our notion of testing, there is no effective test to differentiate between them. Because of Rice's Theorem (Roges 1967), this seems inevitable for any nontrivial notion of semantic equivalence.

When we allow programs to be nondeterministic the situation is immediately more complicated and one has to be more specific about the application of an experiment to a program. In general, when presented with an input, a nondeterministic program might produce one of a number of different outputs or, in addition, it might loop forever. In terms of experiments one can say that such a program P, for a specific experiment $\langle i, o \rangle$, can

i) only fail, i.e. every computation on the input i leads to failure

ii) only succeed, i.e. every computation on the input i leads to success

or iii) both fail and succeed, i.e. there are both failing computations and successful computations on the input i.

Our approach will extract certain of this information as being relevant. We will say:

P *may satisfy* the experiment $\langle i, o \rangle$ if situation ii) or iii) holds

P *must satisfy* the experiment $\langle i, o \rangle$ if situation ii) holds.

Then one may use these relations *may satisfy* and *must satisfy* to define differing notions of semantic equivalence or, more generally, of semantic preorder. The effect of different choices is exemplified using the following four nondeterministic programs. It is assumed that only one branch of an *or* statement can be executed.

P_1:	input x;	P'_1:	input x;
	$x := 1$;		$x := 1$ *or* loop forever;
	output x		output x
P_2:	input x;	P'_2:	input x;
	$x := 1$ *or* $x := 2$;		$x := 1$ *or* $x := 2$ *or* loop forever
	output x		output x.

If we only use the relation *may satisfy*, then P_1 and P'_1 cannot be differentiated (nor indeed can P_2 and P'_2). However P_1 *must satisfy* the experiment $\langle 0, 1 \rangle$ whereas this is not true of P'_1. So a semantic equivalence stated in terms of *must satisfy* will differentiate between them. It will also differentiate between P_1 and P_2. However, a semantic equivalence which only uses *must satisfy* will not differentiate between P'_1 and P'_2

because no experiment is guaranteed of success when applied to either program. However, the relation *may satisfy* differentiates between them since P_2' *may satisfy* the experiment $\langle 0, 2 \rangle$, which is not true of P_1'.

One could envisage more complicated forms of experiment, even for these kinds of programs. For example, the expected output *o* could be replaced by an output predicate Out. Then an application would be successful if the actual output produced satisfied the predicate Out. This form of testing involves some nontrivial active participation by the experimenter: presumably he must do some computation to evaluate the predicate Out on the output produced.

If we enlarge our domain of interest to machines, or processes, even more participation by the experimenter in tests can be contemplated. In general there will be a sequence of interactions between the experimenter and the process under investigation. For example, consider a machine which sorts numbers. At each point in time it can input a number or output a number. Its specification is such that at any point in time it is ready to output the largest number it has received which has not yet been output. An experiment on this machine would consist of a sequence of commands such as

INPUT 6, INPUT 10, INPUT 4, OUTPUT 10?

The application of this experiment involves a series of interactions between the experimenter and the machine. It would be successful if the experimenter gives as input the sequence 6, 10, 4 and when he then demands output receives the number 10. By performing many such experiments a picture of the true behavior of the machine may be constructed.

An experiment demands the continual interaction between the experimenter and the machine being examined. Between any two interactions the experiment could break down (i.e. fail) because the machine does not respond. However, the experimenter can only infer knowledge of the machine by virtue of interacting or communicating with it. So a particular application of an experiment is deemed successful not because the machine reaches a particular configuration, but rather because the experiment induces the experimenter into a particular state. In the above example the experimenter asks for output, receives it, sees that it is 10 and is happy: he deems the experiment a success.

It is this latter very general idea of testing which we wish to formalize. The former notion of test, involving a simple examination of input and subsequent output, is that special case where the number of interactions is limited to at most two.

Let *P*, *E*, be arbitrary sets of processes and experimenters, respectively. These can be interconnected in various ways and we use $\|$ to denote such an interconnection. The nature of this interconnection determines the possible interactions between experimenters and processes. This is represented by a relation

$\rightarrow \subseteq (E \times P) \times (E \times P)$.

We use $\rightarrow^*$ to denote the reflexive transitive closure of $\rightarrow$, and we write $\rightarrow$ in infix notation in the form

$e \parallel p \rightarrow e' \parallel p'$.

This indicates that when e and p interconnected via $\parallel$ they may interact. Moreover, this interaction changes the process p to p' and the experimenter e to e'; so that an experiment affects not only the process being examined but also the experimenter doing the examining. These experimenters may be humans using the processes or, indeed, may be other processes. (One uses a process by interacting or experimenting with it.) An experiment or test is then a sequence of possible interactions between the experimenter and the process. It is represented by a sequence of the form

$$e_0 \parallel p_0 \rightarrow e_1 \parallel p_1 \rightarrow e_2 \parallel p_2 \rightarrow \cdots \rightarrow e_k \parallel p_k \rightarrow \cdots . \qquad (*)$$

Formally we say such a sequence is a *computation* if it is maximal, i.e. it is infinite or it is finite with terminal element $e_n \parallel p_n$ which has the property that $e_n \parallel p_n \rightarrow e' \parallel p'$ for no pair e', p'. In short, it is a computation if it cannot be extended. The success of an experiment depends on the experimenter reaching a particular configuration. So we introduce a subset of E, called *Success*, and stipulate that the computation $(*)$ is *successful* if e_n is in *Success* for some $n \geq 0$. Therefore, a computation is successful if the experimenter passes through a *Success* state. We can now formulate Testing equivalences or preorders as outlined above for nondeterministic machines.

DEFINITION 2.1.1 An *Experimental System* ($\mathscr{E}\mathscr{S}$) is a collection of the form $\langle P, E, \rightarrow, Success \rangle$, where

i) P is an arbitrary set of processes
ii) E is an arbitrary set of experimenters
iii) $\rightarrow \subseteq (E \times P) \times (E \times P)$ is the interacting relation
iv) $Success \subseteq E$ is the success set.

For such an $\mathscr{E}\mathscr{S}$, and e, p in E, P, respectively, we let Comp(e, p) be the set of computations whose initial element is $e \parallel p$. Comp(e, p) is always nonempty; at worst it will contain the computation $\{e \parallel p\}$. We tabulate the possible results of applying e to p by using the symbols

$\top$ to denote a successful computation

$\bot$ to denote an unsuccessful computation.

So let Result$(e, p) \subseteq \{\top, \bot\}$ be defined by:

$\top \in$ Result(e, p) if Comp(e, p) contains a successful computation

$\bot \in$ Result(e, p) if Comp(e, p) contains an unsuccessful computation.

A natural equivalence between processes now suggests itself:

$p \sim p'$ if for every experiment e in E, Result$(e, p) =$ Result(e, p').

However, the conceptual apparatus we have introduced enables us to analyse the equivalence $\sim$ and formulate it in terms of more primitive relations. We write

p *may* e if $\top \in$ Result(e, p)

p *must* e if $\bot \notin$ Result(e, p).

This enables us to define three different preorders between the processes of $\mathcal{ES}$.

DEFINITION 2.1.2 For a given $\mathcal{ES}$, $\langle P, E, \rightarrow, Success \rangle$, let

i) $p \underset{\sim}{\sqsubset}_{MAY} p'$ if, for every e in E, p *may* e implies p' *may* e

ii) $p \underset{\sim}{\sqsubset}_{MUST} p'$ if, for every e in E, p *must* e implies p' *must* e

iii) $p \underset{\sim}{\sqsubset} p'$ if $p \underset{\sim}{\sqsubset}_{MAY} p'$ and $p \underset{\sim}{\sqsubset}_{MUST} p'$.

Notice that $\sim$ can quite easily be reconstituted from these preorders.

LEMMA 2.1.3 $p \sim p'$ if and only if $p \underset{\sim}{\sqsubset} p'$ and $p' \underset{\sim}{\sqsubset} p$.

That is, the natural equivalence can be explained in terms of "*may* satisfy a test" and "*must* satisfy a test" in a very intuitive manner. For the sake of uniformity we will use $\eqsim$ to denote the equivalence generated by $\underset{\sim}{\sqsubset}$, i.e. $p \eqsim q$ if and only if $p \underset{\sim}{\sqsubset} q$ and $q \underset{\sim}{\sqsubset} p$. Similar notation will be applied to $\underset{\sim}{\sqsubset}_{MAY}$ and $\underset{\sim}{\sqsubset}_{MUST}$. Moreover, these three testing preorders are not arbitrary choices but arise naturally from the three different powerdomains of the two-point lattice Θ:

$\top$
|
$\bot$

This represents the natural ordering on the set $\{\top, \bot\}$, success being better than failure. Result associates with each pair (e, p) a nonempty subset of $\{\top, \bot\}$ and there are three different ways of ordering or comparing these subsets, corresponding to the three powerdomain constructions. We examine each in turn.

I *The Hoare Powerdomain* In this case the sets are ordered as

$$\begin{array}{c} \{\top\} = \{\top, \bot\} \\ | \\ \{\bot\}. \end{array}$$

No distinction is made between the sets $\{\top, \bot\}$ and $\{\top\}$, and both are greater than $\{\bot\}$. This ordering reflects the view that possible failure is unimportant: it is only the possibility of success which counts. We represent this ordering between sets by $\underset{\sim}{\sqsubset}_{\mathrm{H}}$. It is very easy to establish that

$p \underset{\sim}{\sqsubset}_{\mathrm{MAY}} p'$ if and only if $\mathrm{Result}(e, p) \underset{\sim}{\sqsubset}_{\mathrm{H}} \mathrm{Result}(e, p')$ for every $e \in E$.

II *The Smyth Powerdomain* This ordering, $\underset{\sim}{\sqsubset}_{\mathrm{S}}$, is given by:

$$\begin{array}{c} \{\top\} \\ | \\ \{\top, \bot\} = \{\bot\}, \end{array}$$

representing the view that failure is catastrophic: one must be assured of success every time the experiment is applied. Once more it is trivial to establish that

$p \underset{\sim}{\sqsubset}_{\mathrm{MUST}} p'$ if and only if $\mathrm{Result}(e, p) \underset{\sim}{\sqsubset}_{\mathrm{S}} \mathrm{Result}(e, p')$ for every $e \in E$.

III *The Egli-Milner Powerdomain* Here the ordering $\underset{\sim}{\sqsubset}_{\mathrm{EM}}$ is given by

$$\begin{array}{c} \{\top\} \\ | \\ \{\top, \bot\} \\ | \\ \{\bot\}. \end{array}$$

Here the ordering represents a more reasoned view of failure, neither ignoring it nor taking it too seriously. Once more one can establish

$p \underset{\sim}{\sqsubset} p'$ if and only if $\mathrm{Result}(e, p) \underset{\sim}{\sqsubset}_{\mathrm{EM}} \mathrm{Result}(e, p')$ for every $e \in E$.

This completes our presentation of the behavioral view of processes or machines. It gives three different operationally motivated preorders on processes which can be used to justify particular choices of semantic interpretations. In part I we deal mainly with $\underset{\sim}{\sqsubset}$ while in parts 2 and 3 we concentrate on $\underset{\sim}{\sqsubset}_{\mathrm{MUST}}$, but throughout we indicate the uniform changes necessary to move from one preorder to another.

Our aim is to embed finite nondeterministic machines $\mathbf{M_l}$ in an $\mathcal{ES}$ and thereby obtain behavioral preorders $\sqsubseteq$, $\sqsubseteq_{MAY}$, $\sqsubseteq_{MUST}$ over machines. We then investigate the fully abstract interpretations as outlined at the end of §1.4. In further chapters we extend the rather trivial language of finite nondeterministic machines so as to encompass more complicated forms of machines, such as the communicating processes, described in the introduction.

2.2 Labeled Transition Systems

A convenient method of modeling the step-by-step operational semantics of a process or machine is by labeled transition systems. Processes evolve by executing actions. The nature of the actions depend on the nature of the processes under consideration. This evolution can be modelled by a binary relation denoted by $\longrightarrow$. We can read

$$p \longrightarrow p'$$

as: the process p may evolve to the process p' by performing some action. Processes may be capable of performing many different kinds of actions and if we wish to discriminate between the different kinds then we can use a different arrow for each possible action. So that if a is an action, then

$$p \xrightarrow{a} p'$$

may be read as: the process p may perform the action a and thereby be transformed into the process p'. This is the intuition underlying labeled transition systems.

Definition 2.2.1 A *labeled transition system* (*lts*) is a triple $\langle P, Act, \longrightarrow \rangle$, where

i) P is an arbitrary set of processes
ii) Act is an arbitrary set of actions
iii) $\longrightarrow$ is a relation in $P \times Act \times P$.

As indicated above, we write $p \xrightarrow{a} p'$ in place of $(p, a, p') \in \longrightarrow$. Moreover, the relations $\xrightarrow{a}$ are extended to relations $\xrightarrow{s}$, for every s in Act^*, in the obvious way:

a) $p \xrightarrow{\varepsilon} p'$ if p' is p
b) $p \xrightarrow{as} p'$ if $p \xrightarrow{a} p''$ for some p'' such that $p'' \xrightarrow{s} p'$.

This means $p \xrightarrow{s} p'$ if p can evolve to p' by performing the sequence of actions s. We also write $p \xrightarrow{s}$ to mean that there exists a p' such that $p \xrightarrow{s} p'$.

Labeled transition systems represent a rather primitive view of the operational nature of processes. However, by using them we can formulate a very useful type of experimental system. Let $\langle P, Act, \longrightarrow\rangle$ be an *lts*. To define an experimental system for the processes in P we need a set of experimenters to interact with them. This interaction affects not only the processes but also the experimenters. They evolve and consequently are endowed with some operational semantics which can also be modeled by an *lts* such as $\langle E, Act, \longrightarrow\rangle$. Notice we assume that experimenters perform the same kinds of actions as processes, i.e. the processes and experimenters are *compatible*. To model success we need to introduce some syntactic method for designating which states are successful. There are many ways of doing this and we choose one which is natural in our framework: we allow experimenters to perform a special action called w. Performing w can be interpreted as reporting success. So we assume that experimenters are modeled by an *lts* $\langle E, Act \cup \{w\}, \longrightarrow\rangle$. The success set *Success* is then $\{e \in E, e \stackrel{w}{\longrightarrow}\}$. To model the interaction between experimenters and processes we need to define a relation $\rightarrow$ between pairs of experimenters and processes. An obvious possibility is to let $\rightarrow$ be the least relation which, for arbitrary a in Act, satisfies

$$e \stackrel{a}{\longrightarrow} e', \quad p \stackrel{a}{\longrightarrow} p' \quad \text{implies} \quad e \parallel p \rightarrow e' \parallel p', \tag{$*$}$$

so that they interact by performing the same action. This gives the experimenter complete control over the process because, during a particular application, the latter can only perform a specific action if the experimenter allows it. Unfortunately this control works both ways since the process also has the same control over the experimenter. To break this symmetry, and allow the experimenter a natural independence from the process it is investigating we allow it to use a special action, 1, which is not constrained as in ($*$). Instead we have

$$e \stackrel{1}{\longrightarrow} e' \quad \text{implies} \quad e \parallel p \rightarrow e' \parallel p. \tag{$**$}$$

The experimenter e can perform this action without p being aware of it. It is an internal action and allows e a certain independence from p in that it does not require its cooperation to perform an internal computation. We summarize this discussion in the following definition.

DEFINITION 2.2.2 Let L_P, L_E, be two compatible *lts*'s, $\langle P, Act, \longrightarrow\rangle$, and $\langle E, Act \cup \{1, w\}, \longrightarrow\rangle$ respectively. Then $\mathscr{ES}(L_P, L_E)$ is the Experimental System $\langle P, E, \rightarrow, \mathit{Success}\rangle$ where $\rightarrow$ is defined by ($*$), ($**$), and $\mathit{Success} = \{e \in E, e \stackrel{w}{\longrightarrow}\}$.

The scenario outlined in the beginning of §2.1 does not constitute an Experimental System in this sense since the processes (i.e. the programs) being examined can compute independently of the will of the user, having obtained the input. We will see examples of such Experimental Systems later.

The Experimental System $\mathscr{E}\mathscr{S}(L_P, L_E)$ induces the three Testing preorders, discussed in the previous section, on the set of processes P of L_P. It is this approach we take to define the Testing preorders on the simple language $\mathbf{M_1}$ of finite nondeterministic machines. We define an *lts* for $\mathbf{M_1}$ which reflects the intuitive interpretations of terms. We use exactly the same *lts* as the experimenters' component of the Experimental System so that the Testing preorders reflect the ability of finite nondeterministic machines to test themselves.

The operational semantics for $\mathbf{M_1}$ is very straightforward and based on two principles:

—the term ap represents a machine which can perform the action a and thereby be transformed into the machine represented by p

—the term $p + q$ represents the machine which can act either like the machine represented by p or that represented by q.

So for each a in A let $\xrightarrow{a}$ be the least relation over $\mathbf{M_1}$ which satisfies

i) $ap \xrightarrow{a} p$

ii) a) $p \xrightarrow{a} p'$ implies $p + q \xrightarrow{a} p'$

b) $q \xrightarrow{a} q'$ implies $p + q \xrightarrow{a} q'$

Note that this is an inductive definition and so we have a powerful proof method for deriving properties of these relations. We will see examples of its use later.

The relations are as one expects. For example

$$(ab + a(c + d)) + c \xrightarrow{a} c + d.$$

However, to prove it from the rules i) and ii) is not completely trivial:

$(ab + a(c + d)) + c \xrightarrow{a} c + d$ by rule ii) a)

because $ab + a(c + d) \xrightarrow{a} c + d$ by rule ii) b)

because $a(c + d) \xrightarrow{a} c + d$ by rule i).

In general to prove $p \xrightarrow{a} p'$ the rules ii) a), ii) b) must be systematically applied until finally rule i) is applicable.

Note that NIL can perform no action, i.e. there exists no p and no a such that $NIL \xrightarrow{a} p$. This is because such a statement cannot be derived from an application of either of the defining rules for the relation $\xrightarrow{a}$.

Using these relations we have that $\langle \mathbf{M_1}(Act), Act, \longrightarrow\rangle$ is an *lts*. The corresponding experimenter's *lts* is $\langle \mathbf{M_1}(Act \cup \{1, w\}), Act \cup \{1, w\}, \longrightarrow\rangle$. Many of these experimenters have very intuitive interpretations. For example *abw* attempts to find out whether a process can perform the sequence of actions *ab*, and $a(bw + cw)$ will be happy if the machine being examined can perform either the sequence *ab* or *ac*. The experimenter *abw* corresponds to the intuitive experiment used in §1.4. However, it is more difficult to interpret experimenters such as $abw + acw$, $abw + c$, $aw + ab$.

EXAMPLES 2.2.3

i. Let p_1, p_2, e_1 be $aNIL$, $a + b$, $aw + b$ respectively. Then p_1 *may* e_1, p_2 *may* e_1, p_1 *must* e_1 but p_2 $mu\not{s}t$ e_1 (i.e. p_2 *must* e_1 is false) because of the unsuccessful computation $a + b \parallel aw + b \rightarrow NIL \parallel NIL$, caused by both processes performing the action b.

ii. Let p_3, p_4, e_2 be $a(b + c)$, $ab + ac$, abw respectively. Then p_3 *may* e_2, p_4 *may* e_2, p_3 *must* e_2 but p_4 $mu\not{s}t$ e_2 because of the unsuccessful computation $(ab + ac) \parallel abw \rightarrow c \parallel bw$. This is essentially the example outlined in §1.4 to show that **PS** is an inadequate model for nondeterministic processes.

iii. Let p_5, p_6, e_3 be $a(b + c) + ad$, $ab + a(c + d)$, $a(cw + dw)$. Then p_5 *may* e_3, p_6 *may* e_3, p_5 *must* e_3 but p_6 $mu\not{s}t$ e_3 because of the unsuccessful computation $ab + a(c + d) \parallel a(cw + dw) \rightarrow b \parallel cw + dw$.

iv. Let p_7, p_8, e_4 be $aNIL$, ab, $a(1w + b)$ respectively. Then p_7 *may* e_4, p_8 *may* e_4, p_7 *must* e_4, but p_8 $mu\not{s}t$ e_4 because of the unsuccessful computation $ab \parallel a(1w + b) \rightarrow b \parallel 1w + b \rightarrow NIL \parallel NIL$.

v. Let p_9, p_{10}, e_5 be NIL, $aNIL$, $1w + a$ respectively. Then p_9 *may* e_5, p_{10} *may* e_5, p_9 *must* e_5 but p_{10} $m\not{u}st$ e_5 because of the unsuccessful computation $a \parallel 1w + a \rightarrow NIL \parallel NIL$.

We use $\mathscr{ES}(\mathbf{M_1})$ as an abbreviation for the experimental system $\mathscr{ES}(\langle \mathbf{M_1}(Act), Act, \longrightarrow\rangle, \langle \mathbf{M_1}(Act \cup \{1, w\}), Act \cup \{1, w\}, \longrightarrow\rangle)$. Then $\mathscr{ES}(\mathbf{M_1})$ induces three Testing preorders on $\mathbf{M_1}$, which we denote by $\underset{\sim}{\sqsubset}_{MAY}$, $\underset{\sim}{\sqsubset}_{MUST}$, $\underset{\sim}{\sqsubset}$, respectively. Strictly speaking, these symbols should be superscripted by $\mathscr{ES}(\mathbf{M_1})$ but for convenience it is omitted.

EXAMPLES 2.2.4

i) $a \underset{\sim}{\sqsubset}_{MAY} a + b$, $a + b \not\underset{\sim}{\sqsubset}_{MAY} a$, $a \not\underset{\sim}{\sqsubset}_{MUST} a + b$.

ii) $a(b + c) \eqsim_{MAY} ab + ac$, $a(b + c) \not\underset{\sim}{\sqsubset}_{MUST} ab + ac$.

iii) $a(b + c) + ad \eqsim_{MAY} ab + a(c + d)$, $a(b + c) + ad \not\underset{\sim}{\sqsubset}_{MUST} ab + a(c + d)$.

iv) $a \underset{\sim}{\sqsubset}_{MAY} ab$, $ab \not\underset{\sim}{\sqsubset}_{MAY} a$, $a \not\underset{\sim}{\sqsubset}_{MAY} ab$.

v) $NIL \underset{\sim}{\sqsubset}_{MAY} a$, $a \not\underset{\sim}{\sqsubset}_{MAY} NIL$, $NIL \not\underset{\sim}{\sqsubset}_{MUST} a$.

In this series of examples the justifications for the negative statements about $\underset{\sim}{\sqsubset}_{MUST}$ are given in the corresponding parts of examples 2.2.3. We leave the reader to justify the negative statements about $\underset{\sim}{\sqsubset}_{MAY}$. To show that two processes are not related it is sufficient to exhibit one experimenter which differentiates between them in the appropriate way. However, to show a positive statement one must consider the effect of all possible experimenters. In the *may* case this is not too difficult: satisfying an experimenter amounts to having a specific sequence of actions. So we leave the readers to convince themselves of the statements about $\underset{\sim}{\sqsubset}_{MAY}$ given above. The relation $\underset{\sim}{\sqsubset}_{MUST}$ is much more difficult to analyze, however, as one must consider all possible effects of all possible experimenters.

EXAMPLE 2.2.5 $ab + ac \underset{\sim}{\sqsubset}_{MUST} a(b + c)$.

For suppose $ab + ac$ *must* e. We must show $a(b + c)$ *must* e. If e can perform w immediately, then this is obvious as p *must* e for any p. Otherwise consider the effect of executing any sequences of 1 actions. Either somewhere in that sequence it can perform w or it must eventually perform an a action to become e', say. Now no matter what e' is, it must be able either eventually (after performing 1 actions) to be capable of performing w or perform either b or c to become e'', which must eventually be capable of w. This is sufficient to establish that $a(b + c)$ *must* e.

Such arguments are nontrivial and we need an easier method of establishing whether or not two processes are related via $\underset{\sim}{\sqsubset}_{MAY}$, $\underset{\sim}{\sqsubset}_{MUST}$, and $\underset{\sim}{\sqsubset}$. We give new relations $<<_{MAY}$, $<<_{MUST}$, and $<<$ which actually coincide with the former but which are easier to comprehend. They are given only in terms of the operational semantics of terms themselves, not in terms of how they can affect other processes. Before defining these new relations the reader might like to establish directly whether any of the following pairs are related via any of the Testing preorders.

EXAMPLE 2.2.6

a) $abc + abd$, $a(bc + bd)$

b) $a(b + c + d) + ac$, $a(b + c + d) + ac + a(b + d)$

c) $ab + a(b + c)$, $a(b + c)$

d) $ab + ac$, $ab + ac + a(b + c)$

e) $a(bc + d) + a(be + f)$, $a(be + d) + a(bc + f)$.

The following notation will be very convenient.

DEFINITION 2.2.7 For any *lts* $\langle P, Act, \longrightarrow\rangle$ and $p \in P$, $s \in Act^*$ respectively, let

i) $D(p, s) = \{p' \in P, p \xrightarrow{s} p'\}$, the *s-derivatives* of p

ii) $L(p) = \{s, p \xrightarrow{s}\}$, the *language* of p

iii) $D(p) = \{p', \text{for some } a \in Act, p \xrightarrow{a} p'\}$, the *derivatives* of p

iv) $S(p) = \{a, p \xrightarrow{a}\}$, the *successors* of p

v) $S(p, s) = \{a, p \xrightarrow{s} p' \xrightarrow{a}\}$, the *successors of p after s*

vi) $\mathscr{A}(p, s) = \{S(p'), p \xrightarrow{s} p'\}$, the *Acceptance sets of p after s*.

For example, in the *lts* $\langle \mathbf{M}_1, Act, \longrightarrow\rangle$ if p is $a(b + c) + ad + ab + ca$ then

$S(p) = \{a, c\}$, $\quad S(p, a) = \{b, c, d\}$, $\quad S(p, c) = \{a\}$, $\quad S(p, d) = \emptyset$,

$L(p) = \{\varepsilon, a, c, ab, ac, ad, ca\}$; $\quad \mathscr{A}(p, \varepsilon) = \{\{a, c\}\}$, $\quad \mathscr{A}(p, a) = \{\{b, c\}, \{d\}, \{b\}\}$,

$\mathscr{A}(p, c) = \{\{a\}\}$, $\quad \mathscr{A}(p, b) = \emptyset$ and $\mathscr{A}(p, ab) = \{\emptyset\}$.

$L(p)$, the language generated by p, is always nonempty and prefix-closed. $S(p, s)$ is the set of next possible actions that p can perform, having performed s; note that $S(p)$ is $S(p, \varepsilon)$. $\mathscr{A}(p, s)$ gives some structure to this set of possibilities. Every set S in $\mathscr{A}(p, s)$ is a subset of $S(p, s)$ and can be viewed as a representation of a possible state that p can attain by performing the sequence s. It represents the state in which p can only proceed by performing some action from S. Note that in our particular *lts* $\langle \mathbf{M}_1, Act, \longrightarrow\rangle$, $\mathscr{A}(p, \varepsilon)$ is always $\{S(p)\}$, i.e. the processes are always initially stable or deterministic.

We will use these Acceptance sets to define alternative characterizations of the Testing preorders. If $\mathscr{A}$, $\mathscr{B}$ are Acceptance sets, i.e. collections of sets of actions, we write $\mathscr{A} \subset\subset \mathscr{B}$ if for every $A \in \mathscr{A}$ there exists some $B \in \mathscr{B}$ such that $B \subseteq A$. For example,

$\{\{a\}\} \subset\subset \{\{a\}, \{a, b\}\}$ but not $\{\{a\}\{a, b\}\} \subset\subset \{\{b\}, \{a, b, c\}\}$.

Note that $\mathscr{A} \subset\subset \mathscr{B}$ is trivially true if $\mathscr{A}$ is the empty set.

DEFINITION 2.2.8 In any *lts*

i) $p <<_{\text{MAY}} p'$ if $L(p) \subseteq L(p')$

ii) $p <<_{\text{MUST}} p'$ if $\mathscr{A}(p', s) \subset\subset \mathscr{A}(p, s)$ for every s in Act^*.

iii) $p << p'$ if both $p <<_{\text{MAY}} p'$ and $p <<_{\text{MUST}} p'$.

These relations, which are in fact preorders, are very straightforward. However there is a subtlety in the definition of $<<_{MUST}$. For example, we need only compare $\mathscr{A}(p', s)$ to $\mathscr{A}(p, s)$ for $s \in L(p')$ for otherwise $\mathscr{A}(p', s)$ is empty. Also the next lemma shows that $p <<_{MUST} p'$ implies $L(p') \subseteq L(p)$ so that if $p << p'$ then $L(p) = L(p')$. This in turn implies that $S(p, s) = S(p', s)$ for every s in Act^* since $a \in S(p, s)$ if and only if $sa \in L(p)$.

LEMMA 2.2.9 If $p <<_{MUST} p'$ then $L(p') \subseteq L(p)$.

Proof Suppose s is in $L(p')$, i.e. $p' \xrightarrow{s} r$ for some r. Then $\mathscr{A}(p', s)$ is nonempty. From the definition of $<<_{MUST}$ it follows that $\mathscr{A}(p, s)$ is nonempty, i.e. s is in $L(p)$. □

The relations $<<_{MAY}$, $<<_{MUST}$ are much easier to manipulate than the corresponding $\sqsubseteq_{MAY}$, $\sqsubseteq_{MUST}$. For example, the following is straightforward.

LEMMA 2.2.10 Both $<<_{MAY}$ and $<<_{MUST}$ are Σ^1-preorders.

Proof We leave the reader to check $<<_{MAY}$. We consider $<<_{MUST}$ only. It is straightforward to prove that it is a preorder. We show that it is preserved by the operators in Σ^1. The case of the nullary operator is vacuous so that we have two to consider, prefixing and $+$.

a) Suppose $p <<_{MUST} p'$. We show $ap <<_{MUST} ap'$. Suppose S is in $\mathscr{A}(ap', s)$. Then s must be either ε or of the form as' where s' is in $L(p')$. In the former case S can only be $\{a\}$ and the required element of $\mathscr{A}(ap, \varepsilon)$ is also $\{a\}$. In the latter case S must be in $\mathscr{A}(p', s')$ and the result follows from the fact that $p <<_{MUST} p'$.

b) Suppose $p <<_{MUST} p'$, $q <<_{MUST} q'$. We must show that $p + q <<_{MUST} p' + q'$. Suppose S is in $\mathscr{A}(p' + q', s)$. If s is not ε then S must be in either $\mathscr{A}(p', s)$ or $\mathscr{A}(q', s)$. Without loss of generality we can suppose the former, and the result follows from the fact that $p <<_{MUST} p'$. This leaves the case when s is ε. Then S can only be $S(p') \cup S(q')$. Since $p <<_{MUST} p'$, $q <<_{MUST} q'$ we have that $S(p) \subseteq S(p')$ and $S(q) \subseteq S(q')$ from which the result follows since $S(p) \cup S(q)$ is the only set in $\mathscr{A}(p + q, \varepsilon)$. □

We now proceed to prove that these two preorders coincide with the Testing preorders $\sqsubseteq_{MAY}$ and $\sqsubseteq_{MUST}$. In doing so we will establish an interesting result about the power of experimenters. The Testing preorders are defined relative to a set of experimenters. As this set changes the resulting preorders change, at least in principle. However, some experimenters are completely useless in that they have no distinguishing power; examples are processes such as $abNIL$ which contain no occurrences of the reporting action w. What kind of experimenters may we drop and still retain the power of discrimination of the complete language? What experimenters are essential? We let $\sqsubseteq^T$, $\sqsubseteq^T_{MAY}$, $\sqsubseteq^T_{MUST}$ denote the relativized preorders. For example

$p \sqsubseteq_{\text{MUST}}^{T} q$ if for every $e \in T$ p *must* e implies q *must* e.

In establishing that $<<$ and $\sqsubseteq$ coincide (and similarly for the other pairs) we will also prove that $\sqsubseteq$ coincides with $\sqsubseteq^{T}$ for a rather small set of experimenters T.

One essential type of test is of the form

$$1w + b_1(1w + \cdots + b_n(1w + a)\ldots).$$

(If $n = 0$ this is simply $1w$.) We use $e(s, a)$ to represent it, where s is the sequence $b_1 \ldots b_n$. It is easy to check that

$$p \textit{ must } e(s, a) \quad \text{if and only if} \quad a \notin S(p, s).$$

Another essential type of test is of the form

$$1w + b_1(1w + b_2(1w + \cdots + b_n(a_1 w + \cdots + a_k w)\ldots).$$

This we denote by $e(s, A)$, where A is the set of actions $\{a_1, \ldots, a_k\}$. Let E denote the set of all experiments of the form $e(s, a)$ or $e(s, A)$.

PROPOSITION 2.2.11 $p \sqsubseteq_{\text{MUST}}^{E} p'$ implies $p <<_{\text{MUST}} p'$.

Proof We must show $\mathscr{A}(p', s) \subset\subset \mathscr{A}(p, s)$, and it is sufficient to consider s such that $s \in L(p')$. First note that $s \in L(p)$. If $s = \varepsilon$ this is immediate; otherwise s has the form $s'a$ and $s \notin L(p)$ implies p *must* $e(s', a)$. It would follow that p' *must* $e(s', a)$ which contradicts the fact that $s \in L(p')$. So we may assume that $\mathscr{A}(p, s)$ is not empty; let $S_1, \ldots, S_k$ be an arbitrary enumeration. Now suppose $\mathscr{A}(p', s) \subset\subset \mathscr{A}(p, s)$ is false; we derive a contradiction. If it is false there exists some $A \in \mathscr{A}(p', s)$ such that for every $1 \leq i \leq k$ there exists some $a_i \in S_i$ and $a_i \notin A$. Then p *must* $e(s, A)$, where A is $\{a_1, \ldots, a_k\}$ and p' *mu$t* $e(s, L)$ because of the unsuccessful computation

$$e(s, L) \parallel p \rightarrow^{*} (a_1 w + \cdots + a_k w) \parallel r,$$

where $p \xrightarrow{s} r$ and $S(r) = A$. However, this contradicts the fact that $p \sqsubseteq_{\text{MUST}} p'$. □

The converse is implied by the more general characterization theorem:

THEOREM 2.2.12 (Alternative Characterization of Testing Preorders) For every p, p' in $\mathbf{M}_1$

a) $p \sqsubseteq_{\text{MAY}} p'$ if and only if $p <<_{\text{MAY}} p'$

b) $p \sqsubseteq_{\text{MUST}} p'$ if and only if $p <<_{\text{MUST}} p'$

c) $p \sqsubseteq p'$ if and only if $p << p'$.

Proof We leave the reader to establish a). We prove b) from which c) follows immediately.

b) Because of the previous proposition it is sufficient to prove $p <<_{\text{MUST}} p'$ implies $p \sqsubseteq_{\text{MUST}} p'$. So suppose $p <<_{\text{MUST}} p'$ and p *must* e. We must prove p' *must* e. Let

$$e \parallel p' = e_0 \parallel p'_0 \rightarrow e_1 \parallel p'_1 \rightarrow \cdots \rightarrow e_k \parallel p'_k$$

be an arbitrary computation from $e \parallel p'$, which means that it cannot be extended. We must prove that $e_n \in Success$ for some $0 \leq n \leq k$. This computation gives rise to two derivations

$$e \xrightarrow{s'} e_k, \quad p' \xrightarrow{s} p'_k$$

for some s in Act^*. Here we use s' to indicate that the derivation of actions from s may be interspersed with 1 derivations. From lemma 2.2.9 s is in $L(p)$ and so we may use the fact that $p <<_{\text{MUST}} p'$ to obtain some S' in $\mathscr{A}(p, s)$ such that $S' \subseteq S(p'_k)$. Since S' is in $\mathscr{A}(p, s)$ there exists some r such that $p \xrightarrow{s} r$ and $S(r) \subseteq S(p'_k)$. So $e_k \parallel r$ cannot be extended and therefore the two sequences $e \xrightarrow{s'} e_k$, $p \xrightarrow{s} r$ can be combined to give a computation $e \parallel p \rightarrow \cdots \rightarrow e_k \parallel r$. Since p *must* e it follows that $e_n \in Success$ for some $0 \leq n \leq k$. □

It is an immediate corollary that $\sqsubseteq_{\text{MUST}}$ coincides with $\sqsubseteq^E_{\text{MUST}}$; adding experimenters to the basic set E does not improve the discriminating power of the Testing preorder. It is simple to extend this result to show that $\sqsubseteq$ and $\sqsubseteq^E$ also coincide for a slight extension E' of E. Indeed for *may*-testing a very simple form of experiment suffices, those of the form $b_1 \ldots b_k w$. Finally, we should point out that although this theorem is stated for processes in $\mathbf{M_1}$, it is quite general. It remains true in any experimental system generated from *lts*'s, provided the *lts* of experimenters contains at least experimenters which behave the same way as the processes in E. For example, it will remain true when we extend the language $\mathbf{M_1}$.

With this characterization theorem the examples 2.2.6 can be reexamined and now deciding whether or not the pairs of processes are related should be straightforward. For example, consider the third pair, $ab + a(b + c)$ and $a(b + c)$. Since they generate the same language $ab + a(b + c) \eqsim_{\text{MAY}} a(b + c)$: also $ab + a(b + c) \sqsubseteq_{\text{MUST}} a(b + c)$ since $ab + a(b + c) <<_{\text{MUST}} a(b + c)$. To establish the latter it is sufficient to compare $\mathscr{A}(ab + a(b + c), s)$ and $\mathscr{A}(a(b + c), s)$ for $s = \varepsilon, a, ab, ac$. They are identical except for the case $s = a$ in which case they are $\{\{b\}, \{b, c\}\}$ and $\{\{b, c\}\}$ respectively. Also $a(b + c) \not\sqsubseteq_{\text{MUST}} ab + a(b + c)$ since $\{b\}$ is in $\mathscr{A}(ab + a(b + c), a)$ but the only element in $\mathscr{A}(a(b + c), a)$ is $\{b, c\}$ which is not contained in $\{b\}$.

As an immediate corollary of this semantic characterization we have that the three

Testing preorders are in fact Σ^1-precongruences. By the remarks at the end of chapter 1 we have that $\mathbf{M_1}/\underset{\sim}{\sqsubset}_{MAY}$, $\mathbf{M_1}/\underset{\sim}{\sqsubset}_{MUST}$, and $\mathbf{M_1}/\underset{\sim}{\sqsubset}$ are fully abstract with respect to $\underset{\sim}{\sqsubset}_{MAY}$, $\underset{\sim}{\sqsubset}_{MUST}$, $\underset{\sim}{\sqsubset}$ respectively. This does not shed much light on these preorders unless we can find more illuminating representations of these interpretations. The *may* case is very simple for it is easy to establish that $\mathbf{M_1}/\underset{\sim}{\sqsubset}_{MAY}$ is essentially the string model **PS**, endowed with the subset ordering. As a corollary the proof system $\mathbf{DED}(E_1)$ is sound and complete for $\eqsim_{MAY}$.

We wish to establish similar results for $\underset{\sim}{\sqsubset}_{MUST}$ and $\underset{\sim}{\sqsubset}$ and we concentrate on the latter. We give a *Σ-po* algebra **fAT** (finite Acceptance Trees), isomorphic to the fully abstract interpretation $\mathbf{M_1}(A)/\underset{\sim}{\sqsubset}$ which is very intuitive and simple to understand. Later we will see that it is initial for a set of inequations, which also gives a sound and complete proof system for $\underset{\sim}{\sqsubset}$. The preorder $\underset{\sim}{\sqsubset}_{MUST}$ can be treated in an analogous fashion.

2.3 The Interpretation fAT

In this section we design an interpretation which is fully abstract with respect to the Testing preorder $\underset{\sim}{\sqsubset}$. It lies somewhere between the interpretations **PS** and **RT**. It should associate with every process just sufficient information to determine it with respect to $\underset{\sim}{\sqsubset}$ and the key to the definition is the alternative relation $<<$. For emphasis we reiterate an equivalent form of its definition:

$p << p'$ if i) $L(p) = L(p')$
ii) for every s in $L(p')$, if $S' \in \mathcal{A}(p', s)$ then there exists some $S \in \mathcal{A}(p, s)$ such that $S \subseteq S'$.

So intuitively **fAT**, the interpretation we construct, associates with every process the language it generates and with every sequence in this language some "*Acceptance set*." Elements of **fAT** can be viewed as certain kinds of rooted trees or as annotated sets of sequences. We develop both these views simultaneously.

Viewed as rooted trees both the branches and the nodes are labeled. The branches are labeled by actions from *Act*. The tree

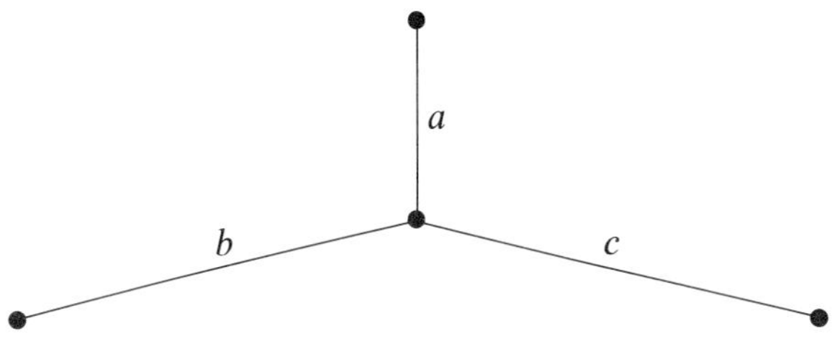

represents a process which can perform the action a and then perform either b or c. We view this machine as being *deterministic*. At each point in time its behavior depends entirely on the user. Initially it can only perform a. Then if requested to perform b it must do so. Similarly if requested to perform c it must do so. Trees of the form

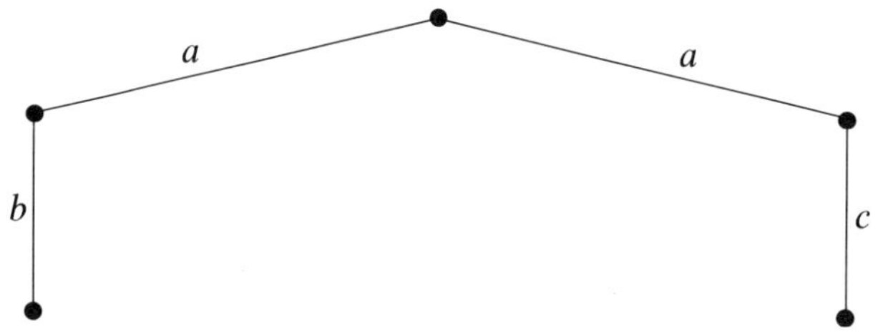

are *not* allowed although such trees can be used to represent machines with inherently nondeterministic behavior. Instead we stipulate that

> R1 For every action a, every node in the tree has at most one successor branch labeled by a.

Because of this requirement, every node in the tree is uniquely determined by a string in Act^*. We denote this set by $L(t)$ and if $s \in L(t)$ we use $t(s)$ to denote the node identified by s. Note that $L(t)$ is prefix-closed, i.e. if $s \in L(t)$ and s' is a prefix of s, then $s' \in L(t)$ also. The set of actions labeling the successor branches of a node n is called its *successor set* and is denoted by $S(n)$. This notation has an analogue in the view of a tree as a set of strings. For any $L \subseteq Act^*$ and s in Act^* let $S(L, s) = \{a, sa \in L\}$. Then it is easy to check that for any tree t, $S(t(s)) = S(L(t), s)$. Moreover, as we shall see, this notation of successor sets is deliberately chosen so as to facilitate the comparison between the behavior of a machine and its denotation in **fAT**. We put a natural restriction on successor sets.

> R2 (*Finite Branching*) For every s in $L(t)$, $S(L(t), s)$ is finite.

Informally speaking, every node has a finite number of successors. This is justified by the fact that every machine in $\mathbf{M}_1$ can perform, at any one time, a finite number of different actions. Formally $S(p, s)$ is always finite. To model the nondeterministic behavior of processes we label the nodes of trees by nonempty subsets of Act. The tree

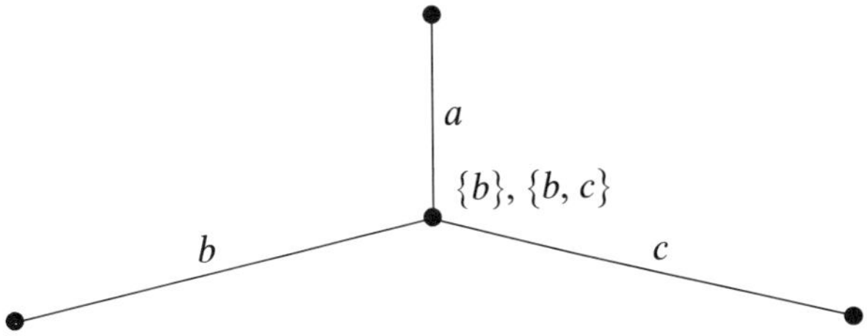

is a typical example. It represents a process which can perform the sequence of actions *ab* or *ac*. After performing the action *a* it can be in one of two internal states represented by the sets $\{b\}$, $\{b, c\}$, respectively. If the process is in the state represented by $\{b, c\}$ then when requested to perform *b* it must perform it, and similarly for *c*. If on the other hand it is in the state represented by $\{b\}$, then it can only perform the action *b*; it can not perform *c*. As we shall see, it is the interpretation of the process $a(b + c) + ab$ from $\mathbf{M_1}$. These sets labeling the nodes, called *Acceptance sets*, must satisfy certain criteria. For $S \subseteq Act$ an *S-set* $\mathscr{A}$ is a nonempty set of subsets of *Act*, which satisfies

i. for every K in $\mathscr{A}$, $K \subseteq S$
ii. for every a in S, there is some K in $\mathscr{A}$ such that a is in K
iii. ($\cup$-closed) if K_1, K_2 are in $\mathscr{A}$ then $K_1 \cup K_2$ is also in $\mathscr{A}$
iv. (convex-closed) if K_1, K_2 are in $\mathscr{A}$ and $K_1 \subseteq K \subseteq K_2$ then K is also in $\mathscr{A}$.

Note that if $\mathscr{A}$ is an S-set and S is finite, then S is in $\mathscr{A}$. This follows from requirements ii) and iii). Indeed, an alternative definition of S-set would be any nonempty set of subsets of S, $\mathscr{A}$, which satisfies

i′) $S \in \mathscr{A}$
ii′) $X \subseteq Y \subseteq S$, $X \in \mathscr{A}$ implies $Y \in \mathscr{A}$.

However, we find its former definition slightly more illuminating. If $\mathscr{A}$ is a set of subsets we say it is *saturated* if it is an S-set for some S. With regard to the interpretation **fAT**, we require that the acceptance set associated with the node n, $\mathscr{A}(n)$, satisfy

R3 $\mathscr{A}(n)$ is an $S(n)$-set.

Because of R2 this immediately implies that $\mathscr{A}(n)$ is a finite set of finite sets and moreover it must contain $S(n)$. If $S(n)$ is the only set in $\mathscr{A}(n)$ then the tree is *deterministic* at that node. When describing trees we omit these labels from the deterministic nodes. This makes the trees more readable. We also abbreviate $\mathscr{A}(t(\varepsilon))$ to $\mathscr{A}(t)$ and $S(t(\varepsilon))$ to $S(t)$.

DEFINITION 2.3.1 $\mathbf{fAT}_{Act}$, the set of finite acceptance trees over Act, is the set of finite rooted trees whose branches are labeled by elements of Act, whose nodes are labeled by subsets of Act^*, and which satisfies the requirements R1, R2, R3.

For convenience we render $\mathbf{fAT}_{Act}$ as **fAT**; the Act will always be clear from the context. Examples of elements from **fAT** are given in figure 2.1.

We now endow **fAT** with a partial order. Intuitively $t \leq_{\mathbf{fAT}} t'$ if they denote more or less the same process except that t' is "more deterministic than" t. Formally we define $t \leq_{\mathbf{fAT}} t'$ if

i) $L(t) = L(t')$
ii) for every s in $L(t')$, $\mathscr{A}(t'(s)) \subseteq \mathscr{A}(t(s))$.

So, to make a tree more deterministic, one omits some of the Acceptance sets, thereby eliminating some of the internal states. The following lemma is trivial.

LEMMA 2.3.2 $\langle \mathbf{fAT}, \leq_{\mathbf{fAT}} \rangle$ is a partial order.

Examples of $\leq_{\mathbf{fAT}}$ are given in figure 2.1; counterexamples are given in figure 2.2. It is worth remarking that the first condition of the definition of $\leq_{\mathbf{fAT}}$ could be replaced by $L(t) \subseteq L(t')$ as the converse is implied by the second condition.

Next we define functions over **fAT** for every function symbol in Σ^1.

1. Let $NIL_{\mathbf{fAT}}$ denote the trivial tree $\bullet\{\emptyset\}$ which consists of one node, labeled by the set consisting of the empty set, and which has no branches. Using the convention about deterministic nodes it can be rendered as $\bullet$.

2. The function $a_{\mathbf{fAT}}$ is straightforward. It maps the tree

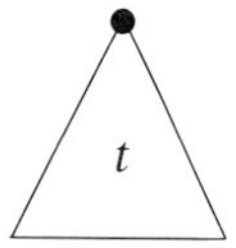

to the tree

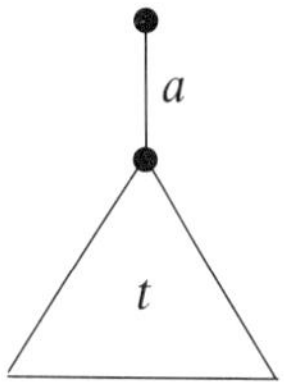

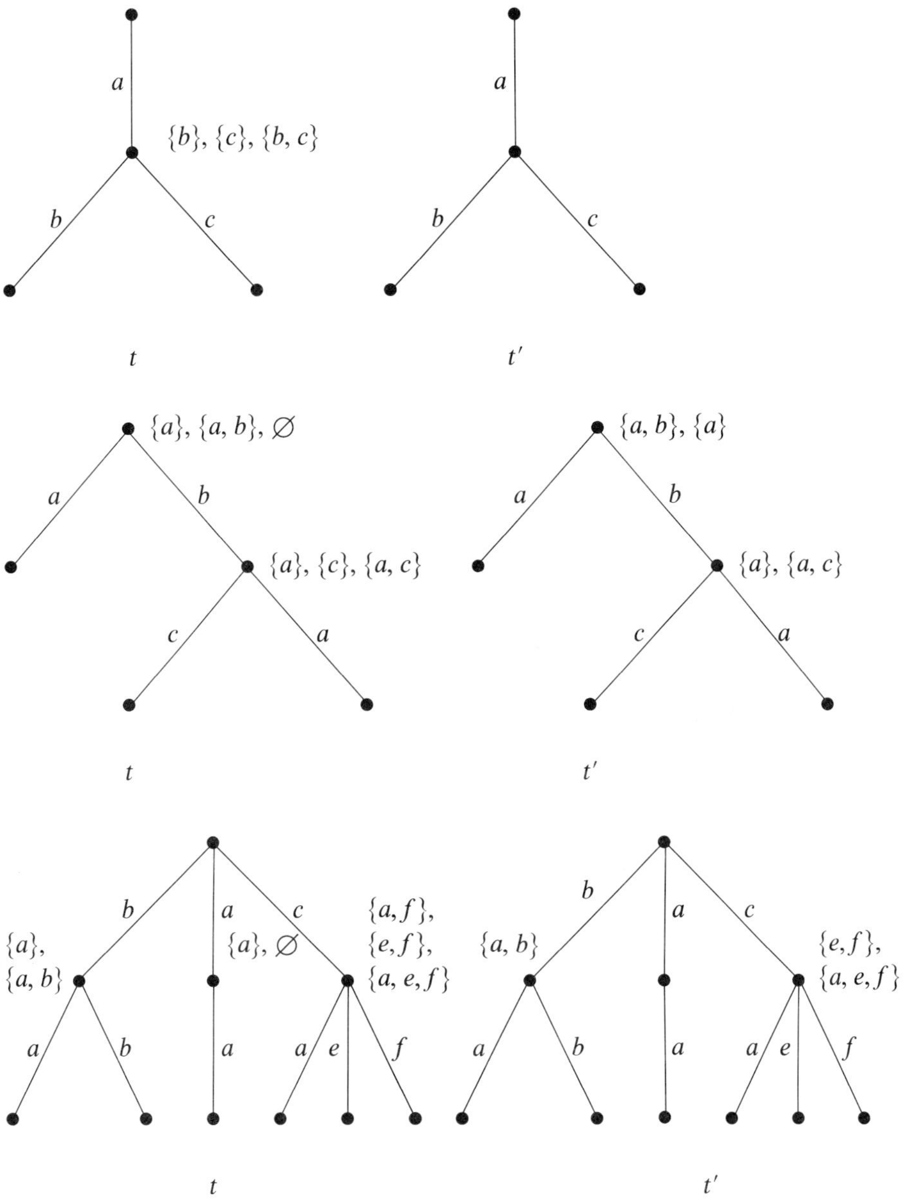

Figure 2.1
Examples from **fAT**. In each case $t \leq_A t'$.

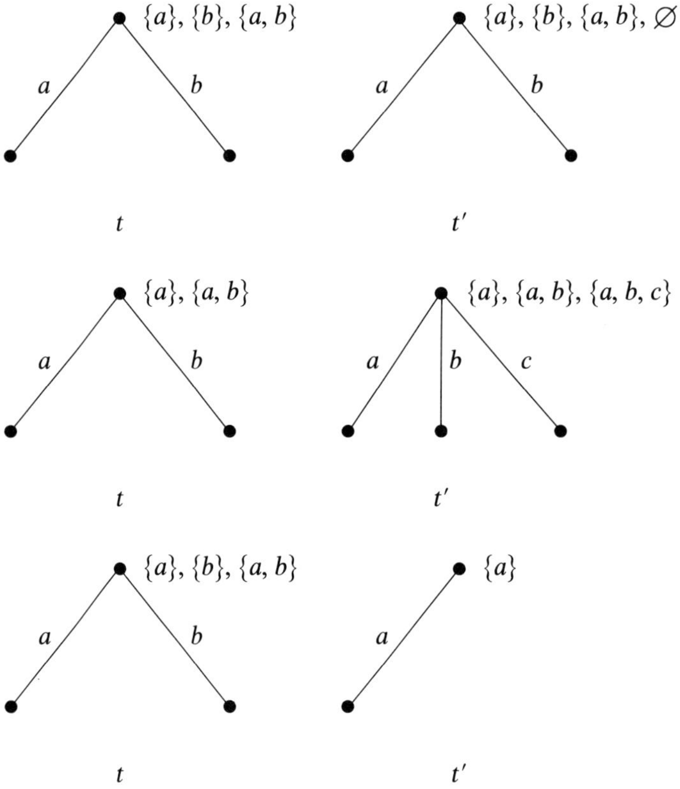

Figure 2.2
Counterexamples from **fAT**. In each $t \not\leq t'$.

Formally if t is in **fAT** let $a_{\mathbf{fAT}}(t)$ be the tree t' determined by

i) $L(t') = \{\varepsilon\} \cup \{as, s \in L(t)\}$

ii) $\mathscr{A}(t'(\varepsilon)) = \{\{a\}\}$, $\quad \mathscr{A}(t'(as)) = \mathscr{A}(t(s))$.

3. The function $+_{\mathbf{fAT}}$ essentially joins two trees together at the root:

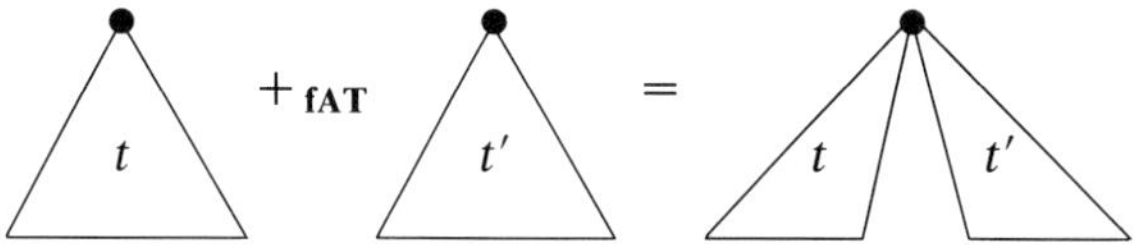

The acceptance set at the root is taken to be the collection of sets of the form $K \cup K'$ where K, K' are in $\mathscr{A}(t)$, $\mathscr{A}(t')$ respectively. For example, the pair

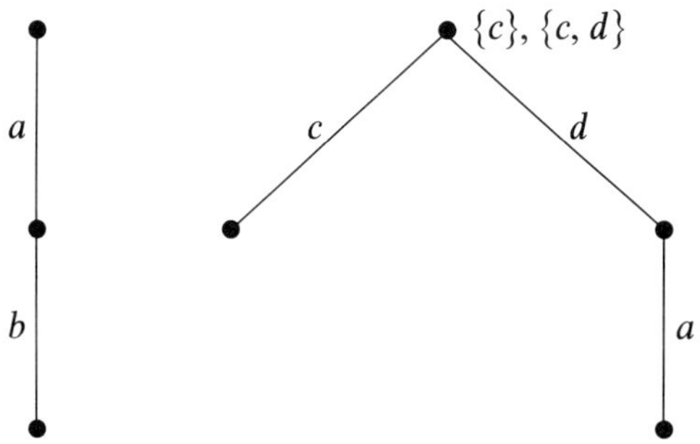

is mapped to

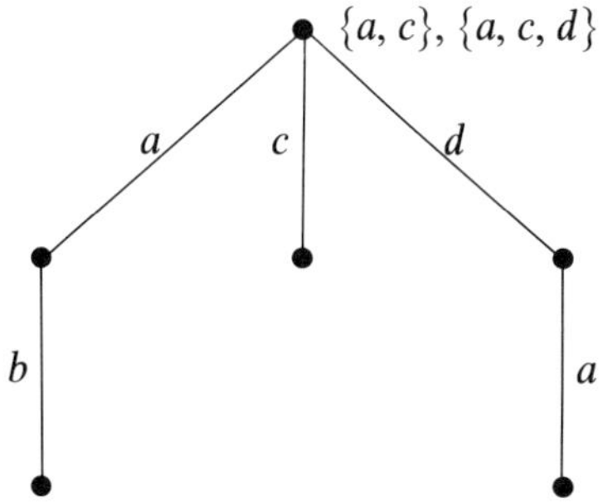

If, however, both roots have successor branches labeled by the same action, then care must be taken. For example, the pair t, t',

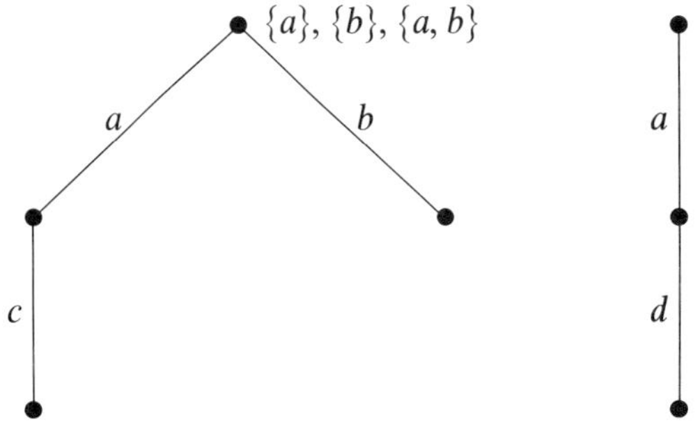

is mapped to

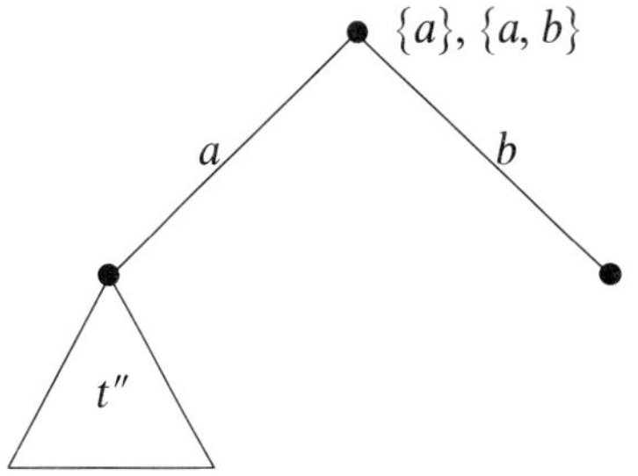

where t'' is obtained in some way from the pair of successor trees

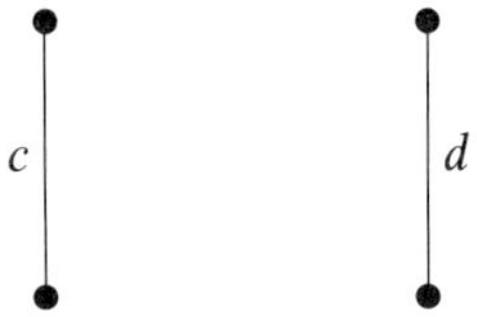

Intuitively when the process denoted by $t +_{\mathbf{fAT}} t'$ performs an a action it can either be the process denoted by

or the process denoted by

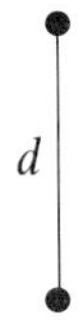

In one case it cannot perform a c action and in the other it cannot perform a d action. So intuitively t'' should be the nondeterministic tree

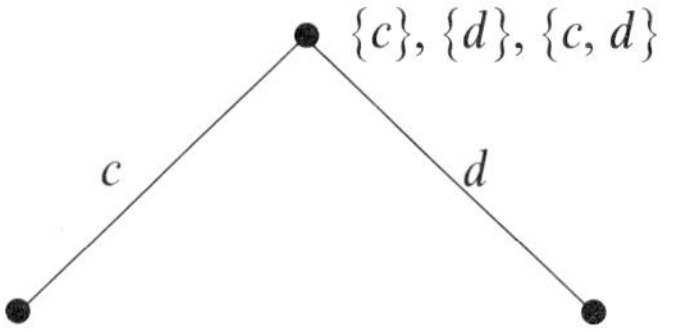

We now introduce some notation which enables us to define $+_{\mathbf{fAT}}$ formally. Let $\mathscr{B}$ be any finite collection of finite subsets of Act. For such a $\mathscr{B}$ let $A(\mathscr{B})$ be the set of actions which appear in $\mathscr{B}$, i.e. $\{a \in Act, a \in B$ for some $B \in \mathscr{B}\}$. In general $\mathscr{B}$ is not an $A(\mathscr{B})$-set. For example, if $\mathscr{B}$ is $\{\{a\}, \{b\}, \{a, b, c\}\}$ then $A(\mathscr{B})$ is $\{a, b, c\}$ and $\mathscr{B}$ is not an $\{a, b, c\}$-set. However, it is always possible to extend $\mathscr{B}$ to become an $A(\mathscr{B})$-set. Let $c(\mathscr{B})$ be the least set such that

a) $\mathscr{B} \subseteq c(\mathscr{B})$
b) $X, Y \in c(\mathscr{B})$ implies $X \cup Y \in c(\mathscr{B})$ – $c(\mathscr{B})$ is $\cup$-*closed*
c) $X, Y \in c(\mathscr{B})$, $X \subseteq Z \subseteq Y$ implies $Z \in c(\mathscr{B})$ – $c(\mathscr{B})$ is *convex-closed.*

So that $B \in c(\mathscr{B})$ if and only if the statement "$B \in c(\mathscr{B})$" can be derived from the three rules

i) $B \in \mathscr{B}$ implies $B \in c(\mathscr{B})$
ii) $X, Y \in c(\mathscr{B})$ implies $X \cup Y \in c(\mathscr{B})$
iii) $X, Y \in c(\mathscr{B})$, $X \subseteq Z \subseteq Y$ implies $Z \in c(\mathscr{B})$

For example, $c(\{a\}, \{b\}, \{a, b, c\})$ is $\{\{a\}, \{b\}, \{a, b\}, \{a, c\}, \{b, c\}, \{a, b, c\}\}$.

LEMMA 2.3.3 a) $c(\mathscr{B})$ is an $A(\mathscr{B})$-set.
b) $c(\mathscr{B})$ is the least saturated set containing $\mathscr{B}$.

Proof a) Referring to the definition of an $A(\mathscr{B})$-set we see that $\mathscr{B}$ satisfies requirements iii)–iv) by construction. Also by the definition of $A(\mathscr{B})$, for every b in $A(\mathscr{B})$ there is a B in $\mathscr{B}$ such that b is in B. So, *a fortiori*, $c(\mathscr{B})$ satisfies condition ii). It remains to show that condition i) is satisfied. Let K be in $c(\mathscr{B})$. We must prove $K \subseteq A(\mathscr{B})$. If K is in $c(\mathscr{B})$ we may prove so using the rules i), ii), iii) above. We used induction on the length of its proof. There are three cases according to whether the last rule used in the proof of "$K \in c(\mathscr{B})$" was i), ii) or iii).

i) Then K is in $\mathscr{B}$ and by definition of $A(\mathscr{B})$, $K \subseteq A(\mathscr{B})$.

ii) Then $K = K_1 \cup K_2$ where $K_1, K_2 \in c(\mathscr{B})$. Moreover, the length of the proof of each of the statements $K_1 \in c(\mathscr{B})$, $K_2 \in c(\mathscr{B})$ is less than that of $K \in c(\mathscr{B})$. So, by induction we can assume $K_1, K_2 \subseteq A(\mathscr{B})$. Therefore $K = K_1 \cup K_2 \subseteq A(\mathscr{B})$.

iii) Then $K_1 \subseteq K \subseteq K_2$ where $K_1, K_2 \in c(\mathscr{B})$. Once more by induction we can assume that $K_1, K_2 \subseteq A(\mathscr{B})$ and therefore $K \subseteq A(\mathscr{B})$.

b) Let $\mathscr{D}$ be any saturated set which contains $\mathscr{B}$. Then $\mathscr{D}$ is also $\cup$-closed and convex-closed. By definition $c(\mathscr{B})$ is the least such set which contains $\mathscr{B}$ and therefore $c(\mathscr{B}) \subseteq \mathscr{D}$. □

The operator c is used extensively in the sequel and consequently it is convenient to list some of its more useful properties here.

c1. if $\mathscr{B}$ is an S-set then $c(\mathscr{B}) = \mathscr{B}$

c2. if $\mathscr{B} \subseteq \mathscr{B}'$ then $c(\mathscr{B}) \subseteq c(\mathscr{B}')$

c3. $c(\mathscr{B}) = c(c(\mathscr{B}))$

c4. $c(\mathscr{B}_1 \cup c(\mathscr{B}_2 \cup \mathscr{B}_3)) = c(\mathscr{B}_1 \cup \mathscr{B}_2 \cup \mathscr{B}_3)$

c5. $c(\mathscr{B}_1 \cup \mathscr{B}_2) = c(c(\mathscr{B}_1) \cup c(\mathscr{B}_2))$

c6. if $A \in c(\mathscr{B})$ then $B \subseteq A$ for some $B \in \mathscr{B}$.

Each of these properties can be proven either using the technique of the previous lemma or by simple calculations which we leave to the reader. We will often be taking the pointwise union of sets of sets so we now introduce a special notation for it. If $\mathscr{B}_1, \mathscr{B}_2$ are sets of sets let

$$\mathscr{B}_1 \mathrel{u} \mathscr{B}_2 = \{B_1 \cup B_2, B_1, B_2 \in \mathscr{B}_1, \mathscr{B}_2, \text{respectively}\}.$$

This operator is used in the definition of $+_{\mathrm{fAT}}$ and so it is useful to derive some of its properties. It is trivial to see that

du1 $(\mathscr{B}_1 \mathrel{u} \mathscr{B}_2) \mathrel{u} \mathscr{B}_3 = \mathscr{B}_1 \mathrel{u} (\mathscr{B}_2 \mathrel{u} \mathscr{B}_3)$

du2 $\mathscr{B}_1 \mathrel{u} \mathscr{B}_2 = \mathscr{B}_2 \mathrel{u} \mathscr{B}_1$

du3 $\mathscr{B} \mathrel{u} \mathscr{B} = \mathscr{B}$, if $\mathscr{B}$ is saturated.

Its interaction with set-theoretic union is given by the following properties, which are also simple to derive:

du4 $\mathscr{B}_1 \; u \; (\mathscr{B}_2 \cup \mathscr{B}_3) = (\mathscr{B}_1 \; u \; \mathscr{B}_2) \cup (\mathscr{B}_1 \; u \; \mathscr{B}_3)$

Its interaction with the operator c is given by:

c7. If $\mathscr{B}_1, \mathscr{B}_2$ are saturated, $\mathscr{B}_1 \; u \; \mathscr{B}_2$ is saturated

c8. $c(\mathscr{B}_1 \; u \; \mathscr{B}_2) = c(\mathscr{B}_1) \; u \; c(\mathscr{B}_2)$.

The proof of c7 is not immediate; it is more easily seen if the alternative definition of an S-set is used. We leave the reader to check c8. One direction requires induction on the inductive definition of the closure operator c.

We are now ready to define the operator $+_{\mathbf{fAT}}$. For $t, t' \in \mathbf{fAT}$ we let $t +_{\mathbf{fAT}} t'$ be the tree t'' determined by:

i) $L(t'') = L(t) \cup L(t')$

ii) $\mathscr{A}(t'') = \mathscr{A}(t) \; u \; \mathscr{A}(t')$

iii) $\mathscr{A}(t''(s)) = c(\mathscr{A}(t(s)) \cup \mathscr{A}(t'(s)))$, $s \neq \varepsilon$ with the convention that if s is not in $L(t)$ then $\mathscr{A}(t(s))$ is $\varnothing$.

An example of $+_{\mathbf{fAT}}$ is given in figure 2.3. For $+_{\mathbf{fAT}}$ to be a well-defined operation on **fAT** we must prove that t'' defined above is indeed an element of **fAT**. This we leave to the reader, as it is straightforward. The only nontrivial part is to show that t'' satisfies R3.

PROPOSITION 2.3.4 $\langle \mathbf{fAT}, \leq_{\mathbf{fAT}}, \Sigma^1_{\mathbf{fAT}} \rangle$ is a Σ-*po* algebra.

Proof We must show that for every operator symbol *op* in Σ^1, $op_{\mathbf{fAT}}$ is monotonic. The only nontrivial case is $+_{\mathbf{fAT}}$.

Suppose $t' \leq_{\mathbf{fAT}} t''$. We must show $t +_{\mathbf{fAT}} t' \leq_{\mathbf{fAT}} t +_{\mathbf{fAT}} t''$, i.e.

a) $L(t +_{\mathbf{fAT}} t') = L(t +_{\mathbf{fAT}} t'')$

b) $\mathscr{A}((t +_{\mathbf{fAT}} t'')(s)) \subseteq \mathscr{A}((t +_{\mathbf{fAT}} t')(s))$ for every s in $L(t +_{\mathbf{fAT}} t'')$.

Part a) is trivial and we concentrate on b). There are two cases:

i) $s = \varepsilon$.

$\mathscr{A}(t +_{\mathbf{fAT}} t'') = \mathscr{A}(t) \; u \; \mathscr{A}(t'')$ by definition

$\subseteq \mathscr{A}(t) \; u \; \mathscr{A}(t')$ since $\mathscr{A}(t'') \subseteq \mathscr{A}(t')$ and u is monotonic with respect to $\subseteq$

$= \mathscr{A}(t +_{\mathbf{fAT}} t')$.

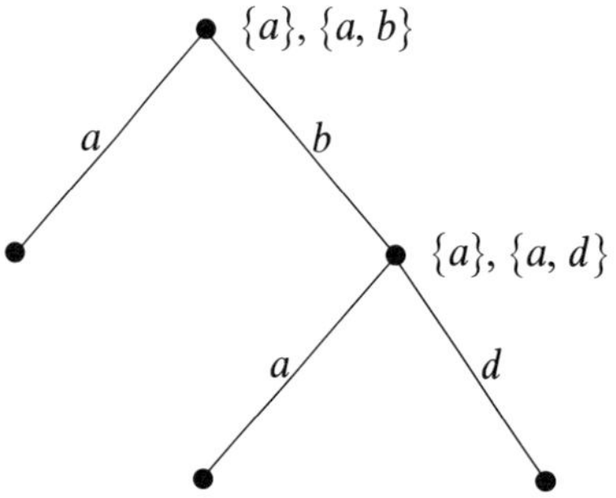

t_1

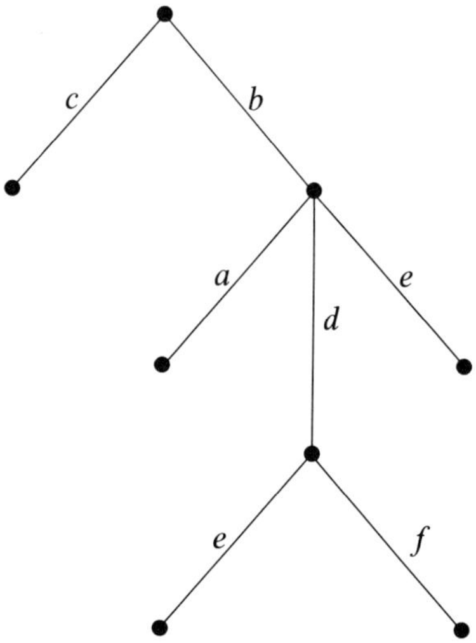

t_2

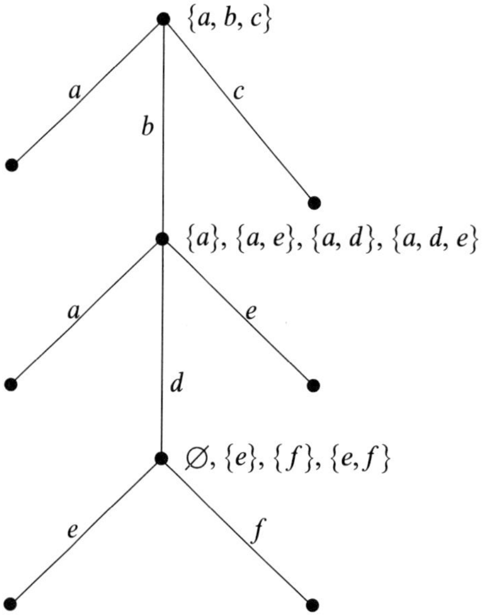

$t_1 + t_2$

Figure 2.3
Example of $+_{\mathbf{fAT}}$.

ii) $s \neq \varepsilon$. Then

$$\mathscr{A}((t +_{\mathbf{fAT}} t'')(s)) = c(\mathscr{A}(t)(s) \cup \mathscr{A}(t'')(s))$$

$$\subseteq c(\mathscr{A}(t)(s) \cup \mathscr{A}(t')(s)) \text{ since } c \text{ is monotonic with respect to } \subseteq, \text{ by c2}$$

$$= \mathscr{A}((t +_{\mathbf{fAT}} t')(s)). \qquad \square$$

The remainder of this section is devoted to this particular Σ^1-*po* algebra. We show that it is fully abstract with respect to $\underset{\sim}{\sqsubset}$. For convenience we abbreviate $\mathbf{fAT}[\![p]\!]$ by $[\![p]\!]$. Also notice that we have introduced notation which applies both to syntax (i.e. elements in $\mathbf{M_1}$) and semantics (i.e. elements of **fAT**). For example,

i) $L(p)$ – the set of sequences of actions that the process p can perform
$L(t)$ – the set of paths through the tree t.

ii) $S(p, s)$ the successors of p after s
$S(t(s))$ the successors of the node identified by s

iii) $\mathscr{A}(p, s)$ the Acceptance set of p after s.
$\mathscr{A}(t(s))$ the Acceptance set of the node identified by s.

In general $\mathscr{A}(p, s)$ is not saturated. For example, if p is $ab + ac$ $\mathscr{A}(p, a) = \{\{b\}, \{c\}\}$. However, we use saturated sets in the model so as to replace the use of the nonstandard relation $\subset\subset$ with the more usual subset inclusion $\subseteq$. The close relation between the denotation of a process and its operational behavior is underscored in the next lemma.

LEMMA 2.3.5 For every p in $\mathbf{M_1}$

a) $L(p) = L([\![p]\!])$

b) $c(\mathscr{A}(p, s)) = \mathscr{A}([\![p]\!](s))$ for every s in $L(p)$.

Proof By structural induction of p. We leave a) to the reader since it is straightforward and we prove part b).

i) p is NIL. Then $L(p) = \{\varepsilon\}$ and $\mathscr{A}(p, \varepsilon)$ is $\{\varnothing\}$. This is also $\mathscr{A}([\![NIL]\!])$ from the definition of $NIL_{\mathbf{fAT}}$.

ii) p is aq. There are two cases:

a) s is ε. Then $\mathscr{A}(p, \varepsilon)$ is $\{\{a\}\}$ which coincides with $\mathscr{A}([\![p]\!])$.

b) s in as'. Then $\mathscr{A}(p, as')$ is $\mathscr{A}(q, s')$ and the result follows by structural induction.

iii) p is $q + r$. Once more, there are two cases:

a) s is ε. Then $\mathscr{A}(p, \varepsilon) = \{S(p)\} = \{S(q) \cup S(r)\} = \mathscr{A}(q, \varepsilon) \mathrel{u} \mathscr{A}(r, \varepsilon)$. So

$$
\begin{aligned}
c(\mathscr{A}(p, \varepsilon)) &= c(\mathscr{A}(q, \varepsilon) \mathrel{u} \mathscr{A}(r, \varepsilon)) \\
&= c(\mathscr{A}(q, \varepsilon)) \mathrel{u} c(\mathscr{A}(r, \varepsilon)) && \text{from condition c8} \\
&= \mathscr{A}([\![q]\!]) \mathrel{u} \mathscr{A}([\![r]\!]) && \text{by induction} \\
&= \mathscr{A}([\![p]\!]) && \text{by definition.}
\end{aligned}
$$

b) $s \neq \varepsilon$. Then $\mathscr{A}(p, s) = \mathscr{A}(q, s) \cup \mathscr{A}(r, s)$. So

$$
\begin{aligned}
c(\mathscr{A}(p, s)) &= c(\mathscr{A}(q, s) \cup \mathscr{A}(r, s)) \\
&= c(c(\mathscr{A}(q, s)) \cup c(\mathscr{A}(r, s))) && \text{from condition c5} \\
&= c(\mathscr{A}[\![q]\!](s) \cup \mathscr{A}[\![r]\!](s)) && \text{by induction} \\
&= \mathscr{A}[\![p]\!](s) && \text{by definition.}
\end{aligned}
$$

□

Recall that in the definition of $<<_{\text{MUST}}$ Acceptance sets are compared using the nonstandard relation $\subset\subset$. This explains the use of saturated sets in the model: under certain conditions $\subset\subset$ on saturated sets coincides with subset inclusion. More generally, comparing collections of sets with $\subset\subset$ is equivalent to comparing their closures with subset inclusion.

LEMMA 2.3.6 If $A(\mathscr{B}) = A(\mathscr{A})$ then $\mathscr{B} \subset\subset \mathscr{A}$ if and only if $c(\mathscr{B}) \subseteq c(\mathscr{A})$.

Proof Suppose $\mathscr{B} \subset\subset \mathscr{A}$. We show $c(\mathscr{B}) \subseteq c(\mathscr{A})$. By definition $c(\mathscr{B})$ is the least $\cup$-closed and convex-closed set which contains $\mathscr{B}$. Also $c(\mathscr{A})$ is $\cup$-closed and convex-closed and therefore to prove $c(\mathscr{B}) \subseteq c(\mathscr{A})$ it suffices to show $\mathscr{B} \subseteq c(\mathscr{A})$. This follows immediately from $\mathscr{B} \subset\subset \mathscr{A}$ and the fact that $A(\mathscr{B})$ is in $c(\mathscr{A})$ since it coincides with $A(\mathscr{A})$.

Conversely, suppose $c(\mathscr{B}) \subseteq c(\mathscr{A})$. Then $\mathscr{B} \subseteq c(\mathscr{A})$ and $\mathscr{B} \subset\subset \mathscr{A}$ follows from property c6. □

These two lemmas combine to give the full abstraction of **fAT** with respect to Testing.

THEOREM 2.3.7 (Full Abstraction for **fAT**) If $p, q \in \mathbf{M}_1$ then $p \sqsubseteq q$ if and only if $\mathbf{fAT}[\![p]\!] \leq \mathbf{fAT}[\![q]\!]$.

Proof Follows directly from the two previous lemmas when $\sqsubseteq$ is replaced by $<<$. □

2.4 Algebraic Characterization of fAT

In this section we show that **fAT** is initial in the class of algebras which satisfy a particular set of inequations, thereby giving a sound and complete proof system for $\sqsubseteq$.

These inequations cannot be expressed in the signature Σ^1 alone. There is a very simple reason why this cannot be the case. For if D is a Σ-*po* algebra which is initial in an equational class, then the unique Σ-homomorphism i_D from T_Σ to D is surjective. But $i_{\mathbf{fAT}}$ is *not* surjective. To see this it is sufficient to remark that for every p in T_{Σ^1} $\mathbf{fAT}[\![p]\!]$ is a tree whose root is labeled by a single Acceptance set, $S(p)$. This is easily proven by structural induction on p. So, for example, the tree

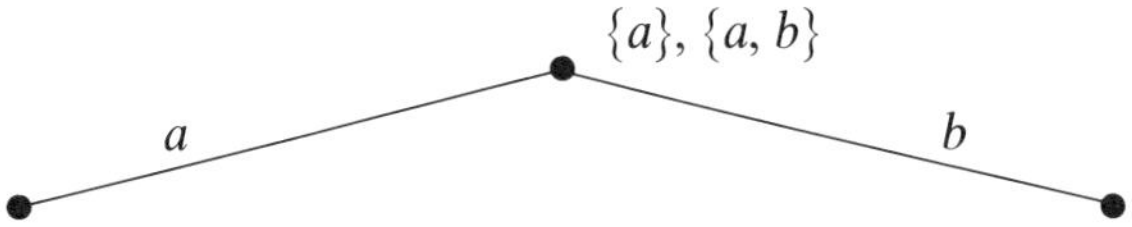

is not in the range of $i_{\mathbf{fAT}}$.

The new operator we introduce is a binary operator, very similar to $+$. The latter treats the roots of its operands differently than the other nodes. The new operator $\oplus_{\mathbf{fAT}}$ treats each node uniformly, in exactly the same way as $+$ treats nonroot nodes. If t, t' are trees in **fAT** let $t \oplus_{\mathbf{fAT}} t'$ be the tree t'' determined by

$$L(t'') = L(t) \cup L(t')$$

$$\mathscr{A}(t''(s)) = c(\mathscr{A}(t(s)) \cup \mathscr{A}(t'(s))) \quad \text{for every } s \text{ in } L(t''), \text{ with the convention that } \mathscr{A}(t(s)) = \varnothing \text{ if } s \notin L(t).$$

It is simple to establish that $\oplus_{\mathbf{fAT}}$ is well-defined and monotonic on **fAT**. So that if we let $\Sigma^2 = \Sigma^1 \cup \{\oplus\}$ then $\langle \mathbf{fAT}, \leq_{\mathbf{fAT}}, \Sigma^2_{\mathbf{fAT}} \rangle$ is a Σ^2-*po* algebra. When writing terms in T_{Σ^2}, prefixing has higher precedence than $\oplus$ which in turn has higher precedence than $+$. We also use $\mathbf{M_2}(Act)$, or simply $\mathbf{M_2}$, in place of T_{Σ^2} to emphasize, as before, that we view elements of T_{Σ^2} as syntactic representations of machines or processes. The operational difference between $+$ and $\oplus$ is quite subtle and we postpone the discussion of how to extend the Testing preorders to the extended language until the next section. It suffices at this moment to say that $+$ is a form of external nondeterminism while $\oplus$ represents a form of internal nondeterminism. In the denotational model the difference between $+$ and $\oplus$ is easily seen when applied to the pair

of trees

The results are, respectively,

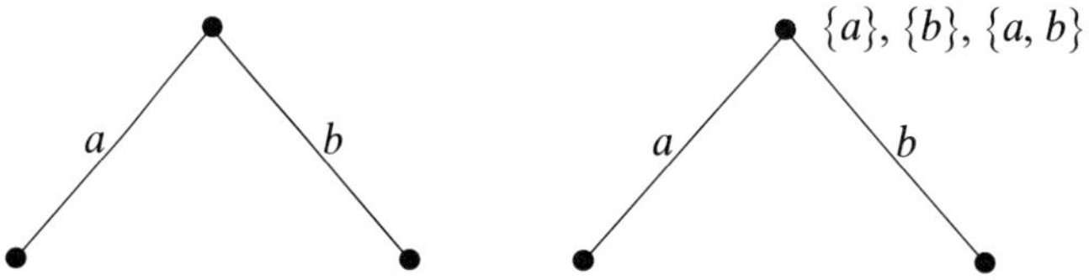

It is important to note that the only difference occurs, if at all, at the root. Moreover, the result of applying $\oplus$ always gives a tree which is more nondeterministic than the result of applying $+$. In short, **fAT** satisfies the inequation

$$x \oplus y \leq x + y \qquad +\oplus 1$$

This is a very simple consequence of the definitions and the remark that $\mathscr{A} \mathbin{\mathit{u}} \mathscr{B} \subseteq c(\mathscr{A} \cup \mathscr{B})$. Also, the effect of applying $+$, $\oplus$ is identical for certain kinds of operands. For example, **fAT** satisfies the equation

$$ax + ay = ax \oplus ay \qquad +\oplus 2$$

for every action a. This is because $\{\{a\}\} \mathbin{\mathit{u}} \{\{a\}\} = \{\{a\}\} = c(\{\{a\}\} \cup \{\{a\}\})$. However, there is a simple equation which $\oplus$ satisfies in **fAT** and $+$ does not satisfy

$$ax \oplus ay = a(x \oplus y). \qquad \oplus 4$$

This can be checked from the definitions of $\oplus$ and a in **fAT**. The corresponding equation for $+$,

$$ax + ay = a(x + y),$$

is *not* true in **fAT**. For example, with x, y replaced by $bNIL$, $cNIL$, respectively, the terms in the equation evaluate to

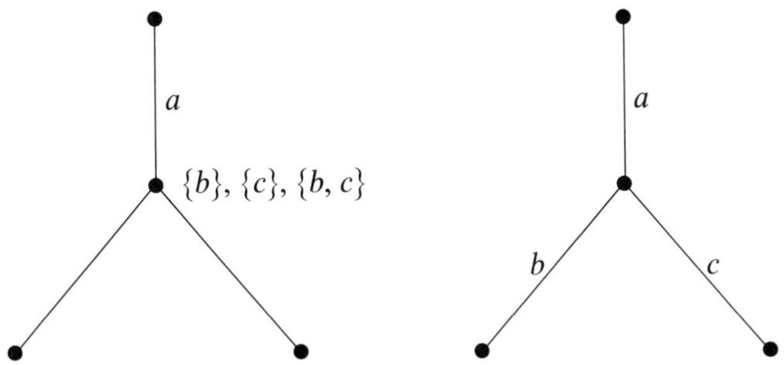

However, note that combining the equations $+\oplus 2$ and $\oplus 4$ we obtain the derived equation

$$ax + ay = a(x \oplus y).$$

Thus the operators $+$, $\oplus$ are very similar but subtly different. In fact they have many properties in common. For example, they are *associative*, *commutative*, and *absorptive*. These properties are expressed by the equations:

$x \oplus (y \oplus z) = (x \oplus y) \oplus z$	associativity for $\oplus$	$\oplus 1$
$x \oplus y = y \oplus x$	commutativity for $\oplus$	$\oplus 2$
$x \oplus x = x$	absorption for $\oplus$	$\oplus 3$
$x + (y + z) = (x + y) + z$	associativity for $+$	$+1$
$x + y = y + x$	commutativity for $+$	$+2$
$x + x = x$	absorption for $+$	$+3$

Each of the these equations is a simple consequence of the definitions. (In the cases of associativity we need property c4 of the operator *c*.) Associativity allows us to omit many brackets from terms without ambiguity. For any sequence of terms $p_1, \ldots p_k$ we may use $p_1 + p_2 + \cdots + p_k$ to denote any one of the terms obtained by bracketing correctly, such as $(p_1 + (p_2 + (.. + p_k) \ldots)$: because $+$ is associative it does not matter how we bracket the terms.

Commutativity allows us to introduce some further notational convenience. If P is a finite set of terms $p_1, \ldots, p_k$ let $\sum \{p, p \in P\}$ denote the term

$$p_1 + p_2 + \cdots + p_k.$$

Because $+$ is associative and commutative it does not matter how the set P is enumerated. If P is in fact empty, then $\sum\{p, p \in P\}$ simply denotes NIL. This convention is consistent since **fAT** satisfies the equation

$$x + NIL = x. \qquad +4$$

However the corresponding equation for $\oplus$,

$$x \oplus NIL = x,$$

is not true in **fAT**. For example $aNIL \oplus NIL$, $aNIL$ evaluate to the respective trees

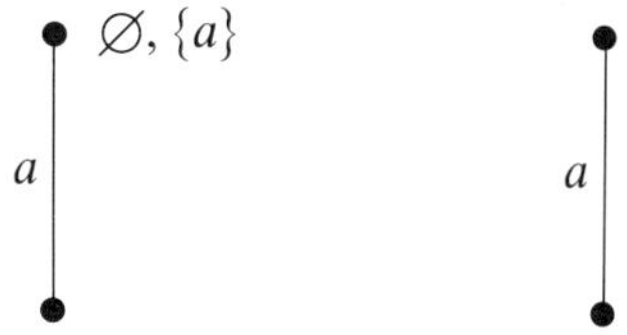

Consequently the corresponding convention for $\oplus$ can only be used for nonempty P. We use $\bigoplus\{p, p \in P\}$ to denote the term $p_1 \oplus p_2 \oplus \cdots \oplus p_k$. In addition to these properties both operators distribute over each other:

$$x + (y \oplus z) = (x + y) \oplus (x + z) \qquad + \text{ distributes over } \oplus \qquad +\oplus 3$$

$$x \oplus (y + z) = (x \oplus y) + (x \oplus z) \qquad \oplus \text{ distributes over } + \qquad +\oplus 4$$

To prove **fAT** satisfies these equations we need the properties

$$\mathscr{A} \mathrel{u} c(\mathscr{B}_1 \cup \mathscr{B}_2) = c((\mathscr{A} \mathrel{u} \mathscr{B}_1) \cup (\mathscr{A} \mathrel{u} \mathscr{B}_2))$$

$$c(\mathscr{A} \cup (\mathscr{B}_1 \mathrel{u} \mathscr{B}_2)) = c(\mathscr{A} \cup \mathscr{B}_1) \mathrel{u} c(\mathscr{A} \cup \mathscr{B}_2)$$

for saturated sets $\mathscr{A}$, $\mathscr{B}_1$, $\mathscr{B}_2$. These are simple to establish using the properties developed in the previous section.

We have now discussed all of the inequations we require to characterize **fAT**. They are collected together in figure 2.4 and we refer to them as the set **A2**.

PROPOSITION 2.4.1 **fAT** is in $\mathscr{C}$(**A2**).

Proof We must show that **fAT** satisfies every inequation in **A2**. In the discussion above we have given sufficient guidelines for the reader to carry out the necessary calculations. □

We show that **fAT** is initial in $\mathscr{C}$(**A2**). To do so it is sufficient, because of corollary 1.4.12, to show

$$x \oplus (y \oplus z) = (x \oplus y) \oplus z \qquad \oplus 1$$

$$x \oplus y = y \oplus x \qquad \oplus 2$$

$$x \oplus x = x \qquad \oplus 3$$

$$x + (y + z) = (x + y) + z \qquad +1$$

$$x + y = y + x \qquad +2$$

$$x + x = x \qquad +3$$

$$x + NIL = x \qquad +4$$

$$x \oplus y \leq x + y \qquad +\oplus 1$$

$$ax + ay = ax \oplus ay \qquad +\oplus 2$$

$$ax \oplus ay = a(x \oplus y) \qquad \oplus 4$$

$$x + (y \oplus z) = (x + y) \oplus (x + z) \qquad +\oplus 3$$

$$x \oplus (y + z) = (x \oplus y) + (x \oplus z) \qquad +\oplus 4$$

Figure 2.4
The set of inequations **A2**.

i) the mapping $i_{\mathbf{fAT}}$ from $\mathbf{M_2}$ to **fAT** is surjective
ii) (completeness) $\mathbf{fAT}[\![p]\!] \leq \mathbf{fAT}[\![q]\!]$ implies $p \leq_{\mathbf{A2}} q$.

The first is straightforward, the second requires some investigation of the power of the proof system **DED**(**A2**). As in §2.3 we abbreviate $\mathbf{fAT}[\![p]\!]$ to $[\![p]\!]$.

LEMMA 2.4.2 $i_{\mathbf{fAT}} : \mathbf{M_2} \rightarrow \mathbf{fAT}$ is surjective.

Proof Let $t \in \mathbf{fAT}$. We must show that there exists some $p \in \mathbf{M_2}$ such that $[\![p]\!] = t$. The proof is by induction on the depth of the tree t.

a) Suppose t has depth 0. Then t must be the tree $\bullet\{\emptyset\}$ and the required term is NIL.

b) Suppose t has depth greater than 0. Then for each action a in $S(t)$ there is a subtree of t, called t_a, obtained by taking as root the node $t(a)$. The depth of t_a is less than that of t and so by induction we may assume that there are terms p_a such that $[\![p_a]\!] = t_a$. For each $A \in \mathscr{A}(t)$ let p_A be the term $\sum \{ap_a, a \in A\}$. Then t is denoted by the term $\bigoplus\!\!\!\!\!\sum \{p_A, A \in \mathscr{A}(t)\}$. □

To prove statement ii), completeness, we use a general technique involving *normal forms*. The idea is to isolate a particular subclass of terms, called normal forms, whose denotations in the model are reflected directly in their syntactic structure. This makes completeness straightforward to prove for this restricted subclass. Completeness for arbitrary terms will follow if we can show that every term can be reduced to a normal form using the equations. Therefore the properties we require of normal forms are:

i) every term has a normal form, i.e. for every $p \in \mathbf{M_2}$ there exists a normal form $nf(p)$ such that $p =_{\mathbf{A2}} nf(p)$

ii) the completeness proof for normal forms, i.e. $[\![n]\!] \leq [\![n']\!]$ implies $n \leq_{\mathbf{A2}} n'$, should be straightforward.

Of course the latter is a rather informal requirement which is difficult to state more precisely.

In our particular case we use as normal forms those terms which were needed in the previous lemma to prove that $i_{\mathbf{fAT}}$ is surjective.

DEFINITION 2.4.3

i) NIL is a *normal form*
ii) if $\mathscr{A}$ is a saturated set and for every a in $A(\mathscr{A})$ there is a normal form $n(a)$ then $\bigoplus\!\!\!\!\sum \{(n(A), A \in \mathscr{A}\}$ is a *normal form*, where $n(A)$ represents $\sum \{an(a), a \in A\}$.

The following are examples of normal forms:

$aNIL \oplus NIL$, $abNIL \oplus (abNIL + cNIL)$

$(a(bNIL + cNIL) + bNIL) \oplus bNIL \oplus a(bNIL + cNIL)$

The following are *not* normal forms:

$aNIL + NIL$, $\quad abNIL + acNIL$, $\quad acNIL \oplus bcNIL$, $\quad acNIL \oplus (bNIL + adNIL)$.

If n is a normal form then, for every $a \in Act$, $D(n, a)$ contains at most one element, called $n(a)$, which is also a normal form. (Recall that $D(n, a)$ is $\{q, n \xrightarrow{a} q\}$, the set of derivations of n after a.) Moreover, the denotation of n is easily described from the syntactic form of n. If n denotes $\bigoplus\!\!\!\!\sum \{n(A), A \in \mathscr{A}\}$, as in the definition, then $\mathscr{A}([\![n]\!])$ is simply $\mathscr{A}$. Moreover, for every $a \in A(\mathscr{A})$, the subtree of $[\![n]\!]$ obtained by taking as root the node $[\![n]\!](a)$ is the denotation of the syntactic term $n(a)$. These properties make the completeness proof for normal forms relatively straightforward.

LEMMA 2.4.4 If n, m are normal forms then $[\![n]\!] \leq [\![m]\!]$ implies $n \leq_{\mathbf{A2}} m$.

Proof If either n or m is NIL then both are and the result is immediate. So we may assume n, m have the form $\bar{\sum}\{n(A), A \in \mathscr{A}\}$, $\bar{\sum}\{m(B), B \in \mathscr{B}\}$ respectively. From the above remarks and the fact that $[\![n]\!] \leq [\![m]\!]$ we can assume $\mathscr{B} \subseteq \mathscr{A}$ and $[\![n(a)]\!] \leq [\![m(a)]\!]$ for every $a \in A(\mathscr{A})$. Applying induction to the latter we obtain $n(a) \leq_{\mathbf{A2}} m(a)$. So systematically replacing each occurrence of $n(a)$ by $m(a)$ in n we have

$$n \leq_{\mathbf{A2}} \bar{\sum}\{m(A), A \in \mathscr{A}\}.$$

Because $\mathscr{B} \subseteq \mathscr{A}$ this may be written as

$$n \leq_{\mathbf{A2}} m \oplus \bar{\sum}\{m(C), C \in \mathscr{A} - \mathscr{B}\}.$$

By systematically applying the axiom $+\oplus 1$ we obtain

$$n \leq_{\mathbf{A2}} m \oplus \sum \{cm(c), c \in C\} \quad \text{where } C \text{ is } A(\mathscr{A} - \mathscr{B}).$$

By using the distributive equation $+\oplus 4$, repeatedly if necessary, we obtain

$$n \leq_{\mathbf{A2}} \sum \{m \oplus cm(c), c \in C\}.$$

To complete the proof we establish for every c in C,

$$m \oplus cm(c) \leq_{\mathbf{A2}} m. \tag{$*$}$$

Since $A(\mathscr{A}) = A(\mathscr{B})$ c must appear in some B from $\mathscr{B}$. For this particular B, $m(B)$ may be written as $q + cm(c)$ for some term q, whose structure is of no concern. Then

$$\begin{aligned} m(B) \oplus cm(c) &\leq_{\mathbf{A2}} q + cm(c) + cm(c) && \text{using } +\oplus 1 \\ &=_{\mathbf{A2}} q + cm(c) && \text{using } +2, +3 \end{aligned}$$

i.e. $m(B) \oplus cm(c) \leq_{\mathbf{A2}} m(B)$

which is sufficient to establish ($*$). □

We now show that every term can be transformed using the inequations into a normal form. More specifically, for every p there is a normal form, $nf(p)$ such that $p =_{\mathbf{A2}} nf(p)$. The required series of transformations is not straightforward. It requires two significant steps. The first is based on the two derived equations

$$x \oplus y = x \oplus y \oplus (x + y) \qquad \text{Der1}$$

$$x \oplus (x + y + z) = x \oplus (x + y) \oplus (x + y + z) \qquad \text{Der2}$$

Der1 is justified as follows:

$$
\begin{aligned}
x \oplus y \oplus (x + y) &= (x \oplus y) \oplus (x + y) \\
&= (x \oplus y \oplus x) + (x \oplus y \oplus y) && \text{using } +\oplus 4 \\
&= x \oplus y && \text{using } \oplus 1\text{–}\oplus 3,\ +1\text{–}+3
\end{aligned}
$$

We leave the reader to justify Der2, which requires some thought.

One significant step in the normalization procedure is encapsulated in the following lemma. As in the definition of normal forms, let p_a denote some term of the form $ap(a)$ for every action a in $A(\mathscr{B})$, where $\mathscr{B}$ is a nonempty set of sets. For each A in $\mathscr{B}$ let $p(A)$ then denote the term $\sum \{p_a, a \in A\}$ and let $p(\mathscr{B})$ denote $\bigoplus\!\!\!\!\sum \{p(A), A \in \mathscr{B}\}$.

LEMMA 2.4.5 $p(\mathscr{B}) =_{\mathbf{A2}} p(c(\mathscr{B}))$.

Proof It is sufficient to show that for every B in $c(\mathscr{B})$ that $p(\mathscr{B}) =_{\mathbf{A2}} p(\mathscr{B}) \oplus p(B)$. We use induction on the length of the proof of $B \in c(\mathscr{B})$.

i) $B \in \mathscr{B}$. This case is a trivial application of absorption.
ii) $B = B_1 \cup B_2$ and $B_1, B_2 \in c(\mathscr{B})$.

By induction we may assume $p(\mathscr{B}) =_{\mathbf{A2}} p(\mathscr{B}) \oplus p(B_1)$, $p(\mathscr{B}) =_{\mathbf{A2}} p(\mathscr{B}) \oplus p(B_2)$. Then

$$
\begin{aligned}
p(\mathscr{B}) &=_{\mathbf{A2}} p(\mathscr{B}) \oplus p(B_1) \oplus p(B_2) && \text{using the inductive assumptions} \\
&=_{\mathbf{A2}} p(\mathscr{B}) \oplus p(B_1) \oplus p(B_2) \oplus (p(B_1) + p(B_2)) && \text{using Der1} \\
&=_{\mathbf{A2}} p(\mathscr{B}) \oplus (p(B_1) + p(B_2)) && \text{using the inductive assumptions again.}
\end{aligned}
$$

However $p(B_1 \cup B_2) =_{\mathbf{A2}} p(B_1) + p(B_2)$, by absorption, and so the required result follows.

iii) $B_1 \subseteq B \subseteq B_2$ where $B_1, B_2 \in c(\mathscr{B})$.

Once more by induction we may assume

$p(\mathscr{B}) =_{\mathbf{A2}} p(\mathscr{B}) \oplus p(B_1)$, $p(\mathscr{B}) =_{\mathbf{A2}} p(\mathscr{B}) \oplus p(B_2)$.

Let C, D be such that $B = C \cup B_1$, $B_2 = D \cup B$. Again using the fact that for any sets A, A', $p(A \cup A') =_{\mathbf{A2}} p(A) + p(A')$, we have that

$$
\begin{aligned}
p(\mathscr{B}) &=_{\mathbf{A2}} p(\mathscr{B}) \oplus p(B_1) \oplus (p(B_1) + p(C) + p(D)) \\
&=_{\mathbf{A2}} p(\mathscr{B}) \oplus p(B_1) \oplus (p(B_1) + p(C)) \oplus (p(B_1) + p(C) + p(D)) && \text{using Der2} \\
&=_{\mathbf{A2}} p(\mathscr{B}) \oplus (p(B_1) + p(C)) && \text{using the inductive assumption again} \\
&=_{\mathbf{A2}} p(\mathscr{B}) \oplus p(B).
\end{aligned}
$$

□

The usefulness of this lemma should be apparent. Assuming each $p(a)$ is a normal form the term $p(\mathscr{B})$ is almost a normal form; the only requirement missing is that $\mathscr{B}$ should be saturated. This lemma gives a valid transformation to a normal form, transforming $p(\mathscr{B})$ to $p(c(\mathscr{B}))$. We call this transformation TRcs, closing with respect to saturation. The other significant step in the normalization procedure is the systematic application of the derived equation

$$(ax_1 + z_1) \oplus (ax_2 + z_2) = (ax + z_1) \oplus (ax + z_2), \quad \text{where } x \text{ denotes } x_1 \oplus x_2. \qquad \text{Der3}$$

It is applied in a left-to-right fashion and is used to coalesce different a-derivatives. The term on the left has at least two different a-derivatives, represented by x_1, x_2. That on the right has eliminated this particular duplication. We call such an application of Der3 the transformation TRgd, the transformation for gathering derivatives. For a term to be in normal form it must, among other things, have a unique a-derivative for every action a. By repeated application of this transformation every term can be automatically transformed into a new term with this property.

To derive Der3 we argue as follows:

$$\begin{aligned}
(ax_1 + z_1) \oplus (ax_2 + z_2) &= (ax_1 + z_1) \oplus (ax_2 + z_2) \oplus (ax_1 + ax_2 + z_1 + z_2) \\
&\qquad \text{using Der1} \\
&= (ax_1 + z_1) \oplus (ax_1 + ax_2 + z_1) \oplus (ax_2 + z_2) \\
&\quad \oplus (ax_1 + ax_2 + z_2) \oplus (ax_1 + ax_2 + z_1 + z_2) \\
&\qquad \text{using Der2, } \oplus 1, \oplus 2 \\
&= (ax_1 + z_1) \oplus (ax + z_1) \oplus (ax_2 + z_2) \oplus (ax + z_2) \\
&\quad \oplus (ax_1 + ax_2 + z_1 + z_2) \qquad \text{using } + \oplus 2 \\
&= (ax_1 + z_1) \oplus (ax + z_1) \oplus (ax_2 + z_2) \oplus (ax + z_2) \\
&\qquad \text{using Der1} \\
&= (ax_1 \oplus ax + z_1) \oplus (ax_2 \oplus ax + z_2) \qquad \text{using } + \oplus 3 \\
&= (ax + z_1) \oplus (ax + z_2) \qquad \text{using } \oplus 4, \oplus 3.
\end{aligned}$$

THEOREM 2.4.6 (Normal Form Theorem) For every term p in $\mathbf{M_2}$ there exists a normal form $nf(p)$ such that $p =_{\mathbf{A2}} nf(p)$.

Proof The proof is by induction on the depth of p, i.e. the length of the largest s such that $s \in L(p)$. There are four cases.

a) p is NIL.

Then p is its own normal form.

b) p is aq.

By induction q has a normal form $nf(q)$ and the required $nf(p)$ is $anf(q)$.

c) p is $q + r$.

By induction q, r have normal forms $\sum\{n(A), A \in \mathscr{A}\}$, $\sum\{m(B), B \in \mathscr{B}\}$ respectively. We argue as follows:

$p =_{\mathbf{A2}} \sum\{n(A) + r, A \in \mathscr{A}\}$, by repeated use of $+\oplus 3$

$=_{\mathbf{A2}} \sum\{n(A) + m(B), A \in \mathscr{A}, B \in \mathscr{B}\}$ by repeated use of $+\oplus 3$

Let $\mathscr{C} = \mathscr{A} \, u \, \mathscr{B}$, which is saturated, and for each c in $A(\mathscr{C})$ let $q(c)$ be defined by

$$
\begin{aligned}
q(c) &= n(c) \oplus m(c) && \text{if } c \in A(\mathscr{A}) \cap A(\mathscr{B}) \\
&= n(c) && \text{if } c \in A(\mathscr{A}), c \notin A(\mathscr{B}) \\
&= m(c) && \text{otherwise.}
\end{aligned}
$$

Then with repeated use of $+\oplus 2$ and the derivative gathering transformation TRgd, the above term can be transformed so that

$p =_{\mathbf{A2}} \sum\{q(C), C \in \mathscr{C}\}$.

The depth of each term $q(c)$ is less than that of p and therefore we may apply induction to obtain, for each $q(c)$, a normal form $o(c)$. By replacing each occurrence of $q(c)$ with its normal form in this term we obtain $nf(p)$ because $\mathscr{C}$ is saturated.

d) p is $q \oplus r$.

Suppose q, r have normal forms as in case c). Then

$p =_{\mathbf{A2}} \sum\{n(A), A \in \mathscr{A}\} \oplus \sum\{m(B), B \in \mathscr{B}\}$.

In this case let $\mathscr{C}$ be $\mathscr{A} \cup \mathscr{B}$ and define $q(c)$ as in the previous case for each c in $A(\mathscr{C})$. Again by systematically applying TRgd we can derive

$q =_{\mathbf{A2}} \sum\{q(C), C \in \mathscr{C}\}$.

$\mathscr{C}$ is not necessarily saturated but we may apply the other transformation TRcs to obtain

$$p =_{\mathbf{A2}} \sum \{q(C), C \in c(\mathscr{C})\}$$

and we can now proceed as in the previous case. □

We are now in a position to prove the main result of this section.

THEOREM 2.4.7 (Initiality of **fAT**) **fAT** is initial in the class $\mathscr{C}(\mathbf{A2})$.

Proof Because of lemma 2.4.1, lemma 2.4.2, and corollary 1.4.12 it is sufficient to establish: $[\![p]\!] \leq [\![q]\!]$ implies $p \leq_{\mathbf{A2}} q$.

So suppose $[\![p]\!] \leq [\![q]\!]$. Since **fAT** satisfies all the axioms in **A2**, $[\![p]\!] = [\![nf(p)]\!]$, $[\![q]\!] = [\![nf(q)]\!]$ and therefore we may assume $[\![nf(p)]\!] \leq [\![nf(q)]\!]$. Applying lemma 2.4.4 we obtain $nf(p) \leq_{\mathbf{A2}} nf(q)$ from which it follows immediately that $p \leq_{\mathbf{A2}} q$. □

Let us at this stage recapitulate the preceding sections. The syntax of the language we have investigated is simply the term algebra T_{Σ^1}, or $\mathbf{M_1}$. In the first section we described a general operational notion of Testing preorder for processes and this was applied to $\mathbf{M_1}$ by considering it as a labeled transition system. This gave an operational criterion $\sqsubseteq$ for discriminating between different semantic interpretations. Then a particular interpretation **fAT** was described and was shown to be fully abstract with respect to the chosen operational criterion. Finally, **fAT** was shown to be initial in the class of interpretations generated by the set of inequations **A2**. This gives a sound and complete proof system for the interpretation **fAT**. However the proof system uses a new operator $\oplus$ which is not in the original language. So although we have connected the three different views of processes, denotational semantics, behavioral equivalence and logical proof systems, the necessity for the new operator introduces a slight discontinuity into the overall framework. To rectify this we extend the behavioral view, that embodied in Testing Equivalence, to the augmented language $\mathbf{M_2}$ and show that the results remain valid in this new setting.

2.5 Internal and External Nondeterminism

In this section we examine a new experimental system which enables us to extend the behavioral view of processes to the extended language $\mathbf{M_2}$. To motivate the definition consider again the denotations of the pair of processes

$$ap + bq, \qquad ap \oplus bq.$$

The first has only one acceptance set at the root $\{a, b\}$ and the second three, $\{a\}$, $\{b\}$, $\{a, b\}$. But $\{a, b\}$ is only included in the latter collection to ensure it is saturated; so we concentrate on $\{a\}$, $\{b\}$. They model two different internal states that the process may be in, one in which only a is possible, the other in which only b is possible. So with $ap \oplus bq$ there is a nondeterministic choice to be made between performing either a or b. Moreover this nondeterminism is resolved *internally* because exactly which action can be performed is determined autonomously by the process itself; it depends only on which internal state it is in.

On the other hand, $ap + bq$ has only one internal state in which it can perform either of a or b (but not both). Which action is actually performed is determined, at least to a certain extent, *externally*. If asked to perform a it will perform it and similarly with b. This external resolution of the choice is limited. For example, if asked to perform either of a or b it is not clear which action will actually be performed. Similarly with the process $ap + aq$: a process cannot determine whether p or q will be the result of performing the action a. This is reflected in the axiom

$$ap + aq = ap \oplus aq.$$

Nevertheless it is reasonable to characterize the intuitive difference between $+$ and $\oplus$ as being a matter of whether the nondeterminism is resolved internally or externally.

Labeled transition systems do not contain sufficient information to distinguish between these two different kinds of nondeterminism. To remedy this we add an extra component which gives information on the internal behavior of processes. This is an extra relation, $\rightarrowtail$, whose intuitive meaning is one step of an internal computation: $p \rightarrowtail q$ means that p can evolve to q autonomously, without any intervention by an external process or experimenter. In our language $\mathbf{M_2}$, an instance of such a move is

$$p \oplus q \rightarrowtail p.$$

DEFINITION 2.5.1 An *extended labeled transition system* is a 4-tuple $\langle P, Act, \longrightarrow, \rightarrowtail \rangle$ where

i) $\langle P, Act, \longrightarrow \rangle$ is an *lts*
ii) $\rightarrowtail$ is a binary relation over P, the internal action relation.

In figure 2.5 we give the definition of $\mathbf{M_2}$ as an extended *lts*. The method used is the same as that in §2.2: the relations $\xrightarrow{a}$ and $\rightarrowtail$ are defined inductively as the least relations which satisfy a set of natural clauses. Those for $\xrightarrow{a}$ are as explained in §2.2 and those for $\rightarrowtail$ essentially say that the only internal movement is that generated by internal choices, embodied in $\oplus$. For convenience we drop the prefix "extended,"

i) $ap \xrightarrow{a} p$

ii) $p \xrightarrow{a} p'$ implies $p + q \xrightarrow{a} p'$

$q \xrightarrow{a} q'$ implies $p + q \xrightarrow{a} q'$

i) $p \oplus q \rightarrowtail p$

$p \oplus q \rightarrowtail q$

ii) $p \rightarrowtail p'$ implies $p + q \rightarrowtail p' + q$

$q \rightarrowtail q'$ implies $p + q \rightarrowtail q'$

Figure 2.5
Operational semantics of $\mathbf{M_2}$.

by viewing the labeled transition systems as extended ones in which the internal move relation is empty. The new *lts*'s generate Experimental Systems as before.

DEFINITION 2.5.2 Let L_p, L_E be two compatible *lts*'s, $\langle P, Act, \longrightarrow, \rightarrowtail \rangle$ and $\langle E, Act \cup \{1, w\} \longrightarrow, \rightarrowtail \rangle$. Then $\mathscr{ES}(L_p, L_E)$ is the *Experimental System* $\langle P, E, \rightarrow, Success \rangle$ where

i) *Success* is, as before, $\{e \text{ in } E, e \xrightarrow{w}\}$
ii) $\rightarrow \subseteq (E \times P) \times (E \times P)$ is the least relation which satisfies
a) $e \xrightarrow{a} e', p \xrightarrow{a} p'$ implies $e \parallel p \rightarrow e' \parallel p'$
b) $e \xrightarrow{1} e'$ implies $e \parallel p \rightarrow e' \parallel p$
iii) $e \rightarrowtail e'$ implies $e \parallel p \rightarrow e' \parallel p$
iv) $p \rightarrowtail p'$ implies $e \parallel p \rightarrow e \parallel p'$

Conditions i) and ii) are exactly those used in the previous definition of Experimental System. Conditions iii) and iv) are designed to explain the operator $\oplus$. They state that a process or experimenter can evolve imperceptively by performing an internal action. This internal movement takes place without the partner in the experiment being aware of it.

This new Experimental System $\mathscr{ES}(L_p, L_E)$ induces three Testing preorders on the processes of P. We are interested in the particular case when both L_p and L_E are generated by the example language $\mathbf{M_2}$.

Let $\mathscr{ES}(\mathbf{M_2})$ denote the Experimental System $\mathscr{ES}(\langle \mathbf{M_2}(Act), Act, \longrightarrow, \rightarrowtail \rangle, \langle \mathbf{M_2}(Act \cup \{1, w\}), Act \cup \{1, w\}, \longrightarrow, \rightarrowtail \rangle)$. It induces three Testing preorders on

$\mathbf{M_2}$. We will see later that when restricted to $\mathbf{M_1}$ these coincide with the preorders of §2.2. This justifies using the same notation for them, namely $\sqsubseteq_{\text{MAY}}$, $\sqsubseteq_{\text{MUST}}$, $\sqsubseteq$.

EXAMPLES 2.5.3

1. $a(b + c) \not\sqsubseteq_{\text{MUST}} a(b \oplus c)$. For example, $a(b + c)$ *must* abw whereas $a(b \oplus c)$ *m̸ust* abw because of the computation $abw \parallel a(b \oplus c) \rightarrow bw \parallel b \oplus c \rightarrow bw \parallel cNIL$. However, one can argue, by examining the effects of all possible experiments that $a(b + c) \eqsim_{\text{MAY}} a(b \oplus c)$, $a(b \oplus c) \sqsubseteq_{\text{MUST}} a(b + c)$.

2. One can also argue that $a \oplus b \sqsubseteq_{\text{MUST}} a + b$ but not the converse: $a + b$ *must* aw but $a \oplus b$ *m̸ust* aw because of the computation $aw \parallel a \oplus b \rightarrow aw \parallel bNIL$.

3. $aNIL + (b \oplus c) \not\sqsubseteq_{\text{MUST}} (a + b) \oplus cNIL$ because $aNIL + (b \oplus c)$ *must* aw whereas $(a + b) \oplus cNIL \parallel aw \rightarrow cNIL \parallel aw$ is an unsuccessful computation. We will see later that the converse is true, i.e. $(a + b) \oplus cNIL \sqsubseteq_{\text{MUST}} aNIL + (b \oplus c)$.

We now wish to relate the principal Testing preorder $\sqsubseteq$ over $\mathbf{M_2}$ with both the interpretation **fAT** and the equational proof system. This relies on extending the alternative characterization in theorem 2.2.12. The definitions of $<<$, $<<_{\text{MAY}}$, and $<<_{\text{MUST}}$ remain exactly the same but that of the properties they use must be modified to take the internal actions into consideration.

Let $p \overset{a}{\Longrightarrow} p'$ if $p \rightarrowtail^* q \overset{a}{\longrightarrow} q' \rightarrowtail^* p'$. So $p \overset{a}{\Longrightarrow} p'$ if p can evolve into p' by performing an a action interspersed with an arbitrary, possibly zero, number of internal moves. For example, if p is $(a(q_1 \oplus q_2) + bq_4) \oplus aq_5$, then p' may be any one of the terms q_5, q_1, q_2 or $q_1 \oplus q_2$. It is worth remarking that on $\mathbf{M_1}$ $\overset{a}{\Longrightarrow}$ coincides with $\overset{a}{\longrightarrow}$ and that $\overset{a}{\longrightarrow}$ is a very strong relation on the extended language $\mathbf{M_2}$; as an instance of this, $p \overset{a}{\not\longrightarrow}$ if p is as above.

These generalized action relations may be extended to arbitrary sequences as follows:

i) $p \overset{\varepsilon}{\Longrightarrow} p'$ if $p \rightarrowtail^* p'$
ii) $p \overset{as}{\Longrightarrow} p'$ if $p \overset{a}{\Longrightarrow} q$ and $q \overset{s}{\Longrightarrow} p'$ for some term q.

So $p \overset{s}{\Longrightarrow} p'$ means that p can be transformed into p' by performing the sequence of actions s, possibly interspersed with internal actions. We now redefine the language of a process, its Acceptance sets, etc., using $\overset{s}{\Longrightarrow}$ in place of $\overset{s}{\longrightarrow}$. For $p \in \mathbf{M_2}$ let

$$\text{Im}(p) = \{p', p \rightarrowtail p'\}$$

$$D(p, s) = \{p', p \overset{s}{\Longrightarrow} p'\}$$

$$D(p) = \{p', \text{for some } a \in Act\ p \overset{a}{\Longrightarrow} p'\}$$

$$L(p) = \{s, p \overset{s}{\Longrightarrow} p' \text{ for some } p'\}$$

$$S(p) = \{a, p \overset{a}{\Longrightarrow}\}$$

$$S(p, s) = \{a, p \overset{s}{\Longrightarrow} p' \overset{a}{\Longrightarrow}\}$$

$$\mathscr{A}(p, s) = \{S(p'), p \overset{s}{\Longrightarrow} p'\}.$$

Let $<<_{\text{MAY}}$, $<<_{\text{MUST}}$, $<<$ be as in definition 2.2.8, but using these modified sets. Note that these modifications do not affect processes in the restricted language $\mathbf{M_1}$. We leave the reader to check:

LEMMA 2.5.4 Both $<<_{\text{MAY}}$ and $<<_{\text{MUST}}$ are Σ^2-preorders. □

This is a simple extension of the corresponding result in §2.2. Most of the proof of the alternative characterization theorem also carries over unchanged.

THEOREM 2.5.5 (Alternative Characterization of Testing Preorders) For every p, p' in $\mathbf{M_2}$,

i) $p \sqsubseteq_{\text{MAY}} p'$ if and only if $p <<_{\text{MAY}} p'$
ii) $p \sqsubseteq_{\text{MUST}} p'$ if and only if $p <<_{\text{MUST}} p'$
iii) $p \sqsubseteq p'$ if and only if $p << p'$.

Proof Only slight modifications are required to the corresponding proof in §2.2. We examine those for the case $p <<_{\text{MUST}} p'$ implies $p \sqsubseteq_{\text{MUST}} p'$. When the sequences $e \overset{s}{\Longrightarrow} e_k$, $p \overset{s}{\Longrightarrow} r$ are combined to give

$e \parallel p \rightarrow \cdots \rightarrow e_k \parallel r$,

this may not be maximal. However it can always be extended to a maximal sequence because there exists an r' such that $r \rightarrowtail^* r' \not\rightarrowtail$. Any such r' satisfies $S(r') \subseteq S(r)$ and therefore r' may be used in the proof in place of r. □

We can also extend the connection between the Acceptance set of the denotation of a process and the behavioral Acceptance set, given in lemma 2.3.5.

LEMMA 2.5.6 For every p in $\mathbf{M_2}$

i) $L(p) = L(\llbracket p \rrbracket)$
ii) $c(\mathscr{A}(p, s)) = \mathscr{A}(\llbracket p \rrbracket(s))$ for every s in $L(p)$.

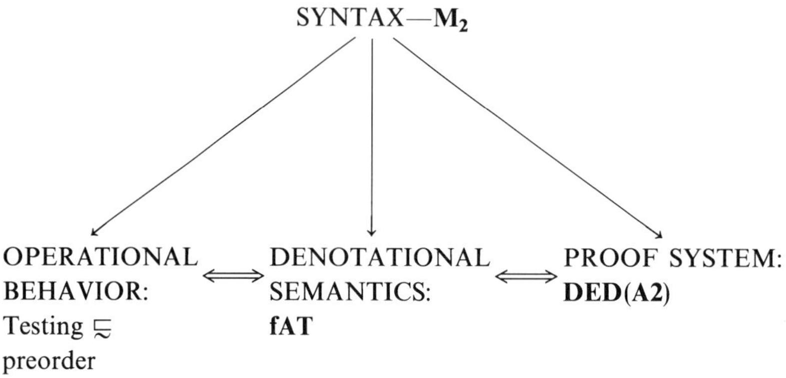

Figure 2.6
Schema of results.

Proof The only difference to the proof of lemma 2.3.5 is in part ii) where there is one extra case to consider, $p = q \oplus r$. The argument used in this case, for arbitrary s is identical to the case $p = q + r$ when $s \neq \varepsilon$. □

As an immediate corollary we have the full-abstraction result:

THEOREM 2.5.7 (Full Abstraction for **fAT**) For p, $q \in \mathbf{M_2}$, $p \sqsubseteq q$ if and only if $\mathbf{fAT}[\![p]\!] \leq \mathbf{fAT}[\![q]\!]$.

Proof Identical to that of the restricted case, theorem 2.3.7, but using the newly established results, theorem 2.5.5 and lemma 2.5.6 in place of the corresponding results, theorem 2.2.12 and lemma 2.3.5. □

The general result, corollary 1.4.14, now implies that **DED(A2)** is both sound and complete with respect to $\sqsubseteq$ over the enriched language $\mathbf{M_2}$. We have thus connected the three different approaches to the semantics of processes for $\mathbf{M_2}$. This is schematized in figure 2.6.

2.6 The Trinity

The theory of testing outlined in §2.1 and 2.2 gives rise in a natural way to *three* operational preorders, $\sqsubseteq$, $\sqsubseteq_{MAY}$, $\sqsubseteq_{MUST}$. In §2.3–2.5 we concentrated on one of these, $\sqsubseteq$, developing for it a fully abstract interpretation and an algebraic characterization. In this final section of this chapter we outline briefly how these sections can be modified so as to treat $\lesssim_{\mathrm{MUST}}$, $\lesssim_{\mathrm{MAY}}$, in place of $\lesssim$. The result will be a uniform treatment of the three preorders.

The MUST case: The two preorders $\lesssim$ and $\lesssim_{\mathrm{MUST}}$ are in general different. For example,

$abNIL + acNIL \lesssim_{\mathrm{MUST}} abNIL$

$abNIL + acNIL \not\lesssim abNIL.$

The latter is true because *abNIL + acNIL may acw* whereas *abw m$\!\!\!/$ay acw*. The former can be checked using theorem 2.2.12.

In general $\lesssim_{\mathrm{MUST}}$ equates more processes than $\lesssim$ and, as we will see, this extra power can be characterized by adding one equation to the set which characterizes $\lesssim$.

We first modify **fAT** so that it models $\lesssim_{\mathrm{MUST}}$ rather than $\lesssim$. The carrier remains the same but the partial order between the objects is different. For $t, t' \in$ **fAT** let $t \leq^{\mathrm{MUST}} t'$ if:

for every s in $L(t')$ $\quad \mathscr{A}(t'(s)) \subseteq \mathscr{A}(t(s))$.

This is essentially the second clause in the definition of $\leq_{\mathbf{fAT}}$, so that $t \leq_{\mathbf{fAT}} t'$ implies $t \leq^{\mathrm{MUST}} t'$ but not necessarily vice versa. Also notice that $t \leq^{\mathrm{MUST}} t'$ implies $L(t') \subseteq L(t)$. For convenience we use $\mathbf{fAT_S}$ (finite Strong Acceptance Trees) to denote $\langle \mathbf{fAT}, \leq^{\mathrm{MUST}}, \Sigma^2_{\mathbf{fAT}} \rangle$.

LEMMA 2.6.1 $\quad \mathbf{fAT_S}$ is a Σ^2-*po* algebra.

Proof i) $\langle \mathbf{fAT}, \leq^{\mathrm{MUST}} \rangle$ is a partial order. The only nontrivial part is to check that if $t \leq^{\mathrm{MUST}} t'$ and $t' \leq^{\mathrm{MUST}} t$ then $t = t'$. This relies on the property: $t \leq^{\mathrm{MUST}} t'$ implies $L(t') \subseteq L(t)$.

ii) The proof that each function in $\Sigma^2_{\mathbf{fAT}}$ preserves the partial order $\leq^{\mathrm{MUST}}$ is essentially given in proposition 2.3.4. □

The proof of the following theorem is also a simple modification of the original theorem in §2.3; all the work has been done in lemmas 2.3.5 and 2.3.6.

THEOREM 2.6.2 (Full Abstractness for $\mathbf{fAT_S}$) $\quad$ If $p, q \in \mathbf{M_2}$, then $p \lesssim_{\mathrm{MUST}} q$ if and only if $\mathbf{fAT_S}[\![p]\!] \leq \mathbf{fAT_S}[\![q]\!]$. □

Finally we modify the algebraic characterization so as to reflect $\lesssim_{\mathrm{MUST}}$. The new interpretation satisfies the extra equation

$x \oplus y \leq x$ **S**

and it is easy to see that it is *not* true in the interpretation of §2.3. This extra equation is the only one required. Let **SA2** denote the set of equations **A2** together with **S**.

THEOREM 2.6.3 (Initiality of $\mathbf{fAT_S}$) $\mathbf{fAT_S}$ is initial in $\mathscr{C}(\mathbf{SA2})$.

Proof i) If $p \leq q$ is an instance of any inequation in **A2** we know that $[\![p]\!] \leq_{\mathbf{fAT}} [\![q]\!]$. Since $\leq_{\mathbf{fAT}}$ implies $\leq^{\mathrm{MUST}}$ it follows that $[\![p]\!] \leq_{\mathbf{fAT_S}} [\![q]\!]$. Simple calculations will show that **S** is also satisfied. Therefore $\mathbf{fAT_S}$ is in $\mathscr{C}(\mathbf{SA2})$.

ii) To show that it is initial in $\mathscr{C}(\mathbf{SA2})$ we use the results of §2.4 to prove $\mathbf{fAT_S}[\![p]\!] \leq \mathbf{fAT_S}[\![q]\!]$ implies $p \leq_{\mathbf{SA2}} q$. The only change necessary is to lemma 2.4.4, in whose place we require:

if n, m are normal forms than $\mathbf{fAT_S}[\![p]\!] \leq \mathbf{fAT_S}[\![q]\!]$ implies $n \leq_{\mathbf{SA2}} m$.

This requires slightly different algebraic manipulations which should provide an interesting exercise for the reader.

The proof now follows that of theorem 2.4.7. □

The MAY case: To obtain the relevant interpretation for the MUST preorder we added the extra axiom $x \oplus y \leq x$ to **A2**. For the MAY case we add its dual:

$$x \leq x \oplus y \qquad \mathbf{W}$$

Let **WA2** denote this set of equations. The resulting proof system is quite strong in that one can derive many theorems relating the operators. For example

$$NIL \leq_{\mathbf{WA2}} x$$

because

$$\begin{aligned} NIL &\leq_{\mathbf{WA2}} x \oplus NIL && \text{from } \mathbf{W} \\ &\leq_{\mathbf{WA2}} x + NIL && \text{from } +\oplus 1 \\ &\leq_{\mathbf{WA2}} x && \text{from } +4. \end{aligned}$$

Applying **W** twice we have

$$\begin{aligned} x + y &\leq_{\mathbf{WA2}} (x \oplus y) + (x \oplus y) \\ &\leq_{\mathbf{WA2}} x \oplus y && \text{from } +3. \end{aligned}$$

It therefore follows (from $+\oplus 1$) that $x + y =_{\mathbf{WA2}} x \oplus y$ so that with this added axiom the distinction between the two nondeterministic operators disappears. It follows, in turn, that prefixing distributes over $+$ because

$$a(x + y) =_{\mathbf{WA2}} a(x \oplus y)$$

$$=_{\mathbf{WA2}} ax \oplus ay \qquad \text{from } \oplus 4$$

$$=_{\mathbf{WA2}} ax + ay.$$

It is therefore not surprising that the model generated by these axioms is essentially that of prefix-closed strings, **PS** in §1.3. However it can be represented in a manner which emphasizes the coherence of the three different viewpoints on operational behavior; we present **PS** as a submodel of **fAT**.

Let $\mathbf{fAT_W}$ (finite Weak Acceptance Trees) denote $\langle D, \leq^{\mathrm{MAY}}, \Sigma^2_{\mathbf{fAT_W}}\rangle$ where

i) D is the set of deterministic trees in **fAT**, i.e. those trees all of whose nodes n are labeled by the acceptance set $\{\{S(n)\}\}$

ii) $t \leq^{\mathrm{MAY}} t'$ if $L(t) \subseteq L(t')$. (Note that this is essentially the first component of the definition of $\leq_{\mathbf{fAT}}$.)

iii) the various operators are defined by

—$a_{\mathbf{fAT_W}}$ coincides with $a_{\mathbf{fAT}}$

—$t +_{\mathbf{fAT_W}} t'$ is the tree t'', determined by

a) $L(t'') = L(t) \cup L(t')$

b) $\mathscr{A}(t''(s)) = \mathscr{A}(t(s)) \mathrel{u} \mathscr{A}(t'(s))$, with the assumption that if $s \notin L(t)$ then $\mathscr{A}(t(s))$ is taken to be $\{\{\varnothing\}\}$

—$\oplus_{\mathbf{fAT_W}}$ coincides with $+_{\mathbf{fAT_W}}$.

We state without proofs the relevant results:

$\mathbf{fAT_W}$ is initial in $\mathscr{C}(\mathbf{WA2})$.

If p, q are in $\mathbf{M_2}$ then $p \sqsubseteq_{\mathrm{MAY}} q$ if and only if $\mathbf{fAT_W}[\![p]\!] \leq \mathbf{fAT_W}[\![q]\!]$.

These various results state that we have a diagram as in figure 2.6, not only for the operational behavior $\sqsubseteq$ but also for the more primitive relations $\sqsubseteq_{\mathrm{MUST}}$, $\sqsubseteq_{\mathrm{MAY}}$. The proof systems for the three relations are closely related; the basic one is determined by the inequations **A2** and the other two are obtained by adding the inequation $x \oplus y < x$ or $x < x \oplus y$. The respective denotational models are also closely related, the basic one being **fAT** and the other two being modifications of it.

Thus there are no inherent mathematical or technical reasons for choosing between these three different notions of operational behavior; they are all equally well-behaved. The choice must be made on other, more philosophical grounds, such as what one believes to be important about the behavior of processes. This is best examined in terms of the extended languages, particularly those which have concurrent and hiding operators such as those discussed in the introduction.

Exercises

Q1 Decide whether or not each of the following pairs of machines in $\mathbf{M}_1$ is related under $\sqsubseteq_{\text{MUST}}$.

i. $a(b + c) + ad, \quad ab + a(c + d)$
ii. $ab(c + d) + abd, \quad abc + a(bc + bd)$
iii. $ab + ac, \quad ab + ac + a(b + c)$
iv. $aNIL + ab, \quad ab$
v. $a(c + be) + a(d + bf), \quad a(c + bf) + a(d + be)$
vi. $a(d + bc) + ab(c + e), \quad a(bc + b(c + e)) + a(d + b(c + e))$
vii. $aNIL + a(b + c + d), \quad aNIL + ab + a(c + d)$.

Q2 Show that **PS** is fully abstract with respect to $\eqsim_{\text{MAY}}$, i.e. $\mathbf{PS}[\![p]\!] = \mathbf{PS}[\![q]\!]$ if and only if $p \eqsim_{\text{MAY}} q$, for every $p, q \in \mathbf{M}_2$.

Q3 For $p, q \in \mathbf{M}_1$ let $p \equiv q$ if $D(p, a) = D(q, a)$ for every $a \in Act$. Prove $p \equiv q$ implies $p \eqsim q$.

Q4 Show $\mathscr{A}$ is an S-set if and only if it satisfies

$A \in \mathscr{A}$ implies $A \subseteq S$

$S \in \mathscr{A}$

$A \subseteq B \subseteq S, \quad A \in \mathscr{A}$, implies $B \in \mathscr{A}$.

Q5 i) Let Seq denote the set of experiments of the form $a_1 \dots a_n w$. Show that $\sqsubseteq_{\text{MUST}}$ and $\sqsubseteq^{Seq}_{\text{MUST}}$ do *not* coincide.

ii) Let D be the set of experiments which do not use the distinguished action symbol 1, i.e. $\mathbf{M}_2(Act \cup \{w\})$. Show that $\sqsubseteq_{\text{MUST}}$ and $\sqsubseteq^{D}_{\text{MUST}}$ do *not* coincide.

iii) Show that $\sqsubseteq_{\text{MUST}}$ and $\sqsubseteq^{B}_{\text{MUST}}$ do not coincide, where B is the set of all experiments of the form $e(s, A)$.

Q6 A process $p \in \mathbf{M_2}$ is *stable* if $p \rangle\!\!\not\longrightarrow$. Let $\mathscr{SA}(p, s) = \{S(p'), p \stackrel{s}{\Longrightarrow} p', p'$ stable$\}$ and $p <\!<\!<_{\text{MUST}} q$ if for every $s \in L(q)$ $\mathscr{SA}(q, s) \subset\!\subset \mathscr{SA}(p, s)$. Prove that $<\!<\!<_{\text{MUST}}$ coincides with $<\!<_{\text{MUST}}$ over $\mathbf{M_2}$.

Q7 For $P \subseteq \mathbf{M_2}$ and finite $L \subseteq Act$ let P *must* L if for every $p \in P$ there exists some $a \in L$ such that $p \stackrel{a}{\Longrightarrow}$. For each $n \geq 0$ define preorder $\leq_n$ on subsets of $\mathbf{M_2}$ by:

a) $P <_0 Q$ if, for every finite $L \subseteq Act$, $D(P, \varepsilon)$ *must* L implies $D(Q, \varepsilon)$ *must* L, where $D(P, s) = \{p', p \stackrel{s}{\Longrightarrow} p'$ for some $p \in P\}$.
b) $P \leq_{n+1} Q$ if, for every $a \in Act$,
   i) $P \leq_0 Q$
   and ii) $D(P, a) \leq_n D(Q, a)$.

Let $P \leq Q$ if $P \leq_n Q$, for every $n \geq 0$.
Alternatively, define $\leq'$ over subsets of $\mathbf{M_2}$ by

$P \leq' Q$ if for every $s \in Act^*$ and finite $L \subseteq Act$
$D(P, s)$ *must* L implies $D(Q, s)$ *must* L.

Prove:

i) $P \leq Q$ if and only if $P \leq' Q$.
ii) $p \sqsubseteq_{\text{MUST}} q$ if and only if $\{p\} \leq \{q\}$.

Q8 (Observational Equivalence, Milner 1980) For any relation $R \subseteq \mathbf{M_1} \times \mathbf{M_1}$ let $D(R) \subseteq \mathbf{M_1} \times \mathbf{M_1}$ be defined by:

$\langle p, q \rangle \in D(R)$ if

$p \stackrel{s}{\longrightarrow} p'$ implies $q \stackrel{s}{\longrightarrow} q'$ for some q' such that $\langle p', q' \rangle \in R$

$q \stackrel{s}{\longrightarrow} q'$ implies $p \stackrel{s}{\longrightarrow} p'$ for some p' such that $\langle p', q' \rangle \in R$.

For each $n \geq 0$ define the equivalence relation $\sim_n$ by:

$\sim_0 = \mathbf{M_1} \times \mathbf{M_1}$
$\sim_{n+1} = D(\sim_n)$

Finally, let $p \sim q$ if $p \sim_n q$ for every $n \geq 0$.

i) Prove $p \sim q$ implies $p \eqsim q$
ii) Show the converse is false, i.e. $p \eqsim q$ does not necessarily imply $p \sim q$.
iii) Find the smallest $n \geq 0$ such that $\sim_n \supsetneq\, \eqsim\, \supsetneq \sim_{n+1}$.

Q9 Use the equations of figure 2.4 to prove $p =_{\mathbf{A2}} q$ for the following pairs:

i) $ab(cd + ce), \quad abcd + abce$
ii) $aNIL + a(b + c + d), \quad aNIL + ab + ac + ad$
iii) $(ab + cNIL) \oplus ad, \quad (ad + cNIL) \oplus ab$
iv) $ab + ac, \quad ab + ac + a(b + c)$
v) $a(c + be) + a(d + bf), \quad a(c + bf) + a(d + be)$.

Q10 Prove that the following are derived equations in **DED(A2)**:

i) $x + y = x + y + (x \oplus y)$
ii) $x + (x \oplus y \oplus z) = x + (x \oplus y) + (x \oplus y \oplus z)$
iii) $\bigoplus\{ax_i, i \in I\} = \sum\{ax_i, i \in I\}$ for every finite I
iv) $\bigoplus\{ax_i, i \in I\} = a\bigoplus\{x_i, i \in I\}$ for every finite I
v) Using the notation of lemma 2.4.5, $\bigoplus\{ap(B), B \in \mathscr{B}\} = \bigoplus\{ap(B), B \in c(\mathscr{B})\}$.

Q11 Show that the axiom $+\oplus 4$ can be derived from Der1, Der2, using the remaining axioms in figure 2.4.

Q12 For any tree t in **fAT** with $a \in S(t)$ let t/a be the subtree of t whose root is the node of t determined by a. For $p \in \mathbf{M_2}$, show $\mathbf{fAT}[\![p]\!]/a = \mathbf{fAT}[\![\bigoplus\{p', p \xrightarrow{a} p'\}]\!]$.

Q13 (Failures; Brookes, Hoare, and Roscoe 1984) For $p \in \mathbf{M_2}$ let failures $(p) = \{\langle s, X\rangle, p \stackrel{s}{\Longrightarrow} p'$ for some p' such that $p' \stackrel{a}{\Longrightarrow}$ for no $a \in X\}$. Intuitively $\langle s, X\rangle \in p$ means that p can perform the sequence of actions s to get into a state where every action in X is refused, i.e. cannot be performed. Prove that $p \lesssim_{\mathrm{MUST}} q$ if and only if failures $(q) \subseteq$ failures (p).

Q14 (Refusal Sets; Brookes, Hoare, and Roscoe 1984) A refusal set R over Act is a set of failures of the form $\langle s, X\rangle$ where $s \in Act^*$ and X is a finite subset of Act, which satisfies

$\langle \varepsilon, \varnothing\rangle \in R$

$\langle st, \varnothing\rangle \in R$ implies $\langle s, \varnothing\rangle \in R$

$X \subseteq Y$ and $\langle s; Y\rangle \in R$ implies $\langle s, X\rangle \in R$

$\langle s, X\rangle \in R$ and $\langle sa, \varnothing\rangle \notin R$ implies $\langle s, X \cup \{a\}\rangle \in R$.

i. Show that for each $p \in \mathbf{M_2}$ failures (p) is a refusal set.
ii. Show $\langle \mathscr{FR}, \supseteq\rangle$ is a partial order where $\supseteq$ is subset inclusion and $\mathscr{FR}$ is the set of finite refusals, i.e. they satisfy $\{s, \langle s, X\rangle \in R$ for some $X)\}$ is finite.

iii. Show $\langle \mathscr{FR}, \supseteq \rangle$ is isomorphic as a partial order with $\mathbf{fAT_S}$

iv. Extend $\langle \mathscr{FR}, \supseteq \rangle$ to a Σ^2-partial order in such a way that the isomorphism in iii. is an isomorphism.

Q15 A process p in $\mathbf{M_2}$ is deterministic if $p \lesssim_{\mathrm{MUST}} q$ implies $p \eqsim_{\mathrm{MUST}} q$. Let $DP(p) = \{d \in \mathbf{M_2}, p \lesssim_{\mathrm{MUST}} d$, and d is deterministic$\}$. Show $p \lesssim_{\mathrm{MUST}} q$ implies $DP(q) \subseteq DP(p)$. Is the converse true?

Q16 (Nondeterministic Processes as Sets of Deterministic Processes) A (finite) deterministic process can be considered to be a finite nonempty prefix-closed subset of Act^*. Let D denote the set of all these objects. Let the set of nondeterministic processes NP be the collection of subsets n of D which satisfy

$d, d' \in n$ implies $d \cup d' \in n$

$d, d', e' \in n$, $d \subseteq e \subseteq d'$ implies $e \in n$.

NP can be made into a Σ^2-algebra by defining $NIL_{NP} = \{\{\varepsilon\}\}$

$a_{NP}(n) = \{ad, d \in n\}$ where $ad = \{as, s \in d\} \cup \{\varepsilon\}$

$n +_{NP} n' = \{d \cup d', d \in n, d \in n\}$

$n \oplus_{NP} n' = c\{n \cup n'\}$ where $c(m)$ is the least set containing m which satisfies the two closure conditions above.

i) Prove that the operations are well-defined and that $\langle NP, \supseteq, \Sigma^2_{NP} \rangle$ is a Σ^2-*po* where $\supseteq$ is subset inclusion.

ii) Show that there exist neither a Σ^2-*po* homomorphism from $\mathbf{fAT_S}$ to NP nor from NP to $\mathbf{fAT_S}$

iii) Show NP is *not* fully abstract with respect to $\lesssim_{\mathrm{MUST}}$

iv) What subset of equations **SA2** does NP satisfy?

v) Find a set of Σ^2-equations E, such that NP is initial in $\mathscr{C}(E)$.

II RECURSIVE PROCESSES

Introduction to Part II

The basic language studied in depth in part I, $\mathbf{M_2}$, is very simple. Essentially it allows one to define nondeterministic processes of a very restricted kind. In part III we will see how the various semantic theories used to explain this language may also be used for more complicated languages. We have in mind a more comprehensive set of combinators, such as those used in the introduction to describe communicating concurrent processes and abstractions. For the moment we continue with the basic set of combinators but extend the language in another dimension. All processes definable in $\mathbf{M_2}$ are strictly finite; there is a definite finite bound on the number of actions they can ever perform. While for most practical purposes these processes are essentially useless, here they allowed us to explain those concepts which form the basis of our approach and the various relationships between them, such as: interpretation, equational deduction system, full abstraction, and Testing equivalence. The aim of part II is to extend these concepts and the results concerning them to languages which support infinite processes.

At the syntactic level this is quite straightforward; innumerable methods have been invented for finitely representing infinite objects. Here we choose a particularly simple one, that of recursive definitions. Using the signature Σ^2,

$$x \Longleftarrow ax + bx + cNIL \qquad (*)$$

is a recursive definition of a process named x. We intuitively ascribe to it the following behavior:

At any point in time it can perform any one of the actions a, b, or c; if it performs c it terminates, i.e. it can perform no more actions; if it performs either of a, b it reconstitutes itself, i.e. it can continue as before.

The behavior of x is determined by that of its definition $ax + bx + cNIL$. This intuition can be captured adequately in the operational semantics by adding one extra rule which allows a process defined in this way to evolve silently to its definition:

If $x \Longleftarrow p$ is a definition then $x \rightarrowtail p$.

This enlarged set of processes now has an operational semantics in terms of a labeled transition system and we can apply the usual definition, of §§2.2, 2.5, to obtain the various Testing preorders.

Purely equational reasoning is not of great help for these extended languages. For example, the processes x_1, x_2 defined by

$$x_1 \Longleftarrow ax_1$$

$$x_2 \Longleftarrow aax_2,$$

have, intuitively, the same behavior but no amount of equational transformations will transform one into the other. Some form of induction is required. We augment the equational proof system **DED**(E) with various forms, such as Recursion Induction and Scott Induction. However the augmented languages, such as **EPL** of the introduction, allow us to define rather complicated processes. In particular we can implement Turing machines. Various undecidable questions can be posed in terms of the Testing preorders and consequently it is not possible to obtain complete proof systems. Nevertheless we shall see that the relationship between the proof systems and the Testing preorders established in part I can be maintained, at least in some sense, in this new setting.

The models also require modification. In an interpretation D it is natural to associate with a process

$$x \Longleftarrow ax$$

the solution of the equation

$$x = a_D(x) \qquad (*)$$

where, as usual, a_D represents the function associated with the unary function symbol a. However, to do so we must ensure that such equations have solutions. For example, in **fAT** the equation $(*)$ has NO solution. The required extension of **fAT** is fairly obvious. We allow the trees to be infinite and the solution to $(*)$ is then the simple tree with one infinite path, each branch of which is labeled by a. However, this solution exists because the proposed extension of **fAT** enjoys certain properties and these properties must be elucidated. Furthermore, these properties must address the problem of equations having more than one solution. For example, the equation

$$x = x + aNIL$$

has an infinite number of solutions in **fAT**.

We take as interpretations Σ-algebras whose carriers are partial orders which enjoy certain continuity constraints. These are sufficient to ensure that the equations associated with recursive definitions always have *least* solutions. These least solutions are taken as the meanings of the associated processes. We demand that the carriers be partial orders $\langle D, \leq \rangle$, where D always contains a "least element $\perp_D$ with respect to $\leq$, and where, essentially, all chains in A:

$$d_0 \leq d_1 \leq d_2 \leq \cdots$$

have limits.

In any such interpretation, the least solution to $(*)$ is then simply the limit of the chain

$$\perp_D \leq a_D(\perp_D) \leq a_D(a_D(\perp_D)) \leq \cdots.$$

Unfortunately there is a price to pay. We have introduced into the semantic domains "partial objects" such as $\perp_D$, the totally undefined object, $a_D(\perp)$, and $a_D(b_D(\perp_D) +_D \perp_D)$. Moreover these play a crucial role in assigning meanings to arbitrary processes. We could try to ignore their presence but this would mean having a very limited understanding of these models. Instead we explain the exact role they play. To do so we introduce a new symbol into our signature, a function symbol of arity 0, or constant symbol Ω. It is the syntactic representation of the totally undefined object, in that we insist every interpretation D is such that Ω_D is the least element in D, $\perp_D$. This allows us to define in our language, in a straightforward manner, the partial objects of the semantic domains.

Having made this extension, we must reconsider both the proof systems and the behavioral concepts. The former is straightforward but the latter does demand attention. How do we define the operational semantics of "partial processes" and apply Testing preorders to them? We will be guided by pragmatics here. Our primary goal is to present a theory of the behavior of "totally defined" processes. To do so we use models which demand the existence of "partial processes." So we assign to these the behavior suggested by the models themselves.

The presentation in part II is somewhat different from that in part I; the mathematical background and the specific application of interest to us are more intertwined. In chapter 3 we present the revised notion of interpretation required for the remainder of the book. These are called Σ-domains, essentially Σ-algebras whose carriers satisfy the continuity constraints discussed above. The theory of this new category of interpretations is developed in chapter 3. In particular we show that there is an initial Σ-domain, CI_E, in the class of Σ-domains which satisfy a given set of (in)equations E, $\mathscr{C}\mathscr{C}(E)$. This chapter ends with particular examples of Σ-domains, called **Acceptance Trees**. These are essentially infinite versions of the finite acceptance trees encountered in part I. However, there are some subtleties in the extension from finite to infinite trees because we require them to be Σ-domains. We also show that they can be characterized equationally; the various varieties, weak and strong, are initial in $\mathscr{C}\mathscr{C}(E)$ for appropriate sets of equations E.

In chapter 4 the language of part I, $T_\Sigma(X)$, is generalized to what we call $REC_\Sigma(X)$, the set of recursively defined terms over the signature Σ. We show how the terms in $REC_\Sigma(X)$ can be interpreted in any Σ-domain A and relate the interpretation in initial Σ-domains, CI_E, to a proof system we call $\boldsymbol{\omega}$**DED**(E), obtained from **DED**(E) by adding a very powerful form of induction. We also examine the specific language $REC_{\Sigma^2}(X)$, where Σ^2 is the signature of part I. We extend the operational semantics of finite processes and thereby obtain an Experimental System for the processes in REC_{Σ^2}. The resulting Testing preorders are investigated. We show that the various interpretations of **Acceptance Trees** are fully abstract with respect to the appropriate Testing preorders. In short, the diagram in figure 2.6 is extended to recursively defined processes.

3 Continuous Algebras

In this chapter we present the basic mathematical structures and their properties which are required to model infinite processes. The first section is an elementary exposition of continuous partial orders and continuous functions. We develop these only to the extent that is required to understand the remainder of the book. This section culminates in the definition of a Σ-domain: a Σ-algebra where the carrier is an algebraic continuous partial order (with a least element) and where each function symbol in Σ is interpreted as a continuous function over the carrier.

Section 3.2, which is brief, considers classes of domains which are characterized by equations. The main result is that $\mathscr{C}\mathscr{C}(E)$, the class of domains which satisfies the equations E, has an initial object. A connection is also established between this initial object and an equational proof system. This generalizes corollary 1.4.12 connecting the initial Σ-algebra I_E with the proof system **DED**(E). Here we use an extension of **DED**(E) obtained by adding one extra axiom $\Omega \leq x$. However, in the continuous setting, the proofs of these results are nontrivial; they occupy all of §3.3 and use a new type of algebra which we call Σ-predomains. They lie between Σ-*po* algebras and Σ-domains, as we require the carriers to be partial orders which contain a least element but do not have any continuity constraints. The initiality result for Σ-predomains is easily obtained by modifying the corresponding result in part I for simple Σ-*po* algebras. This is then "lifted" to Σ-domains using the technique of ideal completion.

In the section 3.4 we give a pertinent example of an initial domain, Strong Acceptance Trees, $\mathbf{AT_S}$. These are generalizations of the finite version of Strong Acceptance Trees, $\mathbf{fAT_S}$, which we saw in part I. This domain is characterized as the initial Σ^2-domain in the class which satisfies a slight modification of the set of equations **SA2**. In the section 3.5 we emphasize our triple view of the world; modifying the set of equations, as in part I, to obtain **A2**, **SA2**, and **WA2**, we obtain three slightly different kinds of Acceptance Trees which, as we will see, will reflect the three different Testing preorders on infinite processes.

3.1 Basic Definitions and Results

Let $\langle A, \leq_A \rangle$ be a partial order. The element $\bot_A$ in A is the *least element* in A if it satisfies

$\bot_A \leq_A a$ for every a in A.

Let D be a subset of A and a an element of A. Then a is an *upper bound* of D if:

$d \leq_A a$ for every d in D.

a is a *least upper bound* (*lub*) of D if

i) a is an upper bound of D
ii) if d' is an upper bound of D then $a \leq_A d'$.

It follows from the antisymmetry of $\leq_A$ that least upper bounds, if they exist, are unique. We use $\bigvee_A D$ to denote the least upper bound of D in A, when it exists.

D is a *directed* subset of A if it is nonempty and for every pair of elements d_1, d_2 in D the set $\{d_1, d_2\}$ has an upper bound which is also in D. This notion is a mild generalization of a chain. A *chain* in A is a sequence $a_0, a_1, \ldots$ of elements from A such that

$$a_0 \leq_A a_1 \leq_A a_2 \leq_A \cdots \leq_A a_n \leq \cdots.$$

Note that every chain is a directed set but not vice versa.

The partial order $\langle A, \leq_A \rangle$ is a *complete partial order* (*cpo*) if

i) it contains a least element $\perp_A$
ii) every directed subset of A has a *lub*.

Examples of *cpo*'s are given in figure 3.1. In each of these examples if x appears below y then $x \leq y$. The first example is a simple method of turning N into a trivial *cpo*, $N_\perp$. An artificial least element $\perp_N$ is appended and the only nontrivial relationship that holds is that $\perp_N \leq_{N_\perp} x$ for every x in $N_\perp$. The method can be used to turn an arbitrary set S into the *cpo* $S_\perp$. Example b) is another way of turning N into a *cpo*, by adding a least element $\perp_N$ and a greatest element ∞. We denote this by N^∞. The partial order is

$n \leq_{N^\infty} m$ if
i. n is $\perp_N$
or ii. m is ∞
or iii. n is less than or equal to m in the usual numerical sense.

Example c) is obtained by appending two extra infinities, ∞_1, ∞_2. Example d) is a partial order but not a complete partial order. The chain $1 \leq 3 \leq 5 \ldots$ lacks a *lub*. Example e) is a *cpo* obtained by adding an isolated element * to N^∞. When there is no possibility of confusion we will refer to $\leq_A$, $\bigvee_A$ simply as $\leq$, $\bigvee$ respectively. However, sometimes we will insert the subscripts, even when they can be determined by the context, as visual aids to the reader. As with partial orders, $\langle A, \leq_A \rangle$ will be referred to as A when the ordering $\leq_A$ is unimportant or apparent from the context.

If D, D' are directed subsets of a *cpo* A and $D \subseteq D'$ then it is easy to check that $\bigvee D \leq \bigvee D'$. More generally we say D' *dominates* D if for every $d \in D$ there is some $d' \in D'$ such that $d \leq d'$.

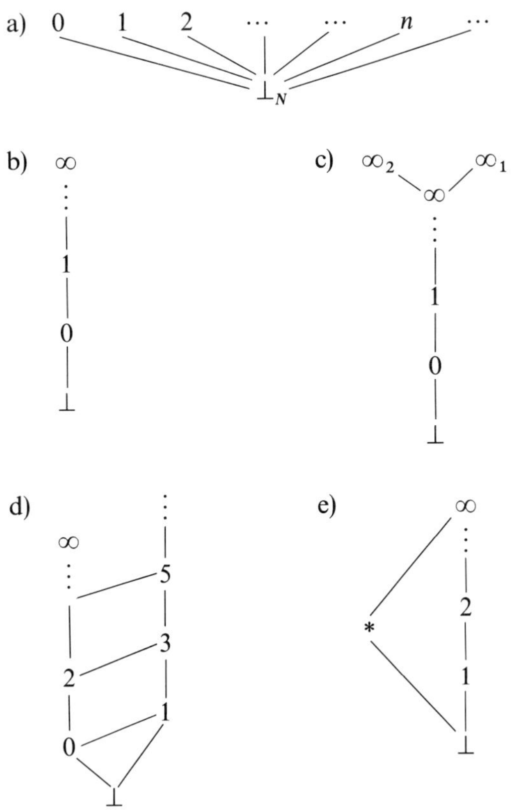

Figure 3.1
Examples of partial orders, *cpo*'s.

LEMMA 3.1.1 If D, D' are directed subsets of a *cpo* A and D' dominates D then $\bigvee D \leq \bigvee D'$.

Proof Let d, d' denote $\bigvee D, \bigvee D'$ respectively. We show $d \leq d'$. Let $a \in D$. Then there is some $a' \in D'$ such that $a \leq a'$. Consequently $a < d'$. This is true for every $a \in D$ and so by the definition of *lub*, $d < d'$. □

We now present a more complicated but equally useful property of directed sets. Let $D = \{d_{ij}, i \in I, j \in J\}$ be a doubly indexed subset of a *cpo* A. Let D_i denote the set $\{d_{ij}, j \in J\}$ for each $i \in I$ and analogously let D^j denote the set $\{d_{ij}, i \in I\}$ for each $j \in J$. Suppose

i. D is directed
ii. each D_i, D^j are directed with *lub* d_i, d^j respectively
iii. the sets $\{d_i, i \in I\}$ and $\{d^j, j \in J\}$ are also directed.

Then we have the following lemma.

LEMMA 3.1.2

a) $\bigvee D = \bigvee \{d_i, i \in I\}$

b) $\bigvee D = \bigvee \{d^j, j \in J\}$.

Proof We prove a) and b) follows by symmetry. First notice that the set $\{d_i, i \in I\}$ dominates D and therefore by the previous lemma $\bigvee D \leq \bigvee \{d_i, i \in I\}$. Also for each $i \in I$, $D_i \subseteq D$ and therefore $d_i \leq \bigvee D$. This is true for each $i \in I$ and so $\bigvee \{d_i, i \in I\} \leq \bigvee D$. □

This lemma has so many premises that it hardly seems worthwhile to state it. However, it will be very useful because many situations arise in which these premises are naturally satisfied. It is most often used in the sequence

$$\bigvee \{\bigvee \{d_{ij}, i \in I\}, j \in J\}$$
$$= \bigvee \{d_{ij}, i \in I, j \in J\}$$
$$= \bigvee \{\bigvee \{d_{ij}, j \in J\}, i \in I\},$$

where all relevant sets are known to be directed.

The Cartesian product of two *cpos* can easily be made into another *cpo*. For any two *cpos* $\langle A_1, \leq_{A_1} \rangle$, $\langle A_2, \leq_{A_2} \rangle$ and $a_1, a_1' \in A_1$, $a_2, a_2' \in A_2$ let

$\langle a_1, a_2 \rangle \leq \langle a_1', a_2' \rangle$

if both $a_1 \leq_{A_1} a_1'$ and $a_2 \leq_{A_2} a_2'$.

LEMMA 3.1.3 $\langle A_1 \times A_2, \leq \rangle$, with $\leq$ defined above, is a *cpo*.

Proof $\langle \bot_{A_1}, \bot_{A_2} \rangle$ is the least element. Let $D \subseteq A_1 \times A_2$ be directed. Define

$D_1 = \{a \in A_1, \langle a, a' \rangle \in D \text{ for some } a' \in A_2\}$

$D_2 = \{a \in A_2, \langle a', a \rangle \in D \text{ for some } a' \in A_1\}$.

Then both D_1 and D_2 are also directed subsets of A_1, A_2 respectively. Let d denote the pair $\langle \bigvee D_1, \bigvee D_2 \rangle$. Then it is easy to check that d is the *lub* of D. □

This construction can be extended in the obvious way to more general products $A_1 \times \cdots \times A_n$.

For any set S and *cpo* A let $(S \longrightarrow A)$ be the set of all functions from S to A. For f, g in $(S \longrightarrow A)$ let

$f \leq g$ if $f(s) \leq_A g(s)$ for every s in S.

LEMMA 3.1.4. $\langle (S \longrightarrow A), \leq \rangle$, with $\leq$ defined above, is a *cpo*.

Proof The least element $\bot$ is defined by

$\bot(s) = \bot_A$ for every s in S.

Let D be a directed set of functions from $(S \longrightarrow A)$. Define f_D in $(S \longrightarrow A)$ by

$f_D(s) = \bigvee_A \{g(s), g \in D\}$.

Note that this is well-defined since D is directed. It is straightforward to show that f_D is in fact the *lub* of D. □

When dealing with structured sets, rather than simply sets with no associated structure, it usually proves fruitful to consider structure-preserving functions in place of arbitrary functions. In the case of partial orders this is captured in the definition of monotonicity; for *cpo*'s it arises from *continuity*.

Let $\langle A, \leq_A \rangle$, $\langle A', \leq_{A'} \rangle$ be two *cpos* and $f\colon A \longrightarrow A'$ function. Then f is *continuous* if for every directed set $D \subseteq A$,

i) $f(D)$ is directed
ii) $f(\bigvee_A D) = \bigvee_{A'} f(D)$.

Here we use $f(D)$ as a notation for the set $\{f(d), d \in D\}$. Condition i) is therefore an abbreviation for

i′) $\{f(d), d \in D\}$ is directed in A'

and condition ii) an abbreviation for

ii′) $f(\bigvee_A D) = \bigvee_{A'}\{f(d), d \in D\}$.

Note that, in general, continuous functions need not preserve the least elements. For example constant functions are continuous: for each a' in A', the function $C_{a'}: A \longrightarrow A'$, defined by $C_{a'}(a) = a'$ for every a in A, is continuous. Also the identity function from A to A is continuous and if $f: A \longrightarrow A'$, $g: A' \longrightarrow A''$ are continuous then their composition $g \circ f: A \longrightarrow A''$ is also continuous. The same is true of monotonic functions.

We now investigate the properties of continuous functions. It will be very convenient to introduce the convention that whenever we write a statement involving $\bigvee D$ to be true we actually mean that the set D is directed and the statement is true of its *lub*. With this convention we avoid many distracting repetitions of phrases such as "the sets D, D' are directed and ... ." For example we may now say that a function $f: A \longrightarrow A'$ is continuous if for every directed set D in A

$f(\bigvee_A D) = \bigvee_{A'} f(D)$.

LEMMA 3.1.5 Continuous functions are monotonic.

Proof Let f be continuous and $a \leq a'$. Then $\{a, a'\}$ is directed with *lub* a'. So $f(a') = \bigvee f(\{a, a'\})$, i.e. $f(a) \leq f(a')$. □

Of course not all monotonic functions are continuous. Consider the function $f: N^\infty \longrightarrow \Theta$, where Θ is the trivial two-point lattice, also a *cpo*:

$\top$
|
$\bot$

defined by

$f(x) = \bot$ if $x \in N$ or $x = \bot_N$

$= \top$ otherwise.

Then f is monotonic but not continuous; N is a directed set in N^∞, with *lub* ∞. However

$f(\bigvee N) = f(\infty) = \top \neq \bot = \bigvee\{\bot\} = \bigvee f(N)$.

However if f is monotonic it is true that $f(D)$ is directed for every directed set D. We also have the following lemma.

LEMMA 3.1.6 $f: A \longrightarrow A'$ is continuous if and only if it is monotonic and for every directed set D in A

$f(\bigvee_A D) \leq \bigvee_{A'} f(D)$.

Proof Obviously if f is continuous it is monotonic and satisfies $f(\bigvee_A D) \leq \bigvee_{A'} f(D)$. So suppose the latter. We first show that since f is monotonic, $f(D)$ is directed—and consequently $\bigvee_{A'} f(D)$ makes sense. Let $f(a), f(a') \in f(D)$. Then $a, a' \in D$ and since D is directed there exists some $a'' \in D$ such that $a \leq a''$, $a' \leq a''$. Now f is monotonic so that $f(a) \leq f(a'')$, $f(a') \leq f(a'')$. So $f(a'')$ is an upper bound in $f(D)$ of $\{f(a), f(a')\}$.

We must now show $f(\bigvee_A D) = \bigvee_{A'} f(D)$, and because of the hypothesis it is sufficient to show

$\bigvee_{A'} f(D) \leq f(\bigvee_A D)$.

Let d denote $\bigvee_A D$. Then $a \leq d$ for every a in D. Since f is monotonic, $f(a) \leq_{A'} f(d)$ for every a in D. Therefore $\bigvee_{A'} f(D) \leq f(d)$. □

If A is a Cartesian product $A_1 \times A_2$ we could demand that a function $f: A \longrightarrow A'$ be continuous in each of its arguments separately. We now show that this coincides with the definition of continuity given above. The function $f: A_1 \times A_2 \longrightarrow A'$ is *left-continuous* if, for every a_0 in A_2 and every directed subset D of $A_1 \times A_2$, all of whose elements are of the form $\langle a, a_0 \rangle$,

$f(\bigvee_{A_1 \times A_2} D) = \bigvee_{A'} f(D)$.

There is a corresponding definition of right-continuous function, using A_1 in place of A_2.

PROPOSITION 3.1.7 $f: A_1 \times A_2 \longrightarrow A'$ is continuous if and only if it is both left-continuous and right-continuous.

Proof Obviously if f is continuous it is both left-continuous and right-continuous. So suppose it is left-continuous and right-continuous. We show f is continuous. Let D be directed in $A_1 \times A_2$. It is easy to see that, because f is both left- and right-continuous, $f(D)$ is directed. As in lemma 3.1.3 let

$D_1 = \{a \in A_1, \langle a, a' \rangle \in D \text{ for some } a' \in A_2\}$

$D_2 = \{a \in A_2, \langle a', a \rangle \in D \text{ for some } a' \in A_1\}$.

Then because D is directed $D_1 \times D_2$ is also directed. Once more because f is both left- and right-monotonic it follows that $f(D_1 \times D_2)$ is directed. Then

$$\begin{aligned} f(\bigvee D) &= f(\langle \bigvee D_1, \bigvee D_2 \rangle) \quad \text{by definition} \\ &= \bigvee \{f(d_1, \bigvee D_2), d_1 \in D_1) \quad \text{since } f \text{ is left-continuous} \\ &= \bigvee \{\bigvee \{f(d_1, d_2), d_2 \in D_2\}, d_1 \in D_1\} \quad \text{since } f \text{ is right-continuous} \\ &= \bigvee f(D_1 \times D_2) \quad \text{by lemma 3.1.2.} \end{aligned}$$

Now it is trivial to establish that $D \subseteq D_1 \times D_2$ and D dominates $D_1 \times D_2$ so that $f(D_1 \times D_2)$ contains $f(D)$ and is dominated by it. It follows from lemma 3.1.1 that

$$\bigvee f(D_1 \times D_2) = \bigvee f(D)$$

and therefore

$$f(\bigvee D) = \bigvee f(D).$$ □

This proposition is often quite useful in showing that particular functions are continuous. It will be very convenient in §3.3 when we consider particular *cpo*'s.

We use $[A \longrightarrow A']$ to denote the set of continuous functions from the *cpo* A to the *cpo* A'. This set can be ordered in the same way as $(A \longrightarrow A')$, the set of all functions from A to A':

$f \leq g$ if for every a in A, $f(a) \leq_{A'} g(a)$.

This ordering if often known as the *induced pointwise ordering.* We will show that $[A_1 \longrightarrow A_2]$ with this ordering is itself a *cpo.* It is convenient to first prove a lemma.

LEMMA 3.1.8 Let $F \subseteq [A \longrightarrow A']$ be a directed set of functions. Define $f: A \longrightarrow A'$ by

$$f(a) = \bigvee_{A'} \{g(a), g \in F\}.$$

Then f is well-defined and continuous.

Proof F is directed and therefore, for each a in A, $\{g(a), g \in F\}$ is directed. So f is well-defined. Moreover for directed $D \subseteq A$,

$$
\begin{aligned}
f(\bigvee D) &= \bigvee\{g(\bigvee D), g \in F\} \\
&= \bigvee\{\bigvee\{g(d), d \in D\}, g \in F\} \\
&= \bigvee\{\bigvee\{g(d), g \in F\}, d \in D\} \quad \text{from lemma 3.1.2} \\
&= \bigvee f(D).
\end{aligned}
$$

□

PROPOSITION 3.1.9 $[A \longrightarrow A']$ is a *cpo* under the induced pointwise ordering.

Proof The function $\perp$ defined, as in lemma 3.1.4, by

$\perp(a) = \perp_A$ for every $a \in A$,

is obviously the least continuous function from A to A'. Let $F \subseteq [A \longrightarrow A']$ be directed. Define $\bigvee F : A \longrightarrow A'$ by

$(\bigvee F)(a) = \bigvee_{A'}\{g(a), g \in F\}.$

From the previous lemma $\bigvee F$ is well-defined and is in $[A \longrightarrow A']$. It is trivial to show that $\bigvee F$ is in fact the *lub* of F. □

Recursively defined processes will be interpreted semantically using least fixpoints. These are now explained and their mathematical properties are derived.

DEFINITION 3.1.10 Let $f \in [A \longrightarrow A]$. The element $a \in A$ is called a *fixpoint* of f if $a = f(a)$. It is called the *least fixpoint* of f if, in addition, $a \leq_A a'$ for every fixpoint, a', of f.

Obviously least fixpoints are unique if they exist. The next proposition shows that they always exist for continuous functions.

PROPOSITION 3.1.11 Every f in $[A \longrightarrow A]$ has a least fixpoint.

Proof Define x_f^n by induction on n:

$x_f^0 = \perp_A$

$x_f^{n+1} = f(x_f^n).$

Then $x_f^0 < x_f^1$ because x_f^0 is the least element in A. Since f is monotonic we have $f(x_f^0) < f(x_f^1)$, i.e. $x_f^1 < x_f^2$. Continuing thus we have the chain

$x_f^0 < x_f^1 < x_f^2 < \cdots < x_f^n < \cdots$

Now A is a *cpo* and so this chain has a limit $\bigvee\{x_f^n, n \geq 0\}$, which we denote by x_f.

i. We show x_f is a fixpoint of f:

$$f(x_f) = \bigvee\{f(x_f^n), n \geq 0\}, \text{ by the continuity of } f$$

$$= \bigvee\{x_f^n, n \geq 1\}$$

$$= x_f.$$

ii. We show x_f is the least fixpoint. Let y be any element of A satisfying $f(y) = y$. Then

a) $x_f^0 \leq y$ because x_f^0 is $\bot_A$.
b) Assume $x_f^n \leq y$.
Then $x_f^{n+1} = f(x_f^n)$
$\leq f(y)$ by induction and the monotonicity of f
$= y$ since y is a fixpoint of f.

Therefore y is an upper bound for the chain $\{x_f^n, n \geq 0\}$ and so $x_f \leq y$. □

The element x_f has in fact a stronger characterization than that of least fixpoint of f. An element $a \in A$ is a *pre-fixpoint* of f if

$$f(a) \leq a.$$

Obviously x_f is a pre-fixpoint of f. Moreover it is the least pre-fixpoint: if b satisfies $f(b) \leq b$ then $x_f \leq a$. This can be proven as in ii. of the last proposition.

The least fixpoints of functions over simple *cpos* such as $N_\bot$ are rather uninteresting. If f is a *strict* function over any *cpo*, i.e. $f(\bot) = \bot$, then its least fixpoint is obviously $\bot$. If $f \in [N_\bot \longrightarrow N_\bot]$ and is not strict then it must be a constant function; if it is of the form C_n i.e. the constant function which always returns n, then its least fixpoint (and unique fixpoint) is n.

The association between a function f and its least fixpoint x_f is itself a mapping. For historical reasons we use Y to denote this mapping. So Y is a function from $[A \longrightarrow A]$ to A, defined by $Y(f) = x_f$. Y should in fact be subscripted by A, but the A in question will always be clear from the context. We now show that Y itself is a continuous function, i.e. $Y \in [[A \longrightarrow A] \longrightarrow A]$.

PROPOSITION 3.1.12 Y is a continuous function from $[A \longrightarrow A]$ to A.

Proof We use lemma 3.1.6.

i. We show Y is monotonic. Let $f \leq g$. Then

a) $x_f^0 \leq_A x_g^0$ since both are equal to $\bot_A$.
b) Assume $x_f^n \leq x_g^n$.
Then $x_f^{n+1} = f(x_f^n)$
$\leq_A f(x_g^n)$ by induction hypothesis and monotonicity of f
$\leq_A g(x_g^n)$ since $f \leq g$
$= x_g^{n+1}$.

Therefore $x_f^n \leq x_g^n$ for every $n \geq 0$. It follows that $x_f^n \leq x_g$ for every $n \geq 0$ since $x_g^n \leq x_g$. So x_g is an upper bound of the set $\{x_f^n, n \geq 0\}$ from which we obtain $x_f \leq_A x_g$, i.e. $Y(f) \leq_A Y(g)$.

ii. Let F be a directed family of functions in $[A \longrightarrow A]$ and let f denote $\bigvee F$. Recall that f is defined by

$$f(x) = \bigvee \{g(x), g \in F\}.$$

We need to show $Y(f) \leq_A \bigvee Y(F)$, i.e.

$$x_f \leq_A \bigvee \{x_g, g \in F\}. \tag{$*$}$$

For convenience let d denote $\bigvee \{x_g, g \in F\}$. We show that $f(d) = d$, from which $(*)$ will follow by the definition of x_f.

Consider the set $\{g'(x_g), g' \in F, g \in F\}$. Because Y is monotonic and each element of F is monotonic it is directed. It contains the set $\{g(x_g), g \in F\}$ and is dominated by it. Therefore

$$\bigvee \{g(x_g), g \in F\} = \bigvee \{g'(x_g), g' \in F, g \in F\}.$$

Then

$$\begin{aligned} d &= \bigvee \{x_g, g \in F\} && \text{by definition} \\ &= \bigvee \{g(x_g), g \in F\} && \text{by definition of } x_g \\ &= \bigvee \{g'(x_g), g' \in F, g \in F\} && \text{from above} \\ &= \bigvee \{\bigvee \{g'(x_g), g' \in F\}, g \in F\} && \text{from lemma 3.1.2} \\ &= \bigvee \{f(x_g), g \in F\} && \text{by definition of } f \\ &= f(d) && \text{since } f \text{ is continuous.} \end{aligned}$$

□

We have assembled all of the facts we need of *cpos* in order to use them as interpretations for recursively defined processes. However we use a very restricted kind of *cpo*.

This class of *cpo*'s, *algebraic cpo*'s, is the last concept to be explained in this section. Let A be a *cpo*. An element $a \in A$ is *compact* or *finite* if whenever $a \leq \bigvee D$, D a directed subset of A, there exists some d in D such that $a \leq d$. A is an *algebraic cpo* if for every a in A

$$a = \bigvee \{d, d \leq a, d \text{ compact}\}.$$

In figure 3.1 the *cpos* in a), b), and c) are algebraic. In example c) both the elements ∞_1 and ∞_2 are compact. The *cpo* in example e) is not algebraic. The element $*$ is not compact because $* \leq \bigvee N$ but $* \leq n$ for no n in N. Therefore $\{d, d \leq *, d \text{ compact}\}$ is $\{\bot\}$ and obviously $* \neq \bigvee \{\bot\}$.

Intuitively the compact elements are the semantic denotations of (syntactically) finite processes. So if an interpretation is algebraic every (recursively defined) process is semantically the limit of a directed set of finite processes. Interpretations such as in example c) will not occur in our theory. These are counterintuitive, as the finite objects ∞_1, ∞_2 dominate the infinite object ∞ which contradicts the informal property of a finite object: that it should contain only a finite amount of information.

We use $\mathrm{Fin}(A)$ to denote the set of finite or compact elements of A and for $a \in A$ let $\mathrm{Fin}(a)$ denote $\{d \in \mathrm{Fin}(A), d \leq a\}$. Then algebraicness may be restated as

$$a = \bigvee_A \mathrm{Fin}(a).$$

Algebraic *cpos* are completely determined by their finite elements. More precisely, if A, A' are algebraic *cpo*'s such that $\langle \mathrm{Fin}(A), \leq_A \rangle$, $\langle \mathrm{Fin}(A'), \leq_{A'} \rangle$ are isomorphic as partial orders, then it is easy to check that A and A' are isomorphic as *cpo*'s.

3.2 Σ-Domains

Let Σ be a signature which contains a distinguished function symbol of arity 0, Ω. In future we always assume that Ω is in every signature. It represents syntactically the least element of the interpretations.

A *Σ-domain* is a triple $\langle A, \leq_A, \Sigma_A \rangle$ where

i. $\langle A, \leq_A \rangle$ is an algebraic *cpo*
ii. for each f in Σ of arity k, there is a continuous function $f_A : A^k \longrightarrow A$ in Σ_A
iii. Ω_A is $\bot_A$.

EXAMPLE 3.2.1 Let Σ be the signature $\{\Omega, \mathit{Zero}, \mathit{Succ}, \mathit{Pred}, \mathit{Plus}\}$ with the usual arities.

1. $\langle N_\perp, \leq, \Sigma_{N_\perp}\rangle$ is a Σ-domain, where $\langle N_\perp, \leq\rangle$ is the *cpo* in figure 3.1 a) and

$$\begin{aligned}
Zero_{N_\perp} &= 0\\
Succ_{N_\perp}(x) &= \perp \text{ if } x = \perp\\
&= x + 1 \text{ otherwise}\\
Pred_{N_\perp}(x) &= \perp \text{ if } x = \perp\\
&= 0 \text{ if } x = 0\\
&= x - 1 \text{ otherwise}\\
Plus_{N_\perp}(x, y) &= \perp \text{ if } x = \perp \text{ or } y = \perp\\
&= x + y \text{ otherwise.}
\end{aligned}$$

2. $\langle N^\infty, \leq, \Sigma_{N^\infty}\rangle$ is a Σ-domain, where $\langle N^\infty, \leq\rangle$ is the *cpo* in figure 3.1 b) and

$$\begin{aligned}
Zero_{N_\perp^\infty} &= 0\\
Succ_{N_\perp^\infty}(x) &= \perp \text{ if } x = \perp\\
&= \infty \text{ if } x = \infty\\
&= x + 1 \text{ otherwise}\\
Pred_{N_\perp^\infty}(x) &= \perp \text{ if } x = \perp\\
&= \infty \text{ if } x = \infty\\
&= 0 \text{ if } x = 0\\
&= x - 1 \text{ otherwise}\\
Plus_{N_\perp^\infty}(x, y) &= \perp \text{ if } x = \perp \text{ or } y = \perp\\
&= \infty \text{ if } x = \infty \text{ or } y = \infty\\
&= x + y \text{ otherwise}
\end{aligned}$$

Homomorphisms between Σ-domains are required to preserve the extra structure. A function $h : A \to B$ is a *Σ-domain homomorphism* from the Σ-domain $\langle A, \leq_A, \Sigma_A\rangle$ to the Σ-domain $\langle B, \leq_B, \Sigma_B\rangle$ if

i. it is continuous
ii. for every f in Σ, $h(f_A(\underline{a})) = f_B(h(\underline{a}))$.

This is a *Σ-domain isomorphism* if in addition there is a Σ-domain homomorphism

$k : \langle B, \leq_B, \Sigma_B \rangle \rightarrow \langle A, \leq_A, \Sigma_A \rangle$

such that

i. $k \circ h = id_A$
ii. $h \circ k = id_B$.

In this case we say that $\langle A, \leq_A, \Sigma_A \rangle$ and $\langle B, \leq_B, \Sigma_B \rangle$ are *isomorphic as Σ-domains.*

Notice that Σ-domain homomorphisms are always strict, i.e. they preserve least elements:

$h(\perp_A) = h(\Omega_A) = \Omega_B = \perp_B$.

There is also a rather simple characterization of isomorphisms. Every Σ-domain is automatically a *Σ-po* algebra because a *cpo* is a partial order and continuous functions are monotonic. To say that two Σ-domains are isomorphic as *Σ-po* algebras means that there are structure-preserving monotonic maps between them, which are inverses of each other. *A priori* this is a weaker requirement than being isomorphic as Σ-domains, since these maps may not be continuous. However this is not the case.

PROPOSITION 3.2.2 The Σ-domains $\langle A, \leq_A, \Sigma_A \rangle$ and $\langle B, \leq_B, \Sigma_B \rangle$ are isomorphic if and only if they are isomorphic as *Σ-po* algebras.

Proof From the hypothesis there exists two monotonic, function-preserving maps

$h : \langle A, \leq_A, \Sigma_A \rangle \rightarrow \langle B, \leq_B, \Sigma_B \rangle$

$k : \langle B, \leq_B, \Sigma_B \rangle \rightarrow \langle A, \leq_A, \Sigma_A \rangle$

satisfying

$k \circ h = id_A, \qquad h \circ k = id_B.$

We must show both h and k are continuous. In fact we only show h is continuous as the argument is symmetric in A, B.

We prove $h(\bigvee_A D) = \bigvee_B h(D)$ for an arbitrary directed set D in A. For convenience let d denote $\bigvee_A D$. Then $h(d)$ is an upper bound of $h(D)$, for if $h(a) \in h(D)$, $h(a) \leq h(d)$ by the monotonicity of h. Moreover it is the *lub* of $h(D)$. For let b be any upper bound of $h(D)$. Then $k(b)$ is an upper bound for D because $h(a) \leq b$ implies $k \circ h(a) \leq k(b)$ by the monotonicity of k, i.e. $a \leq k(b)$. Therefore $d \leq k(b)$ and so $h(d) \leq h \circ k(b)$ by the monotonicity of h, i.e. $h(d) \leq b$. We have established $h(d) = \bigvee_B h(D)$. □

When dealing with Σ-domains we use the usual abbreviations: $\langle A, \leq_A, \Sigma_A \rangle$ will be written as $\langle A, \leq_A \rangle$ when the functions Σ_A are clear from the context and as $\langle A, \leq \rangle$ when the lack of a subscript causes no ambiguity. Moreover we often write simply A for $\langle A, \leq_A, \Sigma_A \rangle$ when both $\leq_A$ and Σ_A are apparent. Finally we abbreviate Σ-domain homomorphism and Σ-domain isomorphism to simply Σ-homomorphism and Σ-isomorphism respectively, when we know that they are mappings between domains.

As we have already stated Σ-domains are also Σ-*po* algebras. So we can apply the definitions of §1.4 to say when a given Σ-domain satisfies a set of equations, or more precisely inequations. For example $\langle N_\perp, \leq, \Sigma_{N_\perp} \rangle$ satisfies the set of equations E_N, given in §1.2, whereas the Σ-domain $\langle N^\infty, \leq, \Sigma_{N^\infty} \rangle$ satisfies the different equations given in example 1.4.7.

Let $\mathscr{C}\mathscr{C}(E)$ denote the class of Σ-domains which satisfy the equations E.

THEOREM 3.2.3 $\mathscr{C}\mathscr{C}(E)$ has an initial object.

Remark By this we mean there is a Σ-domain CI_E which satisfies the equations E, with the property that if A is any Σ-domain which satisfies E there is a unique Σ-homomorphism (i.e. Σ-domain homomorphism)

$i_A : CI_E \longrightarrow A.$

This theorem will be proven in the next section. We leave the reader to check that corollary 1.1.10, which states that initial objects are unique up to isomorphism, remains true in this new setting.

In part I we have two different uses for the corresponding initial objects I_E. The first is to give syntax. If E is the empty set then I_E is (isomorphic to) the term algebra T_Σ. This can be viewed as a simple structural language for defining processes, at least if we choose an appropriate signature Σ. Its main attraction in this role is the unique "meaning"-function $i_A : T_\Sigma \longrightarrow A$; for any Σ-*po* algebra A it assigns a unique meaning in A to every term of T_Σ. We will *not* use the corresponding continuous object in this role. This Σ-domain is often referred to as CT_Σ. Although every element in CT_Σ can be assigned a unique meaning in an arbitrary Σ-domain, it is difficult to look on CT_Σ as a language in the normal sense of the word. There is a relatively natural representation of CT_Σ as finite and infinite words, or trees, (see ADJ 1978), but the infinite nature of these words strains our intuitive understanding of syntactic descriptions. These should at least be finite. Moreover a language should be countable and probably recursive. However CT_Σ, for nontrivial Σ, is uncountable. In the next chapter we introduce a language REC_Σ which will substitute for CT_Σ in its role as syntax.

The second use is to give semantic domains. For specific sets of equations E, we found relatively natural representations of I_E, as sets of trees, which are fully abstract with respect to various Testing preorders. The attraction of I_E in this role is that it has an associated proof system **DED**(E). This proof system is sound and complete with respect to interpretation in I_E and consequently with respect to the various Testing preorders.

The Σ-domains CI_E will play a similar role as semantic domains. The interpretation **fAT** can be extended to a Σ-domains **AT**, which is in fact initial for an equational class. This is the subject of the final section of this chapter. Moreover there is a close link between the interpretations CI_E and proof systems similar to **DED**(E). This is now explained.

For a given set of equations E consider the proof system **DED**(E) defined in §1.4. Since every Σ-domain is a Σ-*po* algebra we have, by lemma 1.4.8,

$\vdash_E t \leq t'$ implies $t \leq_A t'$ for every A in $\mathscr{CC}(E)$.

In particular this is true for the initial Σ-domain CI_E. However, the converse, completeness, will not be true. One reason for this is the presence of Ω, and its special role in interpretations. For example $\Omega \leq_A t$ for any term t and any Σ-domain A. In particular $\Omega \leq_{CI_E} t$ but $\Omega \leq t$ is *not* a theorem of **DED**(E) for any term t. We now modify **DED**(E) to take into consideration the special role of Ω. We add one more rule to the system:

Ω-*rule* $\overline{\Omega \leq t}$ for every t in $T_\Sigma(X)$.

Let **ΩDED**(E) be this new proof system. It is convenient to extend the preorder notation for proof systems in part I to these new proof system; we write $t \leq_{E\Omega} t'$ to mean $t \leq t'$ can be derived in the proof system **ΩDED**(E). We can now derive a strong characterization of initiality in continuous equational classes in terms of **ΩDED**(E), similar to that of corollary 1.4.12.

A Σ-domain A is *finitary* if

i) for every term t in T_Σ, $i_A(t)$ is a finite element in A
ii) for every finite element a of A there exists a term t in T_Σ such that $i_A(t) = a$.

So A is finitary if every (syntactically) finite term is interpreted in A as a finite element and every finite element of A is denotable by a (syntactically finite) term.

THEOREM 3.2.4 The Σ-domain A is initial in $\mathscr{CC}(E)$ if and only if

i) it is finitary
ii) **ΩDED**(E) is sound and complete with respect to $\leq_A$, restricted to T_Σ.

Thus, if we use initial domains as semantic models this theorem automatically gives a sound and complete proof system for finite terms. This is of interest because it can be extended in a standard manner to a sound and complete proof system for the entire syntactic language. This is discussed in §4.2.

These semantic domains are also evaluated with respect to behavioral preorders and as in part I there is a close connection between initiality and full abstraction. However, this connection depends on the behavioral relation satisfying certain continuity constraints and a discussion of these is left until §4.4.

To prove these two central theorems we must introduce a new class of Σ-algebras, which we call Σ-predomains. These are investigated in the next section.

3.3 Σ-Predomains

Σ-domains are obtained from Σ-*po* algebras by adding two extra requirements: the operators should satisfy a continuity constraint and the carrier should contain a least element. If we demand only the latter, we obtain Σ-predomains. Let Σ be a signature which contains the distinguished 0-ary function symbol Ω. A Σ-*predomain* is a triple $\langle A, \leq_A, \Sigma_A \rangle$ where

i) $\langle A, \leq_A \rangle$ is a partial order with a least element $\perp_A$
ii) for each f in Σ of arity k, there is a monotonic function $f_A : A^k \longrightarrow A$
iii) Ω_A is the least element $\perp_A$.

Thus it is a very mild modification of the definition of a Σ-*po* algebra. However it is an important modification because every Σ-domain A which satisfies a simple criterion can be generated in a canonical manner from a Σ-predomain. The Σ-predomain in question is simply the set of finite elements of A. We will use this fact to construct the initial object in $\mathscr{CC}(E)$. Before doing so we examine the relationship between the proof system $\mathbf{\Omega DED}(E)$ and initial Σ-predomains. Σ-predomains are *a priori* Σ-*po* algebras and therefore we may employ the definition of §1.4 to say when a Σ-predomain satisfies a set of inequations E. We let $\mathscr{CP}(E)$ denote the class of Σ-predomains which satisfy E. Of course these inequations may contain occurrences of the special symbol Ω since it is assumed to be part of the signature Σ.

EXAMPLE 3.3.1 Because a Σ-domain is automatically a Σ-predomain both structures in example 3.2.1 can be considered as Σ-predomains. The first one satisfies the usual equations for the signature in question, called E_N in §1.2, while the second satisfies those of example 1.4.7. In addition both satisfy

$$\Omega \leq x$$
$$Succ(\Omega) = \Omega$$
$$Pred(\Omega) = \Omega$$
$$Plus(\Omega, x) = \Omega$$
$$Plus(x, \Omega) = \Omega.$$

The notion of a Σ-predomain homomorphism is borrowed directly from Σ-*po* algebras: it should be monotonic and structure-preserving, i.e. a Σ-homomorphism. In particular it preserves the operator Ω. With this notion of homomorphism the definition of an initial object is: A collection of Σ-predomains, $\mathscr{CP}$ has an *initial object* I if

i) $I \in \mathscr{CP}$
ii) for every $A \in \mathscr{CP}$ there is a unique Σ-predomain homomorphism $h_A : I \longrightarrow A$.

THEOREM 3.3.2 For every set of equations E, $\mathscr{CP}(E)$ has an initial object.

Proof A Σ-predomain is in $\mathscr{CP}(E)$ if and only if it is in $\mathscr{C}(E\Omega)$, i.e. satisfies the inequations $E\Omega$ as a Σ-*po* algebra, where $E\Omega$ denotes the set of inequations E together with the extra equation $\Omega \leq x$. It follows that the initial object in $\mathscr{C}(E\Omega)$, which we know to exist from corollary 1.4.9, is also an initial Σ-predomain in $\mathscr{CP}(E)$. It is worth pointing out that this initial Σ-predomain has the structure $T_\Sigma/\leq_{E\Omega}$. □

In future we will use PI_E to denote the initial object in $\mathscr{CP}(E)$, which is unique up to isomorphism, and recall that the initial object in $\mathscr{C}(E)$ is referred to as I_E.

Experience with Σ-*po* algebras indicate that there is a close relationship between $\mathbf{\Omega DED}(E)$ and the initial object $\mathscr{CP}(E)$. We state the result and refer the reader to the proof of the corresponding result for Σ-*po* algebras, corollary 1.4.12.

COROLLARY 3.3.3 The Σ-predomain $\langle A, \leq_A, \Sigma_A \rangle$ is initial in $\mathscr{CP}(E)$ if and only if it is surjective and $\mathbf{\Omega DED}(E)$ is both sound and complete with respect to $\leq_A$ over T_Σ.

The relation $\leq_{E\Omega}$ over $T_\Sigma(X)$, generated by the proof system $\mathbf{\Omega DED}(E)$, also has a characterization similar to that of the relation $\leq_E$ generated by $\mathbf{DED}(E)$.

PROPOSITION 3.3.4 $\leq_{E\Omega}$ is the least substitution closed Σ-preorder over $T_\Sigma(X)$ which satisfies the inequations E and the additional inequation $\Omega \leq x$.

Although technically there is little difference between Σ-*po* algebras and Σ-predomains the presence of the extra structure in the latter has important consequences. Using the informal terminology of the introduction to part II, the objects in a Σ-*po* algebra can be viewed as totally defined or fully specified. On the other hand in a Σ-predomain we have, in addition to these total objects, what may be viewed as "partial objects." One such partial object is the least element, denoted by Ω. If $a \in \Sigma^0$, $f \in \Sigma^2$ then other partial objects could be denoted by $f(a, \Omega)$, $f(f(a, \Omega), \Omega)$, etc. The presence of what we call "partial objects" has a significant impact on the theoretical development of Σ-predomains. We give three examples which illustrate the difference between the theory of Σ-*po* algebras and that of Σ-predomains. Underlying all of these examples are the presence of "partial objects."

EXAMPLE 3.3.5 The initial Σ-*po* algebra which satisfies the empty set of equations is the term algebra T_Σ. Here $t \leq_{T_\Sigma} t'$ if and only if t and t' are syntactically identical, i.e. the partial order is simply the identity relation.

The initial Σ-predomain which satisfies the empty set of equations has the same carrier, T_Σ, and the functions are defined in the same syntactic manner. However the partial order, which we denote by $\preceq$ is nontrivial. For example $\Omega \preceq t$ for every t, $f(g(c, \Omega), \Omega) \preceq f(g(c, c'), f(\Omega, \Omega))$, etc., assuming f, g, c, c' are in Σ.

We now describe in detail the initial Σ-*po* algebra which satisfies the empty set of equations, $PI_\varnothing$, and thereby lend credence to the remarks in this example. Let $\preceq$ be the least Σ-preorder over $T_\Sigma(X)$ which satisfies

i) $\Omega \preceq t$ for every term t

That is, it is the least relation over $T_\Sigma(X)$ which satisfies i) together with

ii) $t \preceq t$ for each $t \in T_\Sigma$
iii) $t \preceq t'$ $t' \preceq t''$ implies $t \preceq t''$
iv) $\underline{t} \preceq \underline{t}'$ implies $f(\underline{t}) \preceq f(\underline{t}')$ for each $f \in \Sigma$.

Although it is defined over $T_\Sigma(X)$ for the moment we examine only its restriction to closed terms T_Σ. Essentially $t \preceq t'$ if t' can be obtained from t by replacing some occurrences of Ω by new subterms.

The relation $\preceq$ so defined is actually antisymmetric, i.e. it satisfies

$$t \preceq t', \quad t' \preceq t \quad \text{implies} \quad t = t'.$$

This is set as Q11 at the end of the chapter. Therefore $\langle T_\Sigma, \preceq \rangle$ is a partial order. Note also that Ω is the least element of this partial order. Moreover the syntactic functions f_{T_Σ}, mapping the vector of terms $\underline{t}$ to the term $f(\underline{t})$, is monotonic because of condition iv). Therefore $\langle T_\Sigma, \preceq \rangle$ is a Σ-predomain.

This is obviously surjective and it is equally trivial to remark that **ΩDED**(ϕ) is sound and complete with respect to $\preceq$ over T_Σ. By corollary 3.3.3 it follows that $\langle T_\Sigma, \preceq \rangle$ is isomorphic to $PI_\varnothing$.

EXAMPLE 3.3.6 We have seen that $I_\varnothing$ and $PI_\varnothing$ are quite different even though they have the same carrier and functions. In general I_E and PI_E are always different and may even have different carriers. Consider, for example, the signature Σ^2 of part I, augmented by the symbol Ω and the set of equations **A2**. For any action symbol a,

$$\Omega + aa\Omega =_{PI_{\mathbf{A2}}} \Omega + a\Omega + aa\Omega$$

because

$$\Omega + aa\Omega \leq_{\mathbf{A2}} (\Omega + a\Omega) + aa\Omega$$

and

$$\begin{aligned} \Omega + a\Omega + aa\Omega &\leq_{\mathbf{A2}} \Omega + a(a\Omega) + aa\Omega \\ &=_{\mathbf{A2}} \Omega + aa\Omega. \end{aligned}$$

On the other hand one can easily argue that

$$\Omega + aa\Omega \neq_{I_{\mathbf{A2}}} \Omega + a\Omega + aa\Omega$$

because these cannot be transformed into each other without using the inequation $\Omega \leq x$.

In short, for the inequations **A2**, the theory generated by $PI_{\mathbf{A2}}$ is significantly different than that generated by $I_{\mathbf{A2}}$.

EXAMPLE 3.3.7 The introduction of a least element can have drastic effects. A particularly catastrophic example is the theory of groups. A group is an algebra which has a constant symbol 1, called identity, a unary symbol $\bar{}$, called inverse, and a binary infix symbol $\cdot$, often called multiplication and which satisfies the equations

$$x \cdot 1 = 1 \cdot x = x$$

$$x \cdot \bar{x} = \bar{x} \cdot x = 1$$

$$x \cdot (y \cdot z) = (x \cdot y) \cdot z.$$

Let these equations be called G. For example if Σ consists of these operators and say two constant symbols a, b, the initial object in $\mathscr{C}(G)$ is the nontrivial free group over two generators a, b. However if we add Ω to Σ this group collapses to the trivial one point group; the initial object in $\mathscr{C}\mathscr{P}(G)$ consists of one element. This is because we can prove the equation

$$x = \Omega$$

from G and the new inequation $\Omega \leq x$:

$$\begin{aligned} x &= x \cdot 1 \\ &= x \cdot (\bar{\Omega} \cdot \Omega) \\ &\leq x \cdot (\bar{x} \cdot \Omega) \\ &= (x \cdot \bar{x}) \cdot \Omega \\ &= 1 \cdot \Omega \\ &= \Omega \end{aligned}$$

Fortunately, the phenomena seen in this example will not occur in our particular theories.

We now turn our attention to generating Σ-domains from Σ-predomains. For the moment we drop the function symbols and related functions in these structures. For explanatory purposes it is sufficient to consider partial orders and *cpo*'s. Let $\langle A, \leq_A \rangle$ be a partial order. An *ideal* in A, I, is an nonempty subset of A which satisfies

i) $x, y \in I$ implies there exists some $z \in I$ such that $x \leq_A z$ and $y \leq_A z$, i.e. I is *directed*
ii) $x \in I$, $y \leq_A x$ implies $y \in I$, i.e. I is *downwards closed.*

Let $\mathscr{I}(A)$ denote the set of ideals of A.

LEMMA 3.3.8 If A has a least element then $\langle \mathscr{I}(A), \subseteq \rangle$ is an algebraic *cpo*.

Proof It is obviously a partial order and the trivial ideal $\{\perp_A\}$ is the least element since every ideal must contain $\perp_A$. Let I be a directed set of ideals. We must show that I has a limit. Let $J = \bigcup \{I, I \in \mathscr{I}\}$; then J is downwards closed and because $\mathscr{I}$ is directed J is also directed, i.e. it is an ideal. It follows that J is the *lub* of $\mathscr{I}$ and so that $\langle \mathscr{I}(A), \subseteq \rangle$ is a *cpo*.

Let $in : A \longrightarrow \mathscr{I}(A)$ be defined by $in(a) = \{x, x \leq_A a\}$. It is easy to check that $in(a)$ is in fact an ideal. Moreover for any $I \in \mathscr{I}(A)$,

$I = \bigcup \{in(a), a \in I\},$

i.e.

$I = \bigcup \{in(a), in(a) \subseteq I\}.$

So to prove that I is an algebraic *cpo* it remains to show that the compact elements of $\mathscr{I}(A)$ are exactly those of the form $in(a)$ for some $a \in A$.

i) Let $in(a) \subseteq \bigcup \{I, I \in \mathscr{I}\}$. Then $a \in \bigcup \{I, I \in \mathscr{I}\}$ and so $a \in I$ for some $I \in \mathscr{I}$. Therefore $in(a) \subseteq I$. It follows that $in(a)$ is a compact element.

ii) Let I be a compact element in $\mathscr{I}(A)$. Then $I \subseteq \bigcup \{in(a), a \in I\}$ so that there exists some $a \in I$ such that $I \subseteq in(a)$, i.e. $I = in(a)$. Therefore every compact element is of the form $in(a)$ for some $a \in A$. □

This construction of completing a partial order may seem artificial but it can be viewed simply as a way of adding extra points to the partial order $\langle A, \leq_A \rangle$ so as to make it a *cpo*; one new point is added for every directed set in A. This is best seen by identifying a subset of the constructed *cpo* $\langle \mathscr{I}(A), \subseteq \rangle$ with the original partial order $\langle A, \leq_A \rangle$.

THEOREM 3.3.9 $\langle \mathscr{I}(A), \subseteq \rangle$ is the unique algebraic *cpo* (up to isomorphism) whose set of finite elements are isomorphic to $\langle A, \leq_A \rangle$ as partial order.

Proof Consider the function $in : A \longrightarrow \mathscr{I}(A)$ defined in the previous lemma. It can be considered as a mapping to $\mathrm{Fin}(\mathscr{I}(A))$ and as such it is surjective. Moreover it is easy to show that

$in(a) \subseteq in(a')$ if and only if $a \leq_A a'$.

It follows that $\langle A, \leq_A \rangle$ and $\langle \mathrm{Fin}(\mathscr{I}(A)), \subseteq \rangle$ are isomorphic as partial orders. We have already remarked at the end of §3.1 that an algebraic *cpo* is determined by its finite elements, from which the uniqueness follows. □

This theorem enables us to view $\langle \mathscr{I}(A), \subseteq \rangle$ as an extension of $\langle A, \leq_A \rangle$. To emphasize this view we will often refer to $\langle \mathscr{I}(A), \subseteq \rangle$ as the *ideal completion* of $\langle A, \leq_A \rangle$ and refer to it as $\langle A, \leq_A \rangle^\infty$ or simply A^∞. For example, if we complete the partial order described in figure 3.2 a) i), we obtain the *cpo* described in ii), which is isomorphic to that in iii), which we have referred to in the previous section as N^∞. Another example of a completion (up to isomorphism) is given in b). In each case the completion adds an extra limit point for each directed set in the partial order. This addition is carried

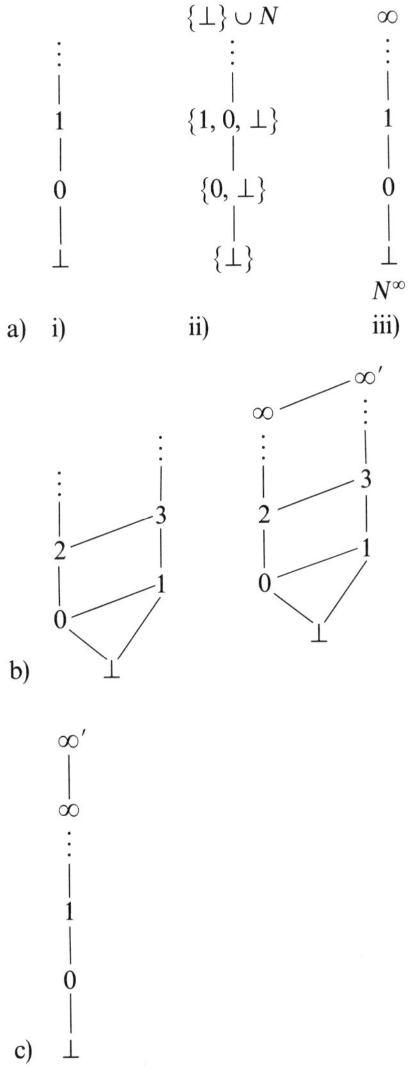

Figure 3.2
Ideal completions.

out in a blind fashion: a new limit point is added regardless of whether or not one exists already. So for example if we complete the partial order N^∞ (which is in fact a *cpo*) the effect is to add a limit point for the chain N, as shown in figure 3.2 c). However, certain directed sets generate the same limit point. Intuitively this happens when they are "tending to the same limit". For example, the two directed sets

$\{0, 2, 4, 6, \ldots\}$

$\{0, 4, 8, 12, \ldots\}$

in figure 3.2 b) are tending to the same limit and so generate the same limit point. Formally this means that they both dominate each other and so generate the same downward closure.

We now argue that this method of completion is canonical in that for a given partial order A, A^∞ is the "least" extension of A to a *cpo*. The most general statement of this fact says that any monotonic function from the partial order A into a *cpo* E can be extended uniquely onto A^∞. This is expressed as a "universal" property of the completion process.

THEOREM 3.3.10 If $k : A \longrightarrow E$ is any monotonic function from the partial order A to the *cpo* E then there exists a *unique* continuous function $ext(k) : A^\infty \longrightarrow E$ such that the following diagram commutes:

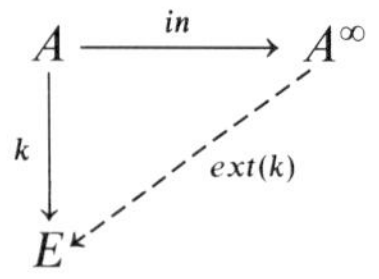

Proof i) We first define $ext(k)$. Informally every element of A^∞ can be viewed as the *lub* of a directed set of elements in A so that intuitively we may define $ext(k)(x)$ to be $\bigvee_E\{k(a), a \leq x\}$. Formally it is defined by

$$ext(k)(I) = \bigvee_E\{k(a), a \in I\}.$$

Since I is directed and k is monotonic this is well-defined. Moreover it is easy to check that it is continuous, using lemma 3.1.2. Finally the diagram above commutes because

$$\begin{aligned} ext(k)(in(a)) &= ext(k)(\{x, x \leq a\}) \\ &= \bigvee_E\{k(x), x \leq a\} \\ &= k(a). \end{aligned}$$

ii) We show that $ext(k)$ is unique. Let h, h' be any two continuous functions from A^∞ to E which make the diagram above commute, i.e. $h(in(a)) = k(a)$ and $h'(in(a)) = k(a)$ for every $a \in A$. We show h and h' are identical, i.e. $h(x) = h'(x)$ for every $x \in A^\infty$. Since every compact element of A^∞ is of the form $in(a)$ we already know that they coincide on $\mathrm{Fin}(A^\infty)$. Then

$$\begin{aligned} h(x) &= h(\bigvee \mathrm{Fin}(x)) \\ &= \bigvee_E \{h(d),\, d \in \mathrm{Fin}(x)\} \\ &= \bigvee_E \{h'(d),\, d \in \mathrm{Fin}(x)\} \\ &= h'(\bigvee \mathrm{Fin}(x)) \\ &= h'(x). \end{aligned}$$ □

Having explained the method of completing a partial order to obtain a *cpo* it is a simple matter to extend it to Σ-predomains. Let $\langle A, \leq_A, \Sigma_A \rangle$ be a Σ-predomain. The carrier A is a partial order with a least element and therefore we can construct the *cpo* A^∞. The functions f_A over A are extended to A^∞ by

$$f_{A^\infty}(\underline{I}) = \{x,\, x \leq f(\underline{a}) \text{ for some } \underline{a} \in \underline{I}\}.$$

LEMMA 3.3.11 $\langle A^\infty, \leq_{A^\infty}, \Sigma_{A^\infty} \rangle$ is a Σ-domain.

Proof It is sufficient to check that each function f_{A^∞} is well-defined and continuous, which is straightforward.

We use $\langle A, \leq_A, \Sigma_A \rangle^\infty$ or simply A^∞ to denote the Σ-domain $\langle A^\infty, \leq_{A^\infty}, \Sigma_{A^\infty} \rangle$ and it has a universal property similar to that expressed in theorem 3.3.10.

THEOREM 3.3.12 If $k : A \longrightarrow E$ is a Σ-predomain homomorphism from the Σ-predomain A to the Σ-domain E then there is a *unique* Σ-domain homomorphism $ext(k) : A^\infty \longrightarrow E$ which makes the following diagram commute:

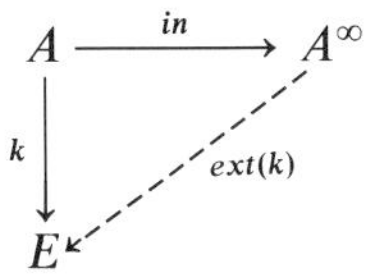

Proof The proof is identical to the corresponding one for partial orders except that one must ensure that $ext(k)$ is in fact a Σ-homomorphism. □

This completion process, together with the universal property of this theorem enables us to establish the main result of this section, namely that $\mathscr{CC}(E)$ has an initial object for any set of inequations E. We have already shown that $\mathscr{CP}(E)$ has an initial object and it will follow that if PI is an initial Σ-predomain in $\mathscr{CP}(E)$ its completion PI^∞ is an initial Σ-domain in $\mathscr{CC}(E)$. Before proving this it is convenient to first prove a lemma.

LEMMA 3.3.13 If $A \in \mathscr{CP}(E)$ then $A^\infty \in \mathscr{CC}(E)$.

Proof Let $\langle t_1, t_2 \rangle \in E$ and ρ_∞ an arbitrary A^∞-assignment. We must show $\rho_\infty(t_1) = \rho_\infty(t_2)$, where we also use ρ_∞ to denote the unique extension of ρ_∞ to $T_\Sigma(X)$. Since ρ_∞ is an A^∞-assignment and every element of A^∞ is the *lub* of a directed set of elements in A, we can find a directed set F of A-assignments such that $\rho_\infty(x) = (\bigvee F)(x)$ for every variable which appears in t_1 or t_2. Let $h: T_\Sigma(X) \longrightarrow A^\infty$ be defined by $h(t) = \bigvee\{\rho_A(t), \rho_A \in F\}$. Then h is a Σ-homomorphism which extends the A^∞-assignment ρ_∞ and therefore by uniqueness $h(t) = \rho_\infty(t)$. The result now follows:

$$\begin{aligned}\rho_\infty(t_1) &= \bigvee\{\rho_A(t_1), \rho_A \in F\}\\ &= \bigvee\{\rho_A(t_2), \rho_A \in F\} \quad \text{because } A \in \mathscr{CP}(E)\\ &= \rho_\infty(t_2)\end{aligned}$$

□

We can now prove theorem 3.2.3, which states that $\mathscr{CC}(E)$ always has an initial object:

PROPOSITION 3.3.14 If PI is initial in $\mathscr{CP}(E)$ then PI^∞ is initial in $\mathscr{CC}(E)$.

Proof From the previous lemma we know at least that $PI^\infty \in \mathscr{CC}(E)$. Let $A \in \mathscr{CC}(E)$. We must show that there exists a unique Σ-domain homomorphism from PI^∞ to A.

A priori A is a Σ-predomain and therefore by the initiality of PI there exists a unique Σ-predomain homomorphism $h: PI \longrightarrow A$. By the universality property of theorem 3.3.12 this can be extended to a Σ-domain homomorphism $ext(h): PI^\infty \longrightarrow A$.

So we know that at least one such homomorphism exists. We now show it is unique. Let $k: PI^\infty \longrightarrow A$ be an arbitrary homomorphism. Define $k_{res}: PI \longrightarrow A$ by $k_{res}(x) = k(in(x))$. This is a Σ-predomain homomorphism and by the initiality of PI this must coincide with h. Therefore both k and $ext(h)$ are extensions of h to PI^∞. Again by the universality property in theorem 3.3.12 such extensions are unique. Therefore k must coincide with $ext(h)$. □

The construction of the initial Σ-domain in $\mathscr{CC}(E)$ is even less illuminating than that of the initial Σ-*po* algebra in $\mathscr{C}(E)$; it consists of ideals of equivalence classes of syntactic terms from T_Σ. The usefulness of initial objects lies in their close connection with equational proof systems. This is particularly true when we can find an intuitive representation of the initial object. This is investigated in the next section for the systems of equations we developed in part I. We end the present section with a proof of the result which underlies the relationship between the initial object in $\mathscr{C}\mathscr{C}(E)$ and the proof system $\mathbf{\Omega DED}(E)$.

THEOREM 3.3.15 The Σ-domain A is initial in $\mathscr{C}\mathscr{C}(E)$ if and only if

i) it is finitary
ii) $\mathbf{\Omega DED}(E)$ is sound and complete with respect to $\leq_A$ over T_Σ.

Proof The proof relies entirely on corollary 3.3.3.

a) If A is initial in $\mathscr{C}\mathscr{C}(E)$ it is isomorphic to I_E^∞ and Fin(A) is isomorphic to I_E. It follows from corollary 3.3.3 that A is surjective. Moreover since A and I^∞ are isomorphic $\leq_A$ and $\leq_{I^\infty}$ coincide as relations over T_Σ. Every term of T_Σ is actually interpreted as a finite element of I^∞. Therefore A is finitary and the soundness and completeness follows from corollary 3.3.3 also.

b) Conversely suppose A satisfies conditions i) and ii). Then Fin(A) is a Σ-predomain and it satisfies the conditions of corollary 3.3.3; therefore it is initial in $\mathscr{C}\mathscr{P}(E)$. Because A is an algebraic *cpo*, A is isomorphic to Fin$(A)^\infty$ and it follows from proposition 3.3.14 that A is initial in $\mathscr{C}\mathscr{C}(E)$. □

3.4 Acceptance Trees

The finite Acceptance Trees of part I were described as a Σ^2-*po* algebra. Here we extend this model to a Σ^2-domain, which is naturally called Acceptance Trees, **AT**. This will be characterized as the initial Σ^2-*cpo* with respect to a set of inequations obtained by extending slightly the set **A2**, which characterizes **fAT** as a Σ^2-partial order.

The development in the previous three sections assumes that every signature contains a 0-ary function symbol Ω, which is always treated in a special way; it is interpreted as the least element of the carrier. So our signature Σ^2 must be extended to include Ω. Rather than write $\Sigma^2 \cup \{\Omega\}$ we will in future always assume that, when appropriate, signatures automatically contain this distinguished symbol. It will be appropriate when we are in the context of Σ-domains or even Σ-predomains. So in particular our signature Σ^2 now consists of the operators

0-ary symbols: NIL, Ω
1-ary symbols: a, for each $a \in Act$
2-ary symbols: $+, \oplus$

Let us first concentrate on defining the carrier of the proposed Σ^2-domain. If it is to be initial then we know from theorem 3.2.4 that it is surjective, i.e. every finite element is denoted by some term in the extended language T_{Σ^2}. Moreover because it is an algebraic *cpo* it is completely determined by these finite elements; two algebraic *cpos* with the same set of finite elements are identical because every element is the limit of the set of finite elements it dominates. So to understand the domain **AT** it is sufficient to concentrate on the finite elements, i.e. the interpretation of the terms in T_{Σ^2}. At worst the nonfinite elements can be viewed as limit points associated with directed sets of finite elements even though in our case they have a natural representation as infinite trees.

The model is an extension of **fAT** so that it contains all of these trees as denotations of the terms in T_Σ which do not contain occurrences of Ω. These can be viewed as total trees, whereas the denotations of the other terms, involving Ω, may be called partial trees. It is the presence of these partial objects (and their associated limit points) which transforms the partial order **fAT** into an algebraic *cpo*. The most important partial tree is the least one: that used as a denotation of the term Ω. This tree is to represent the completely unknown process, a process about which we have absolutely no information. It cannot have any branches since the presence of a branch, labeled a, say, indicates that the process can at least perform the action a. On the other hand, it is different than the trivial tree used to interpret NIL, •. We know everything we need to know about this process: it can perform no actions. To differentiate this from Ω we need to introduce new types of nodes. Ω is interpreted as the trivial tree consisting of a single *open* node:

○

So in general we now have two kinds of nodes: *open*, denoted as ○, and *closed*, denoted as •, which is what has been used up to now. The presence of an open node indicates an "underdefinedness" or, in terms of T_{Σ^2} an occurrence of Ω. For example, $aNIL$, $a\Omega$ are interpreted as (respectively):

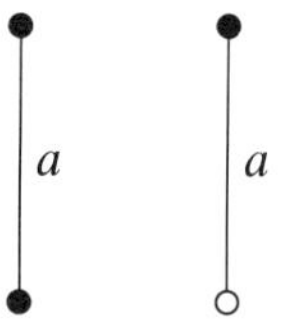

The model is determined by how these two kinds of nodes interact and by how the various operators behave when confronted by the new kind of node. This in turn is a reflection of the operational behavior we associate with processes. We have now extended the set of processes to include "partial processes" such as $a\Omega$, $abNIL + a\Omega$, or indeed, Ω. In part I we gave the operational behavior of the total processes, and partial processes are studied in chapter 4. However to motivate the model we are developing here it suffices to say that Ω represents a process which when tested delays forever without responding; or, in other words, it computes internally indefinitely. So that $a\Omega$ will never pass the test abw ("please do an a followed by a b") and $aNIL + \Omega$ may or may not pass the test aw.

For the sake of variety we consider the strong version of Acceptance Trees, appropriate for the Testing preorder $\sqsubseteq_{MUST}$. Unlike the finite trees in part I there is considerable variation in the trinity of models for the preorders $\sqsubseteq_{MAY}$, $\sqsubseteq_{MUST}$, $\sqsubseteq$. The one we choose is the simplest and most intuitive.

DEFINITION 3.4.1 $\mathbf{AT_S}$, the set of *Strong Acceptance Trees*, is the set of rooted trees such that

i) every node is either open (○) or closed (●)
ii) every branch is labeled by an action from Act
iii) every closed node n is labeled by $\mathscr{A}(n)$, a collection of subsets of Act

and which satisfies the requirements

R1	For every action a, every node in the tree has at most one successor branch labeled by a
R2	(*Finite Branching*) For every closed node n $S(n)$ is finite
R3	For every closed node n $\mathscr{A}(n)$ is an $S(n)$-set
R4	If n is open it is a leaf.

$\mathbf{AT_S}$ is actually parametrized on the set of actions Act, but this set is usually apparent from the context. Finite Acceptance Trees are automatically included in this set. They are precisely the finite trees in $\mathbf{AT_S}$, all of whose nodes are closed. The extension is obtained by allowing the trees to be infinite and the nodes to be open. The requirements R1, R2, and R3 are inherited from part I, although R2 has new significance. The processes we wish to model can exhibit nondeterministic behavior but this nondeterminism is finitely bounded: at any point in time there is a finite number of moves it can make. This is a natural assumption to make if we conceive of processes being physically realizable. R2 is a direct consequence of this assumption. For a process must resolve the nondeterminism by, presumably, some internal computation. If the possible outcomes of this computation, stable states, are infinite then the tree detailing all possible evolutions of the internal computations at this point must be infinite. It follows by Konig's lemma (Knuth 1975) that there is an infinite path through the tree, representing an infinite internal computation. This capability to diverge is represented in the model by an open node. In other words, any point in the history of a process at which the nondeterminism can be resolved in an infinite number of ways must be represented in the model by an open node.

The new requirement R4 reflects the fact that we are only interested in *must*-tests. If a process can diverge then we can never be certain that it passes any test. If we are only modeling processes in terms of the tests which they always pass, then once they have the capability of diverging internally their subsequent behavior is of no interest.

We continue to use the notation developed in §2.3 for these trees. For example $L(t)$ denotes the sequences from Act^* along the paths of the tree t. As in part I it is nonempty and prefix-closed but now may be infinite. However $S(s, L(t))$ is still finite for every s in $L(t)$. The only extra notation we introduce is $CL(t)$, to denote the subset of $L(t)$ which determines closed nodes. It is a prefix-closed subset of $L(t)$ and may be empty. We now endow $\mathbf{AT_S}$ with a partial order. Informally $t \leq t'$ means that, viewing the trees as processes, t' *must* pass every test that t *must* pass. This comes about when the behavior of t' is more determined than that of t. One way in which this happens is when t' is more deterministic than t. In the model this is reflected in t' having fewer Acceptance sets. This is exactly the partial order defined in §2.6 on $\mathbf{fAT_S}$, and an example is given in figure 3.3 a). Another way is when t' is obtained by making t more defined, i.e. by grafting on a subtree to an open leaf. An example is given in figure 3.3 b). In general both these aspects of 'more determined' may be present, as in figure 3.3 c). To define formally the order we need some notation which will also be used with respect to transition systems. For any tree t and $s \in Act^*$ we write $t \downarrow s$ to mean:

if $s' \in L(t)$ then $s' \in CL(t)$ for every prefix s' of s.

We emphasize that $t \downarrow s$ does not mean that there is a node in t determined by s.

DEFINITION 3.4.2 For $t, t' \in \mathbf{AT_S}$ let $t \leq t'$ if for every $s \in Act^*$, $t \downarrow s$ implies

i) $t' \downarrow s$
ii) if $s \in CL(t')$ then $\mathscr{A}(t'(s)) \subseteq \mathscr{A}(t(s))$.

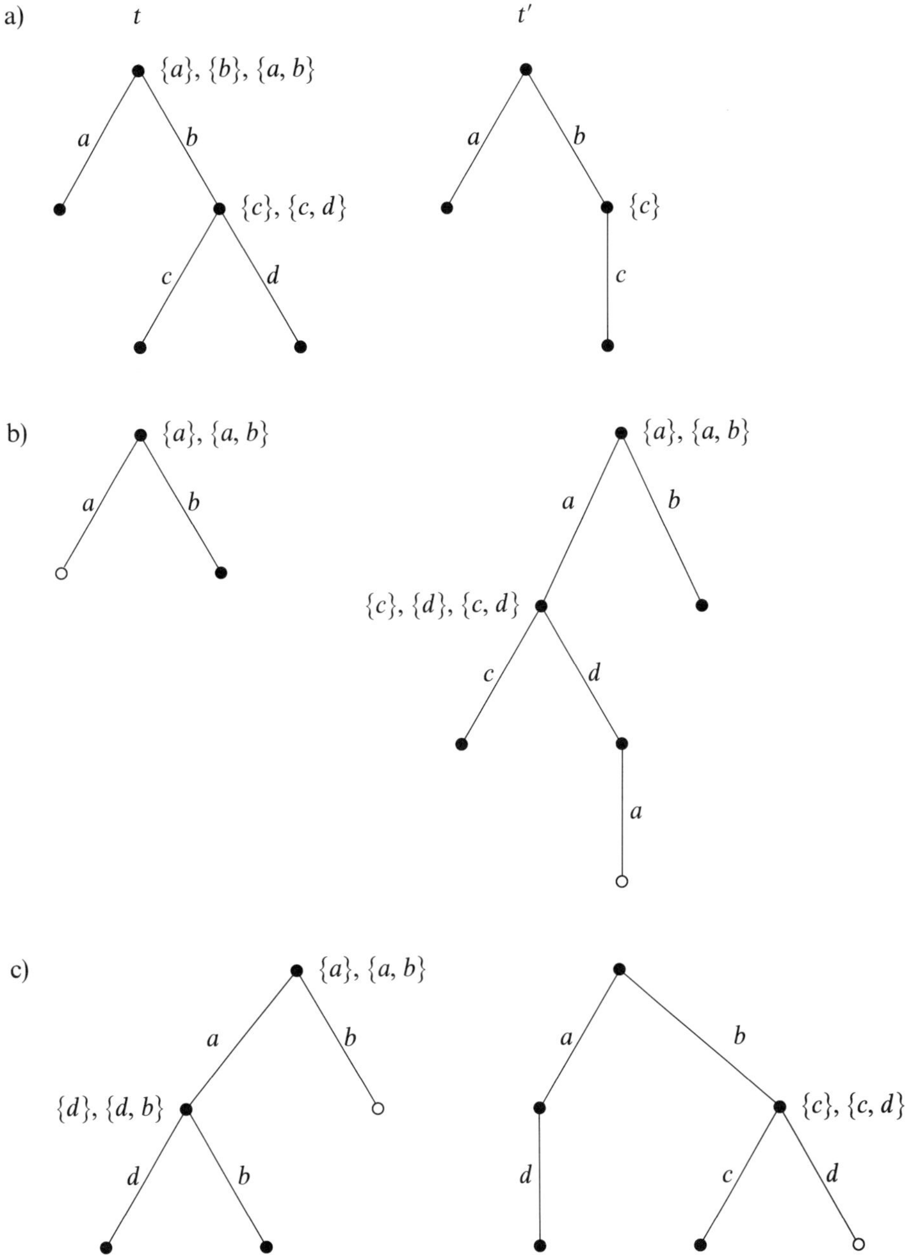

Figure 3.3
Examples in $\mathbf{AT_S}$. In each case $t \leq t'$.

(figure continued on overleaf)

d)

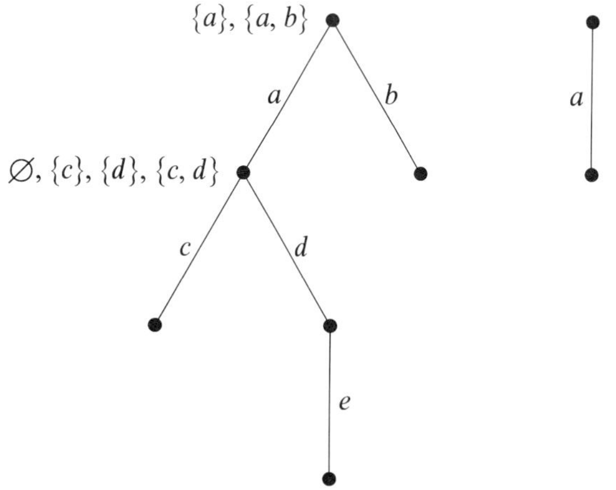

Figure 3.3 (continued)

This definition is somewhat subtle. If we let $O(t)$ denote the strings determining the open leaves of t, i.e. $L(t) - CL(t)$, it implies that if $s \in O(t')$ then some prefix of s is in $O(t)$; in particular if t only has closed nodes then so does t'. For such trees the definition coincides with that of $\leq^{\text{MUST}}$, given in §2.6, one consequence of which is that $CL(t') \subseteq CL(t)$. This is not true in general, as can be seen from the examples in figure 3.3. The net effect is that improving even a fully determined tree may change it structurally. An extreme example is given in figure 3.3 d). This phenomenon did not occur in **fAT** and will not occur in the more general model **AT**. In figure 3.4 we give some counter-examples to $\leq$ which are meant to shed light on it.

PROPOSITION 3.4.3 $\langle \mathbf{AT_S}, \leq \rangle$ is an algebraic *cpo*.

Proof a) It is easy to check from the definition that $\leq$ is a partial order. We prove it is a *cpo*. To do so we use the following terminology: If X is a set we say the property P is true *almost always* in X if for all but a finite number of $x \in X$ P is true of x. So if $x_0, x_1, x_2, \ldots$ is an infinite sequence of elements from X, then there exists some $k \geq 0$ such that P is true of x_n for every $n \geq k$. If X is a directed set in some *po* then it follows that there exists some $x \in X$ such that P is true of x' for every $x' \geq x$.

Let D be a directed set of trees in $\mathbf{AT_S}$. We construct its limit, called t. It is determined by

$$L(t) = \{s, s \in L(t') \quad \text{for almost all } t' \in D\}$$

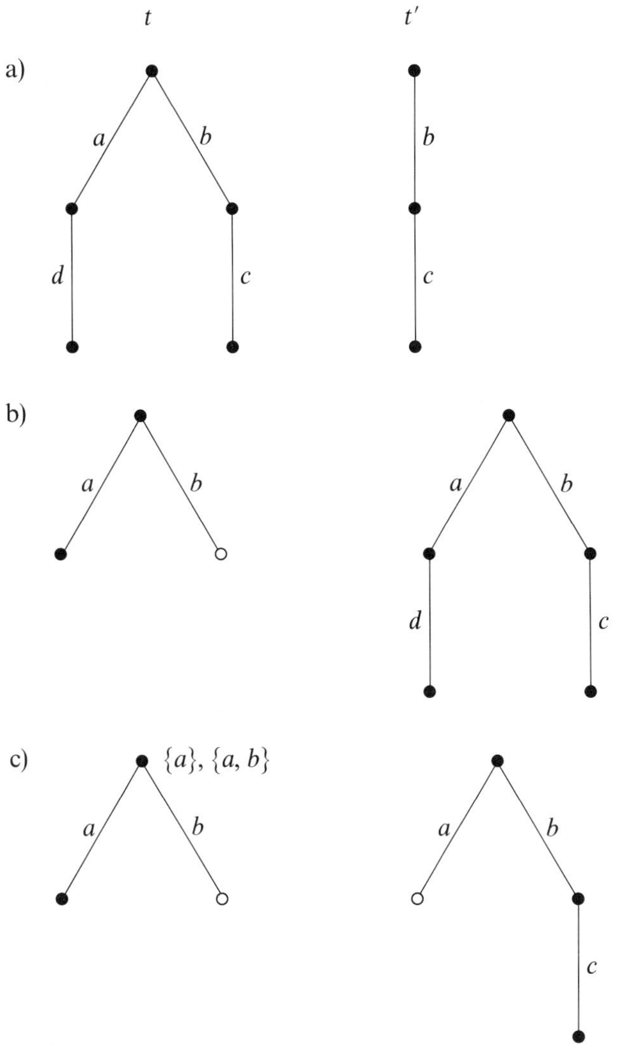

Figure 3.4
Counterexamples in $\mathbf{AT_S}$ In each casc $t \not\lesssim t'$.

$$CL(t) = \{s, s \in CL(t') \quad \text{for almost all } t' \in D\}$$

$$\mathscr{A}(t(s)) = \{A, A \in \mathscr{A}(t'(s)) \quad \text{for almost all } t' \in D\}$$

We must first show that t is actually a tree in $\mathbf{AT_S}$. It is straightforward to show that both $L(t)$ and $CL(t)$ are prefix-closed and $CL(t) \subseteq L(t)$. So we need only to show that the requirements R2–R4 are satisfied. First note that for each s in $CL(t)$ there is a specific t_s in D such that $\mathscr{A}(t_s(s)) = \mathscr{A}(t'(s))$ for every $t' \geq t_s$. For otherwise there would be an infinite sequence

$$\mathscr{A}(t_1(s)) \supsetneq \mathscr{A}(t_2(s)) \supsetneq \cdots$$

However, this is impossible because all these sets are finite. It follows that $\mathscr{A}(t(s)) = \mathscr{A}(t_s(s))$ and $S(t(s)) = S(t_s(s))$ and therefore both R2 and R3 are satisfied. R4 follows directly from the definition of $L(t)$ and $CL(t)$ and the fact that each $t' \in D$ satisfies R4.

This characterization of $\mathscr{A}(t(s))$ is also sufficient to show that $t = \bigvee D$. For example, suppose $t' \leq u$ for every t' in D. We show $t \leq u$.

i) Suppose $s \in O(u)$. We show $s' \in O(t)$ for some prefix s' of s. This means $t \downarrow s$ implies $u \downarrow s$. Since $t' \leq u$ for every t' in D we have that for every such t' there is a prefix of s in $O(t')$. However s has only a finite number of prefixes and therefore there is some prefix s' in $O(t')$ for almost all t' in D. This in turn implies that $s' \in O(t)$.

ii) Suppose $A \in \mathscr{A}(u(s))$ where $t \downarrow s$. Then $t' \downarrow s$ for almost all t' in D. Since $t' \leq u$ for such t', $s \in CL(t')$; hence $s \in CL(t)$. Then $B \in \mathscr{A}(t(s))$ such that $B \subseteq A$ now follows from $t_s \leq t$.

b) Let $fin\mathbf{AT_S}$ be the set of finite trees in $\mathbf{AT_S}$. We show $fin\mathbf{AT_S}$ is precisely the set of compact elements. Let $d \in fin\mathbf{AT_S}$ and $d \leq t$, where $t = \bigvee D$. We must find some $t_d \in D$ such that $d \leq t_d$. Note that $L(d)$, and therefore $CL(d)$ is finite. Consider the finite set $\{t_s, s \in CL(d)\} \subseteq D$, where t_s is the element of D which determines $\mathscr{A}(t(s))$. This set has an upper bound t_d in D since D is directed and it is trivial to establish $d \leq t_d$.

So every element of $fin\mathbf{AT_S}$ is compact.

Conversely, suppose t is compact. We show $L(t)$ is finite and therefore $t \in fin\mathbf{AT_S}$. If $L(t)$ were infinite we could construct a chain $\{t_n, n \geq 0\}$ such that $t = \bigvee\{t_n, n \geq 0\}$ but $t \leq t_n$ for no $n \geq 0$, which would contradict the compactness of t; t_n is defined to coincide with t to depth less than n but all nodes at depth n are open. We leave the reader to fill in the details.

c) Finally, to show $\mathbf{AT_S}$ to be algebraic it is necessary to prove $t = \bigvee\{d \in fin\mathbf{AT_S}, d \leq t\}$. We leave this in the hands of the reader, noting that for each $s \in CL(t)$ we can construct a finite $d \leq t$ such that $\mathscr{A}(d(s)) = \mathscr{A}(t(s))$. □

We now turn our attention to defining continuous operations on the *cpo* $\mathbf{AT_S}$, one for each symbol in Σ^2, thereby viewing it as a Σ^2-domain. The operations are mild extensions of those in part I, defined on the finite model **fAT**; the only problem is to take into consideration the open nodes. This is particularly true for the nondeterministic operators $+$ and $\oplus$.

I. *The Constant Symbols* As we have already seen, $NIL_{\mathbf{AT_S}}$ is the trivial tree • with no branches and $\Omega_{\mathbf{AT_S}}$ is the slightly different tree o which also has no branches.

II. *Prefixing* The function $a_{\mathbf{AT_S}}$ acts in exactly the same way as $a_{\mathbf{fAT}}$. It maps

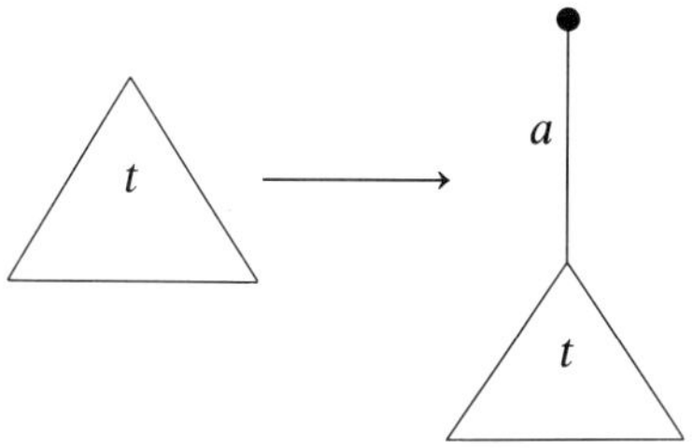

The root of $a_{\mathbf{AT_S}}(t)$ is always closed, which models the fact that the immediate behavior of a process *ap* is totally defined: it can always perform an *a* action, and this is the only action it can perform.

III. *The Nondeterministic Operators* The definition of the two nondeterministic operators is only marginally more interesting. The function $\oplus_{\mathbf{AT_S}}$ operates in exactly the same way as $\oplus_{\mathbf{fAT}}$ until an open leaf is reached in either of its operands. When this occurs an open leaf is produced at the corresponding point of the result. Intuitively the reason for this should be clear. If either p or q can diverge internally then the behavior of $p \oplus q$ is very unstable: we will never be assured of a response to any test. Since the model is only designed to reflect the *must* variety of testing such $p \oplus q$ will be considered no different than Ω. We will see this later when testing is extended to partial processes. The actual definition of $\oplus_{\mathbf{AT_S}}$ is:

$t_1 \oplus_{\mathbf{AT_S}} t_2$ is the tree t determined by

i) $CL(t) = \{s \in L(t_1) \cup L(t_2)$, for every prefix s' of s and for $i = 1, 2$, if $s' \in L(t_i)$ then $s' \in CL(t_i)\}$

ii) $L(t) = \{s \in L(t_1) \cup L(t_2),\ s = s'a$ implies $s' \in CL(t)\}$

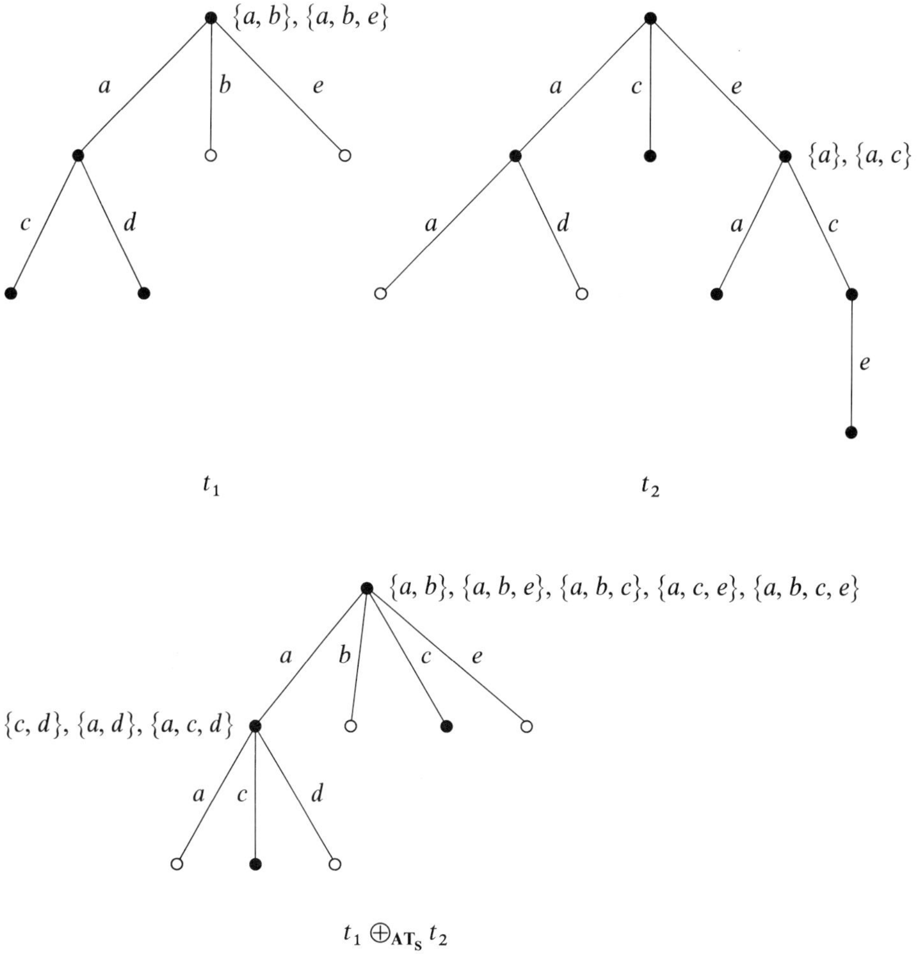

Figure 3.5
Examples of $\oplus_{\mathbf{AT_S}}$, $+_{\mathbf{AT_S}}$.

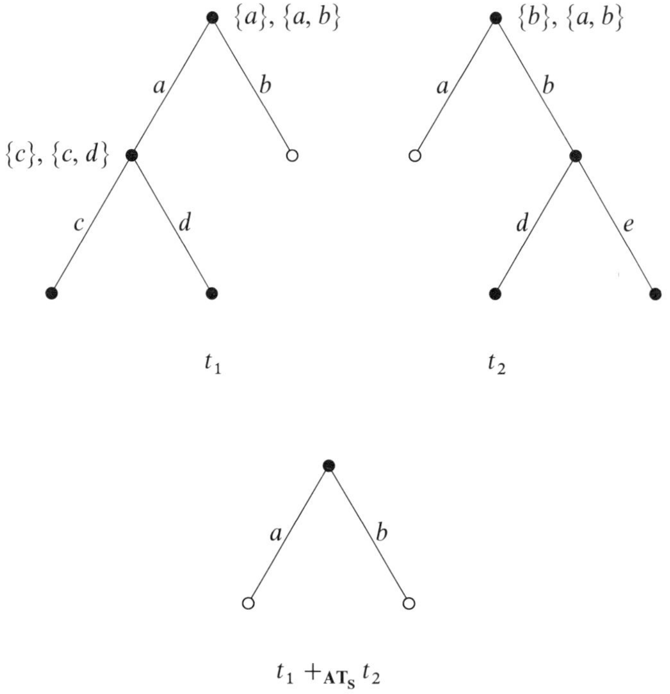

Figure 3.5 (continued)

iii) $\mathscr{A}(t(s)) = c(\mathscr{A}(t_1(s)) \cup \mathscr{A}(t_2(s)))$, for $s \in CL(t)$ where, by convention, $\mathscr{A}(t_i(s)) = \varnothing$ if $s \notin CL(t_i)$.

An example is given in figure 3.5.

The definition of $+_{\mathbf{AT_S}}$ is identical except that the Acceptance set at a closed root is slightly different, as in **fAT**: the tree $t_1 +_{\mathbf{AT_S}} t_2$ is the same as $t_1 \oplus_{\mathbf{AT_S}} t_2$ except clause iii) is replaced by

iii) $\mathscr{A}(t(\varepsilon)) = \mathscr{A}(t_1(\varepsilon)) \mathbin{u} \mathscr{A}(t_2(\varepsilon))$

$\mathscr{A}(t(s)) = c(\mathscr{A}(t_1(s)) \cup \mathscr{A}(t_2(s)))$, with the same convention, for $s \neq \varepsilon$.

PROPOSITION 3.4.4 $\langle \mathbf{AT_S}, \leq_{\mathbf{AT_S}}, \Sigma^2_{\mathbf{AT_S}} \rangle$ is a Σ^2-domain.

Proof It suffices to show that each of the functions defined above is continuous. This is not difficult and we leave it to the reader. The only nontrivial cases are the two nondeterministic operators, for which proposition 3.1.7 is useful. □

We end this section by showing that this intuitive model is actually generated as a Σ^2-domain by a set of equations, i.e. it is initial in the class of Σ^2-domains which satisfies a given set of equations. Since $\mathbf{AT_S}$ is a mild extension of the finite model $\mathbf{fAT_S}$ it is not surprising that it satisfies all of the equations in **SA2**. This consists of all the equations in figure 2.4, **A2**, and the strong axiom $x \oplus y \leq x$. However these are not sufficient to characterize $\mathbf{AT_S}$ as a Σ^2-domain. In addition we need to specify that the nondeterministic operators are *strict*, i.e. when any one of their arguments is the least element $\perp$ the result is also $\perp$.

$$x \oplus \Omega = \Omega \qquad \textbf{(s1)}$$

$$x + \Omega = \Omega. \qquad \textbf{(s2)}$$

In fact **(s2)** can be replaced by

$$x + \Omega \leq \Omega$$

because the converse, $\Omega \leq x + \Omega$, is satisfied by every Σ^2-domain: Ω is always interpreted as the least element. Also, **(s1)** is entirely redundant because $x \oplus \Omega \leq \Omega$ is a simple consequence of the strong axiom $x \oplus y \leq x$, **(S)**. For organizational reasons it is more convenient to consider the slightly weaker axiom

$$x + \Omega \leq x \oplus \Omega \qquad \textbf{(s)}$$

In the context of the strong axiom this implies $x + \Omega \leq \Omega$ and therefore **(s2)**. We refer to the list **A2** plus **(s)** as **A2s** and **A2s** plus **S** as **SA2s**. These equations are reiterated in figure 3.6.

THEOREM 3.4.5 $\mathbf{AT_S}$ is initial in $\mathscr{CC}$(**SA2s**).

The proof of this theorem follows the lines of the related characterization of **fAT** in §2.4 except that here we use theorem 3.2.4: we show $\mathbf{AT_S}$ is finitary and **ΩDED(SA2s)** is both sound and complete with respect to $\leq_{\mathbf{AT_S}}$, restricted to T_Σ. The technical details are simply minor variations on those in §2.4. They are summarized in the following three results.

LEMMA 3.4.6 $\mathbf{AT_S}$ is finitary.

Proof a) It is a simple matter to prove by structural induction that if $t \in T_{\Sigma^2}$, then $\mathbf{AT_S}[\![t]\!]$ is a finite tree.

b) Conversely we show that every finite tree in $\mathbf{AT_S}$ is denoted by some term in T_{Σ^2}. If it has no branches it is denoted by either NIL or Ω. Otherwise we proceed by induction, as in lemma 2.4.2. □

$$x \oplus (y \oplus z) = (x \oplus y) \oplus z \qquad \oplus 1$$

$$x \oplus y = y \oplus x \qquad \oplus 2$$

$$x \oplus x = x \qquad \oplus 3$$

$$x + (y + z) = (x + y) + z \qquad +1$$

$$x + y = y + x \qquad +2$$

$$x + x = x \qquad +3$$

$$x + NIL = x \qquad +4$$

$$x \oplus y \leq x + y \qquad + \oplus 1$$

$$ax + ay = ax \oplus ay \qquad + \oplus 2$$

$$ax + ay = a(x \oplus y) \qquad \oplus 4$$

$$x + (y \oplus z) = (x + y) \oplus (x + z) \qquad + \oplus 3$$

$$x \oplus (y + z) = (x \oplus y) + (x \oplus z) \qquad + \oplus 4$$

$$x + \Omega \leq x \oplus \Omega \qquad \textbf{(s)}$$

$$x \oplus y \leq x \qquad \textbf{(S)}$$

$$x \leq x \oplus y \qquad \textbf{(W)}$$

Figure 3.6
The set of equations **A2s**, **S**, **W**.

LEMMA 3.4.7 **ΩDED(SA2s)** is sound with respect to $\leq_{\mathbf{AT_S}}$ restricted to T_Σ.

Proof It is sufficient to show that $\mathbf{AT_S}$ satisfies all of the equations in **SA2s**, which we leave to the reader. □

To prove completeness of the proof system we use the general technique of normal forms, as outlined in §2.4. The required normal forms are only marginally different.

DEFINITION 3.4.8

i) NIL and Ω are Ω-normal forms.

ii) If $\mathscr{A}$ is a saturated set and for every $a \in A(\mathscr{A})$, there is a normal form $n(a)$ then $\mathbf{\Sigma}\{n(A), A \in \mathscr{A}\}$ is also a Ω-normal form, where as usual $n(A) = \sum\{an(a), a \in A\}$.

PROPOSITION 3.4.9 **ΩDED(SA2s)** is complete with respect to $\leq_{\mathbf{AT_S}}$ over T_{Σ^2}.

Proof Two results are required.

i) for Ω-normal forms n, m, $\mathbf{AT_S}[\![n]\!] \leq \mathbf{AT_S}[\![m]\!]$ implies $n \leq_{\mathbf{SA2s}} m$.

ii) every term has a Ω-normal form, i.e. for every $p \in T_{\Sigma^2}$ there exists a Ω-normal form n such that $p =_{\mathbf{SA2s}} n$.

The proofs follow those of lemma 2.4.4 and theorem 2.4.6 respectively. For example, in i) if n is Ω then the axiom $\Omega \leq x$, which is always available in the proof system **ΩDED**(E), can be applied to prove $n \leq m$. If m is Ω then n is also Ω and the results also follow. Otherwise we can proceed as in lemma 2.4.4. Condition ii) above is proved by structural induction on p. For example, if it has the form $q \oplus r$ then we may assume both q and r have Ω-normal forms $nf(q)$, $nf(r)$ respectively, and $p =_{\mathbf{SA2s}} nf(q) \oplus nf(r)$. If either of these normal forms is Ω then we can apply **(s1)** to obtain $p =_{\mathbf{SA2s}} \Omega$. Otherwise we may proceed as in the case d) of theorem 2.4.6. □

3.5 The Continuous Trinity

This brief section emphasizes that the model we are interested in actually comes in three forms, as in the previous chapter. We will not go into many details and providing the proofs will be left to the exercises at the end of the chapter.

We have already seen the strong model which satisfies the strong axiom **S**:

$x \oplus y \leq x$.

If we drop this axiom the model we obtain is that of Acceptance Trees. **AT**, the set of

Acceptance Trees, is defined in the same manner as $\mathbf{AT_S}$ but with R4 replaced by

R4′ If n is open every descendant of n is also open.

Intuitively open nodes in a tree represent points during the history of the process represented by the tree where possible divergences may occur; or more generally points at which responses to tests can no longer be guaranteed. From this point of view the hereditary nature of open nodes is natural. These open nodes do, however, contain some information, at least from the point of *may*-tests: Acceptance Trees are designed to reflect processes under both forms of testing.

Notice that trees may now be infinite-branching but R3 states that nodes at which this occurs must be open. In the extended syntax for processes introduced in the next chapter these trees will not be denotable. However they are required to act as limit points for chains of simple denotable (and partial) processes.

DEFINITION 3.5.1 For $t, t' \in \mathbf{AT}$ let $t \leq_{\mathbf{AT}} t'$ if

i) $L(t) \subseteq L(t')$
ii) for every $s \in Act^*$,
  $t \downarrow s$ implies a) $t' \downarrow s$
  b) if $s \in CL(t')$ then $\mathscr{A}(t'(s)) \subseteq \mathscr{A}(t(s))$.

This is a generalization of the corresponding definition in **fAT** in §2.3, and the second clause is essentially the partial order on $\mathbf{AT_S}$. It coincides exactly with the definition of $\leq$ on **fAT** when the first tree only contains closed nodes.

In this generalization if $t \leq t'$ then the closed part of t is structurally identical to the corresponding part of t'; only Acceptance sets may be removed. However the open parts of t may be improved by adding on new branches or possibly closing certain open nodes. Examples are given in figure 3.7.

LEMMA 3.5.2 $\langle \mathbf{AT}, \leq_{\mathbf{AT}} \rangle$ is an algebraic *cpo*. □

We leave the reader to verify this and to endow it with operators so that it becomes a Σ^2-domain which is characterized by the set of equations **A2s** i.e. it is initial in $\mathscr{CC}(\mathbf{A2s})$.

AT is the basis of the trinity of models. As we have seen the strong model $\mathbf{AT_S}$ can be obtained by imposing the strong requirement that every node except leaves is closed and limiting the definition of the partial order to clause ii); algebraically the strong equation is added. To obtain the weak model $\mathbf{AT_W}$ we impose the constraint on

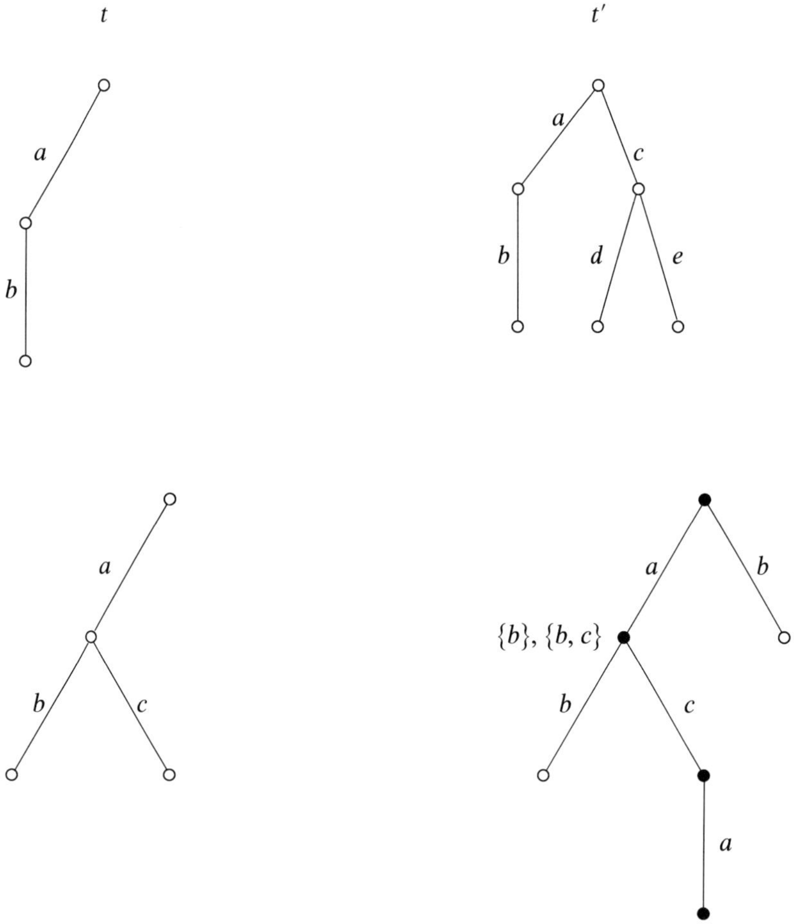

Figure 3.7
Examples in **AT**. In each case $t \leq t'$.

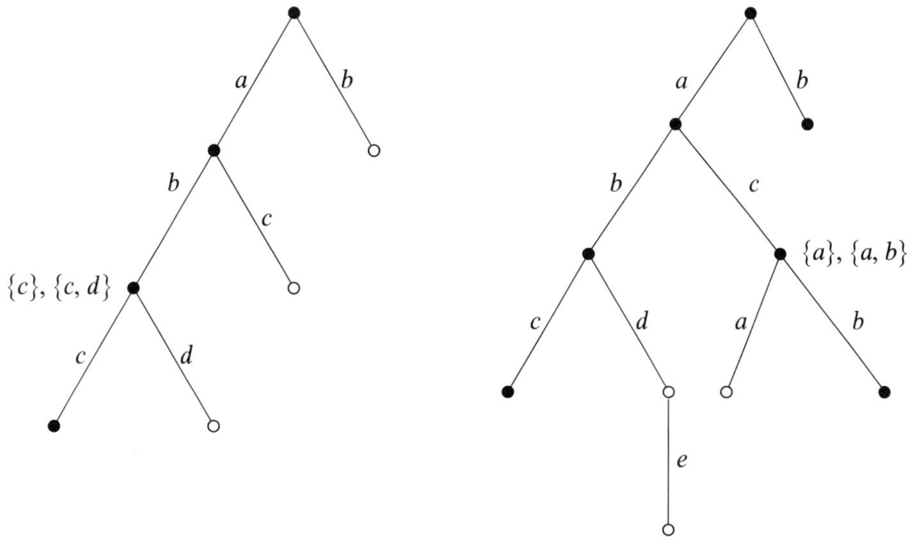

Figure 3.7 (continued)

elements of **AT** that

R4″ every node is open.

This is intuitive since the weak model is not concerned with the ability of processes to react to *must*-tests.

$\mathbf{AT_W}$ is a very simple model. Every element is simply a nonempty prefix-closed set of strings. We leave the reader to check that it is an algebraic *cpo*; the partial order is inherited from **AT** even though only clause i) of the definition has any effect. In terms of prefix-closed strings this partial order is simply set inclusion. We also leave it to the reader to endow it with operators so that it becomes a Σ^2-*cpo*, initial in $\mathscr{C}\mathscr{C}$(**WA2s**) where **WA2s** is simply **A2s** + **W**.

We end this chapter by introducing some convenient notational conventions. As already stated, we now assume that every signature contains the distinguished symbol Ω. Because of this we will abbreviate **ΩDED**(E) to simply **DED**(E), i.e. we always assume the presence of the equation $\Omega \leq x$. Therefore, for any set of equations E, we drop the subscript Ω from $E\Omega$, rendering it simply E. In other words, the presence of Ω in the signature means that our basic framework is that of Σ-predomains rather than Σ-*po*'s and therefore we can transfer the notation developed for the latter to the former.

Exercises

Q1 a) Show that least elements and least upper bounds are unique in partial orders, if they exist.

b) A partial order A is chain complete if every chain in A has a *lub*. Show that if A is countable then this implies every directed set has a *lub*.

Q2 a) Let A, B be two Σ-predomains. Prove that if they are isomorphic as Σ-*po*-algebras they are also isomorphic as Σ-predomains.

b) Let A be a finitary Σ-domain. Prove that Fin(A), with the functions f_A restricted to Fin(A), for each f in Σ, is a Σ-predomain.

c) Prove that for Σ-domains A, B if Fin(A) and Fin(B) are isomorphic as Σ-predomains then A and B are isomorphic as Σ-domains.

Q3 A partial function $f: A \longrightarrow B$ is a mapping which associates with each element of A at most one element of B. Let graph $(f) = \{\langle a, b\rangle \in A \times B, f(a)$ is defined and $f(a) = b\}$. Such an f can be extended to $\varepsilon(f): A_\perp \longrightarrow B_\perp$ by

$$\varepsilon(f)(x) = f(a) \text{ if } x \in A \text{ and } f(a) \text{ defined}$$

$$= \perp \text{ otherwise.}$$

a) Show $\varepsilon(f) \in [A_\perp \longrightarrow B_\perp]$.

b) Show graph $(f) \subseteq$ graph (g) if and only if $\varepsilon(f) \leq \varepsilon(g)$.

Q4 For any $d \in D$, D a *cpo*, let $D_d = \{x \in D, x \leq d\}$.

a) Show that D_d is also a *cpo*.

b) If $k: D \longrightarrow E$ is an isomorphism between two *cpo*'s prove that D_d and $E_{k(d)}$ are also isomorphic as *cpo*'s.

c) If k is as in b) prove that d is finite if and only if $k(d)$ is finite.

Q5 Let Π denote the flat *cpo* {true, false, $\perp$} and Π^ω the set of infinite sequences of elements from Π. This is ordered in the following way:

$$x_0 x_1 \ldots \leq y_0 y_1 \ldots$$

if $x_n \leq y_n$ for every $n \geq 0$.

Show that Π^ω under this ordering is an algebraic *cpo*.

Q6 Let $x \vee y$ denote $lub\{x, y\}$. A *cpo* A is said to be *consistently complete* (*cp*) if whenever $\{x, y\}$ has an upper bound in A $x \vee y$ exists.

a) Prove that if x and y are finite then $x \vee y$ is also finite whenever it exists.

b) Show that $[A \longrightarrow B]$ is consistently complete whenever B is consistently complete.

Q7 Let A, B be two algebraic *cpo*'s. If $d \in A$, $e \in B$, let $d \Longrightarrow e \in (A \longrightarrow B)$ be defined by

$$d \Longrightarrow e(a) = e \quad \text{if } d \leq a$$
$$= \bot \quad \text{otherwise.}$$

a) Prove $d \Longrightarrow e$ is continuous whenever d is finite.

b) Prove $d \Longrightarrow e$ is finite in $[A \longrightarrow B]$ if in addition e is finite.

c) Let F be the collection of all functions of the form $k_1 \vee \cdots \vee k_n$, where each k_i has the form $d \Longrightarrow e$ for some finite elements d, e. Prove $f = \bigvee\{k \in F, k \leq f\}$.

d) Show that $[A \longrightarrow B]$ is an algebraic *cpo* whenever A and B are algebraic and consistently complete.

Q8 A partial order $\langle A, \leq \rangle$ is called a *complete lattice* if every subset of A has a *lub*.

a) If A is a complete lattice show that every subset of A has a greatest lower bound; that is, for every $D \subseteq A$ there exists a $glbD$ in A which satisfies

i) $glbD \leq d$ for every $d \in D$
ii) if $z \leq d$ for every $d \in D$ then $z \leq glbD$.

b) Let $f: A \longrightarrow B$ be a monotonic function between two complete lattices. Define $l(f) = glb\{a, f(a) \leq a\}$. Show $l(f)$ is a pre-fixpoint of f, i.e. $f(l(f)) \leq l(f)$.

c) Show that every monotonic function between complete lattices has a least fixpoint.

Q9 Let $f, g \in [A \longrightarrow A]$ be continuous functions.

a) Prove $f\,Y(g \circ f) = Y(f \circ g)$

b) If $f(\bot) = g(\bot)$ and $f \circ g = g \circ f$ prove $Yf = Yg = Y(f \circ g)$.

Q10 a) Let $\sqsubseteq$ be a Σ-preorder over a Σ-predomain A which satisfies $\Omega \sqsubseteq x$. Prove $A/\!\sqsubseteq$ is a Σ-predomain.

b) Let $\sqsubseteq$ be a Σ-preorder over the predomain T_Σ which satisfies $\Omega \sqsubseteq x$. Prove that $T_\Sigma/\!\sqsubseteq$ is the initial Σ-predomain in the class of Σ-predomains which satisfy $\sqsubseteq$.

Q11 a) Show $\preceq$ is antisymmetric, i.e. $t \preceq t'$, $t' \preceq t$ implies $t = t'$.

b) Show that if a set in the partial order $\langle T_\Sigma(X), \preceq \rangle$ has an upper bound it has a least upper bound.

Q12 Elements of T_Σ can be viewed as trees. For example if $\Sigma_0 = \{a, b\}$, $\Sigma_1 = \{h\}$, $\Sigma_2 = \{f\}$ then the term $f(a, h(b))$ can be represented by the tree

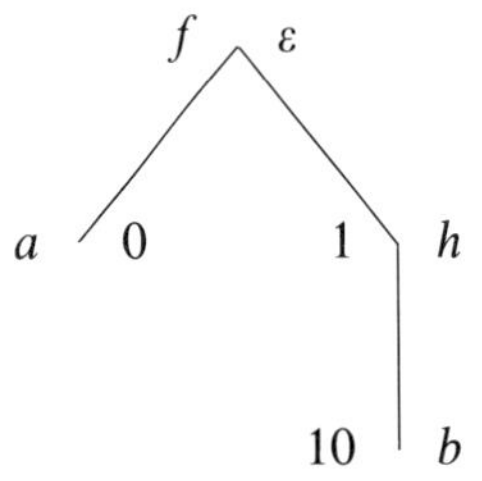

Each node of the tree is labeled by a function symbol: the only constraint is that if the symbol has arity k then the node must have exactly k successors.

Each node in a tree can be uniquely identified by a string in N^* and this gives a formal representation of trees. Let FTR (Finite Tree Representations) be the collection of subsets R of N^* which satisfy

i) R is nonempty, finite and prefix-closed
ii) if $n(k + 1) \in R$ then $nk \in R$.

A labeled tree assigns to each node an element of Σ, i.e. it is a function $t : R \longrightarrow \Sigma$ which satisfies

iii) $ni \in R$ implies $t(n) \in \Sigma_k$ where $i < k$ and $k > 0$
iv) $t(n) \in \Sigma_k$ implies $ni \in R$ for every i, $0 \leq i < k$.

Let FLTR be the set of pairs $\langle t, R \rangle$ which satisfy i)–iv). For each $f \in \Sigma$ define a corresponding function over FLTR so that it is isomorphic as a Σ-algebra to T_Σ.

Q13 a) Here we generalize the representation of T_Σ in the previous question to contain partial objects. Each partial tree is represented as a finite prefix-closed subset of N^*. A Finite Partial Tree Representation is a pair $\langle R, t \rangle$ where $R \subseteq N^*$ satisfies

i) R is finite and prefix-closed
ii) $ni \in R$ implies $t(n) \in \Sigma_k$ where $i < k$ and $k > 0$.

For example, the partial term $f(\Omega, h(b))$, using Σ of the last question is represented by

$R = \{\varepsilon, 1, 10\}$

$t : \varepsilon \longrightarrow f,\ 1 \longrightarrow h,\ 10 \longrightarrow b.$

Endow these trees with functions corresponding to each function symbol in Σ so that the result is a Σ-predomain isomorphic to the initial one described in example 3.3.5.

b) We extend these trees by allowing them to be infinite. Let CT_Σ be the set of pairs $\langle R, t\rangle$ where $R \subseteq N^*$, $t : R \longrightarrow \Sigma$ satisfies

i) R is prefix-closed
ii) $ni \in R$ implies $t(n) \in \Sigma_k$ where $i < k$ and $k > 0$.

These trees are ordered as in part (a) and the functions corresponding to the function symbols are also defined similarly. Show that CT_Σ is initial in the class of all Σ-domains.

Q14 a) Show that **AT** is an algebraic *cpo*

b) For each $f \in \Sigma^2$ let $f_{\mathbf{AT}}$ be defined in the same way as $f_{\mathbf{AT_S}}$, except in the definition of $t_1 \oplus_{\mathbf{AT}} t_2$ and $t_1 +_{\mathbf{AT}} t_2$, $L(t)$ should simply be $L(t_1) \cup L(t_2)$. Show that the two nondeterministic operators are in general not strict.

c) Show that **AT** is a Σ^2-domain which satisfies all the equations in **A2s**.

d) normal forms:

i) Ω, NIL are normal forms

ii) If $\mathscr{A}$ is a saturated set and for each $a \in A(\mathscr{A})$ $n(a)$ is a normal form then $\bigoplus \{n(A), A \in \mathscr{A}\}$ is a normal form

iii) if $n(a)$ is a normal form for every $a \in A$, A a nonempty set then $\Omega + \sum \{n(a), a \in A\}$ is a normal form.

Prove that for every t in $\mathbf{M_2}$ there exists a normal form n such that $t =_{\mathbf{A2s}} n$.

e) Prove that every finite tree is denotable by a normal form.

f) Prove that **AT** is initial in the class of Σ^2-domains which satisfies **A2s**.

Q15 a) Show that $\mathbf{AT_W}$ as outlined in §3.5 in a Σ^2-domain which satisfies all the equations in **WA2s**. The functions corresponding to the function symbols in Σ^2 are defined as in the previous question.

b) weak normal forms:

i) Ω is a weak normal form

ii) if $n(a)$ is a weak normal form for each $a \in A$, then $\Omega + \sum \{n(a), a \in A\}$ is also a weak normal form.

Prove that for every $t \in \mathbf{M_2}$ there exists a weak normal form n such that $t =_{\mathbf{WA2s}} n$.

c) Prove every finite element in $\mathbf{AT_W}$ is denotable by a weak normal form.

d) Prove $\mathbf{AT_W}$ is initial in the class of Σ^2-domains which satisfy **WA2s**.

Q16 a) Let t_1, t_2 be two trees in **AT** which contain no open nodes. They are also in $\mathbf{AT_S}$. Prove $t_1 =_{\mathbf{AT}} t_2$ if and only if $t_1 =_{\mathbf{AT_S}} t_2$.

b) Let $t_1, t_2 \in \mathbf{M_2}$ be two terms not involving the distinguished symbol Ω. Show that $t_1 =_{\mathbf{A2s\Omega}} t_2$ implies $t_1 =_{\mathbf{A2}} t_2$.

Q17 (Bounded Refusal Sets, DeNicola 1985). A refusal set $\mathscr{R}$ is *bounded* if

$$\{a, \langle sa, \varnothing\rangle \in \mathscr{R}\} \text{ is infinite implies } \langle st, X\rangle \in \mathscr{R} \text{ for every } t \in Act^* \text{ and finite } X \subseteq Act.$$

Let $\mathscr{BR}$ be the set of bounded refusal sets and assume *Act* is infinite.

a) Show $\langle \mathscr{BR}, \supseteq\rangle$ is a *cpo*, isomorphic to $\mathbf{AT_S}$.

b) Give direct definitions of functions over $\langle \mathscr{BR}, \supseteq\rangle$ corresponding to each symbol in Σ^2 in such a way that $\mathscr{BR}$ and $\mathbf{AT_S}$ are isomorphic as Σ^2-domains.

4 Recursive Processes

In this chapter we extend the three different views of processes to recursively defined processes. The previous chapter has laid the groundwork for the denotational models; we use Σ-domains and in particular the Acceptance Trees developed in §3.4. Contrary to the result in part I, the general framework of Σ-domains does not lead automatically to a denotational semantics of processes because the syntax of processes cannot simply be the carrier of the initial domain. We need to develop a specific syntax for infinite processes and show how to interpret it in Σ-domains; this syntax is developed in §4.1. Instead of writing recursive definitions such as

$$x \Longleftarrow ax + bx,$$

we introduce constructors of the form *rec x.-*, rendering this definition as the term $rec\, x.ax + bx$. These are called recursive terms. In this extended syntax variables play two distinct roles and this complicates some concepts, such as substitution, which were previously very straightforward. In §4.2 we show how these extended terms may be interpreted in arbitrary Σ-domains. Intuitively the idea is to interpret the term $rec\, x.t$ as the least solution in the domain of the equation

$$x \Longleftarrow t.$$

To state this precisely we use the least-fixpoint operator Y defined in §3.1. We also try to characterize this method of assigning meanings to terms as a very natural extension of that used in part I which only applies to finite terms.

In the next section we turn our attention to proof systems for the extended syntax. Although the system **DED**(E) can be used to characterize indirectly the initial domain CI_E it is not very powerful when applied to recursive terms. We add an infinitary rule called ω-Induction and show that this augmented system completely characterizes the interpretation of closed recursive terms in CI_E. However most of the section is devoted to examining effective forms of induction, such as Recursion Induction and Scott Induction.

In §4.4 we give an operational semantics for a particular recursive language which uses the combinators in Σ^2. It is called **RM** for Recursive Machines. The semantics is a simple extension of that in §2.5; essentially we add one extra rule

$$rec\, x.t \rightarrowtail t[rec\, x.t/x]$$

which says that recursive definitions may expand at any point in time to gain the capabilities of their bodies. This view of **RM** as a labeled transition system induces Testing preorders in the usual way—we also use recursive machines for experiments. The final section relates this behavioral view to the proof-theoretic view, the infinitary proof systems based on the equations of part I. This in turn implies that the Acceptance Trees of chapter 3 are fully abstract with respect to the Testing preorders. The main technical tool here is the alternative characterization of the Testing preorders. We also show that even on recursive terms it is sufficient to use finite tests.

4.1 Syntax

In this section we introduce the syntax for recursively defined processes and examine its properties. The most natural idea is to allow, as in the introduction to part II, definitions such as

$$x \Longleftarrow ax + bx. \qquad (*)$$

However this requires us to consider two different syntactic categories, terms and definitions, to investigate their individual properties and the relationships between them. Instead we stay within the framework of term algebras. Each function symbol in Σ gives a constructor in T_Σ or $T_\Sigma(X)$: f, of arity k, takes k terms $t_1, \ldots, t_k$ and construct the new term $f(t_1, \ldots, t_k)$. We add a new constructor, or more accurately a family of constructors, one for each variable. If t is a term, $rec\, x.t$ is also a term. So for example $(*)$ above will be rendered as

$$rec\, x.ax + bx.$$

This choice of syntax has certain disadvantages. For example the translation of mutually recursive definitions is somewhat opaque. The definition

$$x \Longleftarrow ax + by + cy$$

$$y \Longleftarrow by + x$$

is rendered as the term

$$rec\, x.ax + b\, rec\, y.(by + x) + c\, rec\, y.(by + x).$$

However these are outweighed by the advantages; the uniformity of our syntax makes it very convenient to investigate and develop its properties.

We now define the language in question. Let X be an infinite set of variables which we assume has an enumeration $x_0, x_1, x_2, \ldots$. We use x, y, z, etc., to range over X. For a given signature Σ (which we assume contains the symbol Ω), the set of *recursive terms over* Σ, $REC_\Sigma(X)$, is the least set which satisfies

i. if $x \in X$ then $x \in REC_{\Sigma}(X)$

ii. if $f \in \Sigma$ of arity k and $t_1, \ldots, t_k$ are in $REC_{\Sigma}(X)$, then $f(t_1, \ldots, t_k) \in REC_{\Sigma}(X)$

iii. if $x \in X$ and $t \in REC_{\Sigma}(X)$, then $rec\, x.t \in REC_{\Sigma}(X)$.

Notice that $T_{\Sigma}(X)$ is a subset of $REC_{\Sigma}(X)$ since it is the least set which satisfies i. and ii. above. Throughout this chapter we will emphasize the fact that $REC_{\Sigma}(X)$ plays a role similar to that of $T_{\Sigma}(X)$ in part I. To help emphasize this comparison and to make the notation uniform we will use $FREC_{\Sigma}(X)$ to denote the subset of terms in $REC_{\Sigma}(X)$ which have no occurrences of *rec x.*, i.e. we use $FREC_{\Sigma}(X)$ in place of $T_{\Sigma}(X)$. These will be called *(syntactically) finite* terms.

An alternative method of describing recursive terms is to give a BNF-type description of how they can be formed:

$$t ::= x \mid f(t_1, \ldots, t_k) \mid rec\, x.t.$$

Here t ranges over $REC_{\Sigma}(X)$, x ranges over X and f ranges over Σ. So there are three different ways of forming recursive terms: using a simple variable, using a function symbol as a constructor, or using *rec x.* as a constructor.

We will abbreviate "recursive term" to simply "term" and use, as in part I, symbols such as t, t', u, u', etc., to indicate terms. When writing terms in $REC_{\Sigma}(X)$ we assume *rec x.* has the lowest precedence of all the operators and between different occurrences of *rec x.* the precedence is assigned from left to right in ascending order. The following pairs exemplify this convention:

$rec\, x.ax + b, \quad rec\, x.(ax + b)$

$rec\, x.ax + rec\, y.ay + x, \quad rec\, x.(ax + rec\, y.(ay + x))$

$rec\, x.ax + (rec\, y.ay) + x, \quad rec\, x.(ax + (rec\, y.ay) + x)$

$rec\, x.\, rec\, y.ax + by, \quad rec\, x.(rec\, y.(ax + by)).$

$REC_{\Sigma}(X)$ can be viewed as a Σ-algebra in much the same way as $T_{\Sigma}(X)$. The carrier is $REC_{\Sigma}(X)$ and for each symbol f in Σ of arity k, $f_{REC_{\Sigma}}$ is the function which maps the k terms $t_1, \ldots, t_k$ to the term $f(t_1, \ldots, t_k)$. Viewing $REC_{\Sigma}(X)$ as a Σ-algebra in this way will help in the next section when we generalize the Freeness property of $T_{\Sigma}(X)$ to a relativised Freeness property of $REC_{\Sigma}(X)$.

In the new syntax variables play two roles; one to indicate the presence of recursion, the other as a genuine unknown. For example in

$$(rec\, x.ax + bNIL) + y, \qquad (**)$$

the variable y is a genuine unknown; to assign a meaning to (**) one must first assign a meaning to y. On the other hand the meaning is independent of x since it is only used as part of the syntactic machinery to indicate a recursion. In fact the meaning of (**) should always be the same as the meaning of

$$(rec\, z.az + bNIL) + y.$$

The difference between these two roles is explained in the definition of free and bound occurrences of variables. It is worthwhile giving a detailed definition of these concepts.

DEFINITION 4.1.1

i. a) x occurs free in y if x and y are the same variable,
   b) x occurs free in $f(\underline{t})$ if it occurs free in $\underline{t}$,
   c) x occurs free in $rec\, y.t$ if x and y are different, and x occurs free in t.

ii. a) x does not occur bound in y for any y,
   b) x occurs bound in $f(\underline{t})$ if it occurs bound in $\underline{t}$,
   c) x occurs bound in $rec\, y.t$ if x and y are the same variable or if it occurs bound in t.

We use $FV(t)$ to denote the set of variables which occurs free in t and $BV(t)$ the set which occurs bound. For example if t is

$$a\, rec\, x.aNIL + (rec\, y.ax + y) + (rec\, z.bNIL) + x + z,$$

then

$$FV(t) = \{z\}, \qquad BV(t) = \{x, y, z\}.$$

Note that variables may have both free and bound occurrences in a given term. Also an occurrence may be bound in a term but free in a subterm. For example x is bound in t but occurs free in the subterm $rec\, y.ax + y$. We use REC_Σ to denote the set of *closed* terms, i.e. those terms t in $REC_\Sigma(X)$ which satisfy $FV(t) = \emptyset$. They contain no free variables and they play a role analogous to the elements of the free algebra T_Σ in part I. We also use $FREC_\Sigma$ to denote the set of finite closed terms. We will normally use letters such as p, q to refer to REC_Σ and d, e to refer to $FREC_\Sigma(X)$.

Intuitively the meaning of a term should not depend on the meaning of bound variables. In other words we should be able to rename bound variables without affecting the meaning. For example, $rec\, y.ay + bx$ and $rec\, z.az + bx$ will always have the same meaning; they represent the process p defined recursively by

$$p \Longleftarrow ap + bx.$$

However we cannot make arbitrary substitutions for bound variables. For example if we substitute y for x in $rec\, y.ay + bx$, we obtain $rec\, y.ay + by$ which represents the very different process

$$q \Longleftarrow aq + bq.$$

The problem occurs here because a free variable is transformed into a bound variable. However it is better to postpone this discussion of the renaming of bound variables until we have developed the notion of substitution.

Here a substitution will mean a mapping from X to $REC_\Sigma(X)$ and as in part I we use Greek letters such as ρ as meta-variables. We use $\rho[x \to t]$ to denote the substitution which is identical to ρ except that it maps x to t. Finally I denotes the identity substitution, which maps every variable x to itself.

To define $t\rho$, the result of applying the substitution ρ to t, we have to take into consideration the different roles the variables play. Since bound variables do not contribute to the meaning of terms they should be impervious to substitutions: substitutions should only affect free variables. Furthermore we should make sure that free variables are not captured, i.e. transformed into bound variables, as in the example above. The approach we take follows that of Stoughton 1987, and is a generalization of that given in part I. It makes extensive use of the benign renaming of bound variables, on the understanding that this will not affect the meanings associated with terms. (How we associate meanings with terms will be the subject of the next section.) We require a subsidiary function called *new*. If x is a variable, t a term, and ρ a substitution, let $new\, xt\rho$ denote the least variable y in the enumeration of X, $x_0, x_1, x_2, \ldots$ such that for every $z \in FV(t)$ which is different from x, $y \notin FV(\rho(z))$.

Definition 4.1.2 (Substitution) For $t \in REC_\Sigma(X)$ and ρ a substitution $t\rho$ is defined by structural induction:

i) $x\rho = \rho(x)$

ii) $f(t_1, \ldots, t_n)\rho = f(t_1\rho, \ldots, t_n\rho)$

iii) $rec\, x.t\rho = rec\, y.(t\rho[x \to y])$, where $y = new\, xt\rho$.

The subtle part of the definition is clause iii). To avoid the possibility of a free occurrence of x in $t\rho$ being captured by $rec\, x.\_$ x is renamed to the harmless variable

y. Every term t has a finite number of free variables and $FV(t\rho) = \{x, x \in FV(\rho(z))$ for some $z \in FV(t)\}$. This should help explain the use of the function *new*. The overall effect is that *rec* y._ is guaranteed not to capture any free variable in $t\rho[x \to y]$, other than y, of course. Many substitutions we use are simple substitutions which affect only one variable and leave the remainder untouched. We introduce a special notation for these cases: $t[u/x]$ will be used in place of $t(I[x \to u])$. More generally, if $\underline{x}$ is a vector of distinct variables and $\underline{u}$ is a vector of terms $t[\underline{u}/\underline{x}]$ will denote $t(I[x_1 \to u_1][x_2 \to u_2]\ldots[x_k \to u_k])$. Note that it does not matter in what order these are enumerated, provided $x_1, \ldots, x_k$ are distinct.

Examples 4.1.3

1. If t is $(rec\, x.ax + b) + ax$, where x is any variable, and ρ is any substitution such that $\rho(x) = cx + d$ then $t\rho = (rec\, x_0.ax_0 + b) + a(cx + d)$. Alternatively, $t[cx + d/x] = (rec\, x_0.ax_0 + b) + a(cx + d)$.
2. If t is $rec\, x.ay + x$ then $t[cx_0 + d/y]$ is $rec\, x_1.a(cx_0 + d) + x_1$.
3. If t is $rec\, x.ax + (y + rec\, y.bx + cy)$ then $t[cx_2 + d/y]$ is

   $rec\, x_0.ax_0 + ((cx_2 + d) + rec\, x_1.bx_0 + cx_1)$.
4. For any two distinct variables x and z, $(rec\, z.z)[u/x] = rec\, x_0.x_0$. Thus

   $(rec\, z.z)I = rec\, x_0.x_0$.

From these examples we see that substitution nearly always involves the systematic renaming of bound variables. Nevertheless the purely syntactic instance of the Substitution lemma 1.2.6 extends to this new setting. As in part I if ρ, ρ' are two substitutions $\rho \circ \rho'$ is also a substitution defined by $\rho \circ \rho'(x) = (\rho'(x))\rho$.

Lemma 4.1.4 (Syntactic Substitution Lemma) For every t in $REC_\Sigma(X)$,

$(t\rho)\rho' = t(\rho' \circ \rho)$.

We will not give the proof of this lemma. Instead it will be outlined in the exercises.

Having formalized substitution we return to the discussion of changing bound variables. We define an equivalence between terms, called α-equality, which is slightly weaker than syntactic identity. Intuitively $t =_\alpha t'$ if t and t' are identical apart from benign changes of bound variables. The equivalence is defined to be the least Σ-congruence over $REC_\Sigma(X)$ which satisfies

i) $t[y/x] =_\alpha t'$ and $y \notin FV(t)$ implies $rec\, x.t =_\alpha rec\, y.t'$
ii) $t =_\alpha t'$ implies $rec\, x.t =_\alpha rec\, x.t'$.

The first clause is the most important. As an example of its application we have $tI[x \to y] =_\alpha tI[x \to y]$, because $=_\alpha$ is reflexive, and therefore $rec\, x.t =_\alpha rec\, y.tI[x \to y]$ if $y \notin FV(t)$. The variable x may be changed to y in $rec\, x.t$ provided in the resulting term a free occurrence of y in t is not captured. For example, $rec\, x.ax =_\alpha rec\, y.ay$ for any variables x, y but $rec\, x.ay + bx \neq_\alpha rec\, y.ay + by$.

We will see in the next section that the meaning associated with terms is preserved by this weak form of identity, i.e. if $t =_\alpha t'$ then both t and t' will have the same meaning. In general it is more natural to use $=_\alpha$ when dealing with $REC_\Sigma(X)$, rather than syntactic identity. However we will not actually make much use of it.

In the Introduction to part II we indicated that the meaning in a Σ-domain of a term is taken to be the limit of the meanings of its finite approximations. These approximations are defined by progressively expanding out recursive subterms.

For each $n \geq 0$ and $t \in REC_\Sigma(X)$ let t^n be defined by

a) $t^0 = \Omega$

b) i) $x^{n+1} = x$
ii) $f(\underline{t})^{n+1} = f(\underline{t}^{n+1})$
iii) $(rec\, x.t)^{n+1} = t^{n+1}[(rec\, x.t)^n/x]$.

Note that $t^n \in FREC_\Sigma(X)$ for every $n \geq 0$. These are called the *principal approximations* to t, and t^n is referred to as the nth approximation. Let $\mathrm{App}(t) = \{t^n, n \geq 0\}$. These finite approximations are related via the syntactic preorder $\preceq$.

LEMMA 4.1.5 $n \leq m$ implies $t^n \preceq t^m$.

Proof Consider the proof system **DED**(E) (or more strictly **ΩDED**(E)) for any set of equations E. We show that within this system $\vdash t^n \leq t^{n+1}$. It will follow that $\vdash t^n \leq t^m$ whenever $n \leq m$. This proof system is sound for any model in $\mathscr{CP}(E)$. Taking E to be the empty set of equations and the model to be the initial model $PI_\emptyset$, we obtain $t^n \preceq t^m$ whenever $n \leq m$. The proof of $\vdash t^n \leq t^{n+1}$ is by induction on n. For $n = 0$ it is immediate from the equation $\Omega \leq x$. So we assume that $\vdash t^k \leq t^{k+1}$ for every term t and prove $\vdash t^{k+1} \leq t^{k+2}$. This proof is by structural induction on t and the only nontrivial case is when t has the form $rec\, x.u$. Then

$t^{k+1} = u^{k+1}[t^k/x]$ by definition

$\leq u^{k+1}[t^{k+1}/x]$ by generalized substitution (see Q19, chapter 1) and induction

$\leq u^{k+2}[t^{k+1}/x]$ by structural induction and instantiation

$= t^{k+2}.$ □

As a trivial corollary we have the following.

COROLLARY 4.1.6 For every $t \in REC_\Sigma(X)$, App(t) is directed with respect to $\preceq$.

The approximations to t in App(t) take a very particular form. They represent a uniform unwinding of the recursive subterms. For example if t is $rec\,x.ax + bx$ then App(t) contains

Ω

$a\Omega + b\Omega$

$a(a\Omega + b\Omega) + b(a\Omega + b\Omega)$

etc.

but it does not contain terms such as

$a\Omega + b(a\Omega + b\Omega)$

$a(a\Omega + b\Omega) + b\Omega.$

Nevertheless these also represent finite approximations to t where the recursions are unwound to different depths. In the exercises we explore a more general type of approximations called Fin(t) which include such terms. However the set App(t) will suffice for our purposes.

4.2 Assigning Meaning to Recursive Terms

Let A be a Σ-domain and ρ_A an A-assignment, i.e. a mapping which associates with each variable x in X an element $\rho_A(x)$ of A. The Freeness theorem 1.2.5 says that ρ_A can be extended uniquely to a Σ-homomorphism from $T_\Sigma(X)$ to A. In part I we used $A[\![t]\!]\rho_A$ to denote the unique element of A associated with t by this extension, the evaluation of t with respect to ρ_A. Essentially this evaluates t in A on the basis of the assignment of values to the variables of X. In this section we show how to extend this mapping to $REC_\Sigma(X)$. $REC_\Sigma(X)$ is also a Σ-algebra and the extension will also be a Σ-homomorphism. However it will no longer be the unique extension of the A-assignment ρ_A. Instead it will be the least one, in some sense, satisfying

$A[\![rec\, x.t]\!] = A[\![t[rec\, x.t/x]]\!]$.

Alternatively, it can be characterized as the unique extension satisfying

$A[\![t]\!] = \bigvee_A A[\![\mathrm{App}(t)]\!]$.

To define this extension it is convenient to reformulate the Freeness theorem of part I. Let ENV_A denote the collection of A-assignments. In part I we used ρ_A as a typical A-assignment and the unscripted ρ for a substitution (which in part I is a $T_\Sigma(X)$-assignment.) We continue this practice here although in a context where there are no substitutions under discussion we prefer to abbreviate ρ_A by ρ. Whenever any doubt may occur we will always subscript A-assignments. The notation for modifying substitutions will also apply to assignments: $\rho_A[\underline{a}/\underline{x}]$ is identical to ρ_A except that it maps x_i to a_i. ENV_A can be ordered pointwise via

$\rho \leq \rho'$ if for every x in X, $\rho(x) \leq_A \rho'(x)$.

As we have seen in §3.1, ENV_A is a domain under this ordering, provided A is a domain. In addition to this ordering we will sometimes use a parameterized identity on ENV_A. If Y is a set of variables, $Y \subseteq X$, we write

$\rho_A =_Y \rho'_A$ if $\rho_A(y) = \rho'_A(y)$ for every $y \in Y$.

In the case when $Y = X$ this relation coincides with the usual functional identity on ENV_A. Similarly we write $\rho_A \leq_Y \rho'_A$ to mean $\rho_A(y) \leq \rho'_A(y)$ for every $y \in Y$.

The Freeness theorem 1.2.5 associates with each term t in $T_\Sigma(X)$ and each A-assignment ρ a unique element in A. This unique element is also denoted by $\rho(t)$ although the more suggestive notation $A[\![t]\!]\rho$ was introduced. As stated above, $A[\![t]\!]\rho$ denotes the unique evaluation of the term t in A with respect to the A-assignment ρ. $A[\![\ ]\!]$ can be considered as a function from $T_\Sigma(X)$ to the set $(\mathrm{ENV}_A \longrightarrow A)$. Indeed it has a very simple definition, using structural induction on $T_\Sigma(X)$:

Let $A[\![\ ]\!] : T_\Sigma(X) \longrightarrow (\mathrm{ENV}_A \longrightarrow A)$ be defined by
i. $A[\![x]\!]\rho = \rho(x)$
ii. $A[\![f(\underline{t})]\!]\rho = f_A(A[\![\underline{t}]\!]\rho)$.

We immediately have the following proposition.

Proposition 4.2.1 $A[\![\ ]\!]$ is a function from $T_\Sigma(X)$ to $[\mathrm{ENV}_A \longrightarrow A]$.

Proof We must show that for every t, $A[\![t]\!]$ is a continuous function in $[\mathrm{ENV}_A \longrightarrow A]$.

If t is a variable x then this is immediate. So we may assume it is of the form $f(\underline{t})$ and by induction each $A[\![t_i]\!]$ is continuous. Now let $F \subseteq \mathrm{ENV}_A$ be directed with *lub* ρ. Then

$$\begin{aligned} A[\![f(\underline{t})]\!]\rho &= f_A(A[\![\underline{t}]\!]\rho) \quad \text{by definition} \\ &= f_A(\bigvee A[\![t_1]\!](F), \ldots, \bigvee A[\![t_k]\!](F)) \quad \text{since each } A[\![t_i]\!] \text{ is continuous} \\ &= \bigvee\{f_A(A[\![t_1]\!]\rho^1, \ldots, A[\![t_k]\!]\rho^k), \rho^i \in F\} \quad \text{since } f_A \text{ is continuous} \\ &= \bigvee\{f_A(A[\![t_1]\!]\rho, \ldots, A[\![t_k]\!]\rho), \rho \in F\} \quad \text{because } F \text{ is directed} \\ &= \bigvee A[\![f(\underline{t})]\!](F). \end{aligned}$$

□

Moreover we can recapture the Σ-homomorphisms given in the Freeness theorem from the function $A[\![\ ]\!]$: For any A-assignment ρ let $\rho : T_\Sigma(X) \longrightarrow A$ be defined by $\rho(t) = A[\![t]\!]\rho$.

PROPOSITION 4.2.2

a) $\rho : T_\Sigma(X) \longrightarrow A$ is a Σ-homomorphism.

b) if $h: T_\Sigma(X) \longrightarrow A$ is a Σ-homomorphism which satisfies $h(x) = \rho(x)$ then h coincides with ρ.

c) if ρ, ρ' are A-assignments such that $\rho =_{FV(t)} \rho'$ then $\rho(t) = \rho'(t)$.

Proof We omit the proof as it follows the lines of the proof of the Freeness theorem 1.2.5. □

Part c) of this proposition implies that if $t \in T_\Sigma(X)$ is closed, i.e. $t \in T_\Sigma$, then $\rho(t)$ evaluates to the same element of A for every A-assignment ρ. This element is of course $i_A(t)$, the meaning of t in A, and in part I we always took $A[\![t]\!]$ to mean this element rather than the constant function which always returned it.

We wish to extend the function $A[\![\ ]\!]$ from $T_\Sigma(X)$ to $REC_\Sigma(X)$. To do so we must define $A[\![rec\, x.t]\!]$ for arbitrary x and t. By induction we can assume that we have already defined $A[\![t]\!]$ as a function in $[\mathrm{ENV}_A \longrightarrow A]$.

Let $\rho \in \mathrm{ENV}_A$. We wish to define $A[\![rec\, x.t]\!]\rho$ as an element of A. To do so we associate a function in $[A \longrightarrow A]$ with the pair t, ρ. This function maps $a \in A \longrightarrow A[\![t]\!]\rho[a/x]$. We use the notation $\lambda a.A[\![t]\!]\rho[a/x]$ to denote this function. Then define $A[\![rec\, x.t]\!]\rho$ as the least fixpoint of this function, i.e. $Y\lambda a.A[\![t]\!]\rho[a/x]$. Intuitively this corresponds to taking the least fixpoint of the interpretation in A of the equation

$x = t.$

We now reiterate the definition of the function $A[\![\ ]\!]$.

DEFINITION 4.2.3 If A is a Σ-domain let $A[\![\ ]\!]: REC_\Sigma(X) \longrightarrow [\mathrm{ENV}_A \longrightarrow A]$ be defined by

i. $A[\![x]\!]\rho = \rho(x)$

ii. $A[\![f(\underline{t})]\!]\rho = f_A(A[\![\underline{t}]\!]\rho)$

iii. $A[\![rec\, x.t]\!]\rho = Y\lambda a.A[\![t]\!]\rho[a/x].$

PROPOSITION 4.2.4 $A[\![\ ]\!]$ is well-defined.

Proof We must show that for every t in $REC_\Sigma(X)$, $A[\![t]\!]$ is a continuous function from ENV_A to A. The proof proceeds by structural induction and the only case not covered by proposition 4.2.2 is when t has the form *rec x.u.*

In this case $A[\![t]\!]\rho = Y\lambda a.A[\![u]\!]\rho[a/x]$. The least-fixpoint operator Y can only be applied to continuous functions, so to ensure that $A[\![t]\!]\rho$ is well-defined we must show that, for an arbitrary environment ρ, $\lambda a.A[\![u]\!]\rho[a/x]$ is continuous.

a) Let D be directed in A with *lub* b and for convenience let f denote the function $\lambda a.A[\![u]\!]\rho[a/x]$. Let $F \subseteq \mathrm{ENV}_A$ denote the set $\{\rho[d/x], d \in D\}$, which is easily seen to be directed, with *lub* $\rho[b/x]$.

$$\begin{aligned} f(b) &= A[\![u]\!]\rho[b/x] \\ &= A[\![u]\!](\bigvee F) \\ &= \bigvee_A A[\![u]\!](F) \quad \text{since, by induction, } A[\![u]\!] \text{ is continuous} \\ &= \bigvee_A \{A[\![u]\!]\rho[d/x], d \in D\} \\ &= \bigvee f(D) \qquad \text{by definition.} \end{aligned}$$

b) We now show that $A[\![rec\, x.u]\!]$ is continuous. We need to be more precise than in part a) about our notation. Let $f\rho$ denote the function $\lambda a.A[\![u]\!]\rho[a/x]$ and let $g: \mathrm{ENV}_A \longrightarrow [A \longrightarrow A]$ be defined by $g(\rho) = f\rho$. Then $A[\![rec\, x.u]\!] = Y \circ g$. By proposition 3.1.12 Y is continuous, and composition preserves continuity. Therefore it suffices to show that the function g is continuous.

Let F be directed in ENV_A with *lub* σ. We must show $g(\sigma) = \bigvee g(F)$, i.e. for every a

in A, $g(\sigma)(a) = \bigvee\{g(\rho)(a), \rho \in F\}$. The set $F[a/x] = \{\rho[a/x], \rho \in F\}$ is also directed with *lub* $\sigma[a/x]$. By induction $A[\![u]\!]$ is continuous and therefore

$$A[\![u]\!]\sigma[a/x] = \bigvee_A\{A[\![u]\!]\rho[a/x], \rho \in F\},$$

$$\text{i.e.}\quad f_\sigma(a) = \bigvee\{f_\rho(a), \rho \in F\},$$

which is what is required. □

We now derive some properties of the mapping $A[\![\ ]\!]$. We extend the notation of part I by writing, when convenient, $t \leq_A t'$, $t =_A t'$, to mean $A[\![t]\!] \leq A[\![t']\!]$, $A[\![t]\!] = A[\![t']\!]$ respectively, for arbitrary t, t' in $REC_\Sigma(X)$. The following is straightforward and the proof is left to the reader.

LEMMA 4.2.5 If $\rho \leq_{FV(t)} \rho'$ then $A[\![t]\!]\rho \leq A[\![t]\!]\rho'$.

An immediate corollary of this lemma is that if t is closed, i.e. $FV(t) = \emptyset$, then $A[\![t]\!]\rho$ is independent of ρ, i.e. for every ρ, ρ' $A[\![t]\!]\rho = A[\![t]\!]\rho'$. So $A[\![t]\!]$ is a constant function from $\mathrm{ENV}_A \longrightarrow A$ and, as in part I, we usually take the element $A[\![t]\!]\rho$ of A to be the meaning or interpretation of the closed term t in A.

To derive more substantial properties of $A[\![\ ]\!]$ we need to generalize the Substitution lemma of part I. Here we will indicate an A-assignment by ρ_A and a substitution by ρ. The composition of an A-assignment and a substitution give another A-assignment: $\rho_A \circ \rho(x) = A[\![\rho(x)]\!]\rho_A$.

PROPOSITION 4.2.6 (Substitution Lemma) $A[\![t\rho]\!]\rho_A = A[\![t]\!](\rho_A \circ \rho)$.

That is to say, we may either substitute into the term and then evaluate in an A-assignment or substitute in the A-assignment and then evaluate.

Proof The proof, as usual, is by structural induction on t and the only nontrivial case is when t has the form *rec x.u*. In this case

$$A[\![(rec\,x.u)\rho]\!]\rho_A = A[\![rec\,y.(u\rho[x \to y])]\!]\rho_A \quad \text{where} \quad y = new\,xt\rho$$

$$= Y\lambda a.A[\![u\rho[x \to y]]\!]\rho_A[a/y].$$

By induction we can assume for each particular a that

$$A[\![u\rho[x \to y]]\!]\rho_A[a/y] = A[\![u]\!](\rho_A[a/y] \circ \rho[x \to y])$$

$$= A[\![u]\!](\rho_A \circ \rho)[a/x] \quad \text{by the previous lemma.}$$

The second equality arises because the two A-assignment $\rho_A[a/y] \circ \rho[x \to y]$ and

$\rho_A \circ \rho[a/x]$ agree on the free variables of u. Therefore

$$A[\![(rec\,x.u)\rho]\!]\rho_A = Y\lambda a.A[\![u]\!](\rho_A \circ \rho)[a/x]$$

$$= A[\![rec\,x.u]\!]\rho_A \circ \rho. \qquad \square$$

Remark When applied to simple substitutions of the form $I[x \to u]$ this result may be rewritten as

$$A[\![t[u/x]]\!]\rho_A = A[\![t]\!]\rho_A[A[\![u]\!]\rho_A/x].$$

With this lemma we can show that changing bound variables in a term, provided it is done properly, does not affect its meaning.

COROLLARY 4.2.7 If $y \notin FV(t)$, $A[\![rec\,x.t]\!] = A[\![rec\,y.t[y/x]]\!]$.

Proof For any $a \in A$ and any A-assignment ρ_A,

$$\rho_A[a/y] \circ I[x \to y] =_{FV(t)} \rho_A[a/x] \quad \text{because} \quad y \notin FV(t).$$

Therefore, by lemma 4.2.5,

$$A[\![t]\!]\rho_A[a/y] \circ I[x \to y] = A[\![t]\!]\rho_A[a/y].$$

From the Substitution lemma we have

$$A[\![t[y/x]]\!]\rho_A[a/y] = A[\![t]\!]\rho_A[a/x] \circ I[x \to y],$$

so that we can conclude

$$A[\![t[y/x]]\!]\rho_A[a/y] = A[\![t]\!]\rho_A[a/x].$$

The result now follows:

$$A[\![rec\,x.t]\!]\rho = Y\lambda a.A[\![t]\!]\rho_A[a/x]$$

$$= Y\lambda a.A[\![t[y/x]]\!]\rho_A[a/y] \quad \text{from above}$$

$$= A[\![rec\,y.t[y/x]]\!]\rho. \qquad \square$$

More generally we can prove

COROLLARY 4.2.8 $t =_\alpha t'$ implies $A[\![t]\!] = A[\![t']\!]$.

Proof Because of the manner in which the semantic mapping is defined it is immediate that $=_A$ is a Σ-congruence. We show that it also satisfies the defining conditions of $=_\alpha$. Since $=_\alpha$ is the least such Σ-congruence the result follows.

We must show

i) if $t[y/x] =_A t'$ and $y \notin FV(t)$, then $rec\, x.t =_A rec\, y.t'$
ii) $t =_A t'$ implies $rec\, x.t =_A rec\, x.t'$

The second requirement is obvious and the first follows from the immediately preceding result:

If $t[y/x] =_A t'$ then $rec\, y.t' =_A rec\, y.t[y/x]$

$$=_A rec\, x.t \text{ if } y \notin FV(t). \qquad \square$$

We can also prove that the meaning of $rec\, x.t$ satisfies an appropriate fixpoint condition.

COROLLARY 4.2.9 $A[\![rec\, x.t]\!] = A[\![t[rec\, x.t/x]]\!]$.

Proof $A[\![rec\, x.t]\!]\rho$ is Yf where f is the function $\lambda a.\, A[\![t]\!]\rho[a/x]$. By the definition of Y we have $Yf = f(Yf)$, i.e. $Yf = f(A[\![rec\, x.t]\!]\rho)$. By applying the function f we obtain

$$A[\![rec\, x.t]\!]\rho = A[\![t]\!]\rho[A[\![rec\, x.t]\!]\rho/x]$$

$$= A[\![t[rec\, x.t/x]]\!]\rho \quad \text{by the Substitution lemma.} \qquad \square$$

This corollary tells us that our method of assigning meanings to terms is in harmony with the natural syntactic concepts which were introduced in the previous section. It also enables us to state the characterizing property of the semantic function $A[\![\ ]\!]$: the meaning of a term is the *lub* of the meaning of its finite approximations.

First let us derive a property of these finite approximations. Recall that t^n is the nth principal approximation to t.

LEMMA 4.2.10 For every $n \geq 0$, $A[\![t^n]\!] \leq A[\![t]\!]$.

Proof Let us abbreviate $A[\![u]\!]$ to $[\![u]\!]$. The proof is by induction on n and for $n = 0$ it is immediate. So let us assume that $[\![t^k]\!] \leq [\![t]\!]$ for every term t. We then prove $[\![t^{k+1}]\!] \leq [\![t]\!]$ from which the result will follow by induction. The proof of the inductive step, namely $[\![t^{k+1}]\!] \leq [\![t]\!]$, is itself by induction, this time structural induction on t. As usual the only nontrivial case is when t has the form $rec\, x.u$. Then

$$[\![t^{k+1}]\!]\rho = [\![u^{k+1}[(rec\, x.u)^k/x]]\!]\rho$$

$$= [\![u^{k+1}]\!]\rho[[\![(rec\, x.u)^k]\!]\rho/x] \quad \text{by the Substitution lemma}$$

$$\leq [\![u]\!]\rho[[\![(rec\, x.u)^k]\!]\rho/x] \quad \text{by structural induction}$$

$\leq [\![u]\!]\rho[[\![rec\, x.u.]\!]\rho/x]$ by the outer level induction and lemma 4.2.5

$= [\![u[rec\, x.u/x]]\!]\rho$ again by the Substitution lemma

$= [\![t]\!]\rho$ by corollary 4.2.9. □

In §3.1 we defined a syntactic preorder $\preceq$ between terms and showed that App(t) is directed with respect to it. It is trivial to establish that $A[\![\ ]\!]$ preserves $\preceq$, i.e. $t \preceq t'$ implies $A[\![t]\!] \leq A[\![t']\!]$. Therefore $A[\![\mathrm{App}(t)]\!]$ is directed in $[\mathrm{ENV}_A \longrightarrow A]$. The characteristic property of $A[\![\ ]\!]$ is given by the following theorem.

THEOREM 4.2.11 (Finite Approximation) For every t in $REC_\Sigma(X)$, $A[\![t]\!] = \bigvee[\![\mathrm{App}(t)]\!]$.

Proof As in the previous lemma we abbreviate $A[\![t]\!]$ to $[\![t]\!]$. From this result we know $\bigvee[\![\mathrm{App}(t)]\!] \leq [\![t]\!]$ and we now prove $\bigvee[\![\mathrm{App}(t)]\!] = [\![t]\!]$ using structural induction on t.

a) t is x. Immediate.

b) t is $f(\underline{u})$.

$[\![t]\!]\rho = f_A([\![\underline{u}]\!]\rho)$ by definition

$= f_A(\bigvee_A\{[\![u_1^n]\!]\rho, n \geq 0\}, \ldots, \bigvee_A\{[\![u_k^n]\!]\rho, n \geq 0\})$ by induction

$= \bigvee_A f([\![\underline{u}^n]\!]\rho)$ since f is continuous

$= \bigvee_A\{[\![f(\underline{u}^n)]\!]\rho, n \geq 0\}$.

c) t is $rec\, x.u$ This is the only nontrivial case.

Let l denote $\bigvee_A\{[\![t^n]\!]\rho, n \geq 0\}$. We already know $l \leq [\![t]\!]\rho$. Now $[\![t]\!]\rho$ is Yf where f is the function $\lambda a.[\![u]\!]\rho[a/x]$. So to show $[\![t]\!]\rho \leq l$, and therefore $[\![t]\!]\rho = l$, it is sufficient to prove $f(l) = l$. Now

$f(l) = [\![u]\!]\rho[l/x]$

$= \bigvee_A\{[\![u]\!]\rho[[\![t^n]\!]\rho/x], n \geq 0\}$ since $[\![u]\!]$ is a continuous function.

For convenience we denote $\rho[[\![t^n]\!]\rho/x]$ by ρ^n. So

$f(l) = \bigvee_A\{[\![u]\!]\rho^n, n \geq 0\}$.

We may now apply structural induction to u, to obtain

$f(l) = \bigvee_A\{\bigvee\{[\![u^m]\!]\rho^n, m \geq 0\}, n \geq 0\}$

$= \bigvee_A\{[\![u^m]\!]\rho^n, m, n \geq 0\}$.

Now $\{[\![u^{n+1}]\!]\rho^n, n \geq 0\}$ dominates $\{[\![u^m]\!]\rho^n, m, n \geq 0\}$. So by lemma 3.1.1,

$$f(l) = \bigvee_A \{[\![u^{n+1}]\!]\rho^n, n \geq 0\}$$

$$= \bigvee_A \{[\![t^{n+1}]\!]\rho, n \geq 0\} \quad \text{by the Substitution lemma}$$

$$= l.$$

□

Let us now return to the discussion at the beginning of this section. There we promised an extension of the Freeness theorem for $T_\Sigma(X)$ to our extended syntax $REC_\Sigma(X)$. Let ρ be any A-assignment, an association of elements of the domain A with variables in X. This association can be extended, as in proposition 4.2.2, by defining $\rho : REC_\Sigma(X) \longrightarrow A$ via $\rho(t) = A[\![t]\!]\rho$.

It is easy to see that this is indeed an extension of the A-assignment ρ. Viewing $REC_\Sigma(X)$ as a syntactic Σ-algebra it is also a Σ-homomorphism. Moreover because of the theorem above it satisfies

$$\rho(t) = \bigvee_A \rho(\text{App}(t)). \qquad (*)$$

We now prove it is the unique such Σ-homomorphism.

Proposition 4.2.12 Let $h : REC_\Sigma(X) \longrightarrow A$ be a Σ-homomorphism which satisfies

i. it is an extension of the A-assignment ρ, i.e. $h(x) = \rho(x)$
ii. for every t in $REC_\Sigma(X)$, $h(t) = \bigvee_A h(\text{App}(t))$.

Then $h = \rho$.

Proof The proof is trivial. As in the proof of the Freeness theorem we can prove by structural induction that $h(d) = \rho(d)$ for every d in $FREC_\Sigma(X)$. Then for arbitrary t in $REC_\Sigma(X)$,

$$h(t) = \bigvee_A h(\text{App}(t))$$

$$= \bigvee_A \rho(\text{App}(t)) \quad \text{from the above remark}$$

$$= \rho(t).$$

□

Informally this result says that our method of assigning meanings to terms, using the "meaning function" $A[\![\ ]\!]$ is essentially unique if we demand that

i. meaning-functions are structurally induced
ii. the meaning of a recursive term is the limit of the meanings of its finite unwindings.

We can also give an alternative characterization of this method of assigning meanings to terms, based on the property that recursive terms are interpreted as fixpoints.

A function $h: REC_\Sigma(X) \longrightarrow [\mathrm{ENV}_A \longrightarrow A]$ is *reasonable* if

i. $h(rec\, x.t) = h(t[rec\, x.t/x])$

ii. $h(t[u/x])\rho = h(t)\rho[h(u)\rho/x]$.

Condition i. says that h interprets recursive terms as fixpoints and condition ii. says that h is well-behaved with respect to substitution. In particular it handles the interpretation of free and bound variables in a reasonable manner. $A[\![\ ]\!]$ is a reasonable Σ-homomorphism and is in fact the least one.

PROPOSITION 4.2.13 If $h: REC_\Sigma(X) \longrightarrow [\mathrm{ENV}_A \longrightarrow A]$ is a reasonable Σ-homomorphism then $A[\![t]\!] \leq h(t)$ for every t in $REC_\Sigma(X)$.

Proof Because of theorem 4.2.10 it is sufficient to prove $A[\![t^n]\!] \leq h(t)$ for every $n \geq 0$. For $n = 0$ this is obvious. On the assumption that it is true for $n = k$ we prove it true for $n = k + 1$, using structural induction. As usual the only nontrivial case is when t has the form $rec\, x.u$. In this case

$$A[\![t^{k+1}]\!]\rho = A[\![u^{k+1}[t^k/x]]\!]\rho$$

$$= A[\![u^{k+1}]\!]\rho[A[\![t^k]\!]\rho/x] \quad \text{by the Substitution lemma}$$

$$\leq A[\![u^{k+1}]\!]\rho[h(t)\rho/x] \quad \text{by induction on } n$$

$$\leq h(u)\rho[h(t)/x] \quad \text{by structural induction}$$

$$= h(u[t/x])\rho \quad \text{by condition ii.}$$

$$= h(t)\rho \quad \text{by condition i..}$$

□

We hope that these two propositions convince the reader that $A[\![\ ]\!]$ is a reasonable extension of the meaning function for $T_\Sigma(X)$ obtained by the Freeness theorem.

We end this section with some examples of this semantic mapping where Σ is the signature Σ^2 and the Σ-domains are those described in the previous chapter. Let r be the term $rec\, x.ax + bx$. Then $\mathbf{AT}[\![r]\!]$ is described in figure 4.1 a). It is a very regular tree. Each node has the single Acceptance set $\{a, b\}$ and exactly two branches, one labeled by a, the other by b. Note that this tree duplicates itself after both these branches. For this reason it is a fixpoint of the function

$$t \longrightarrow \mathbf{AT}[\![ax + bx]\!]\rho[t/x].$$

a) **AT**$[\![rec\ x.ax + bx]\!]$

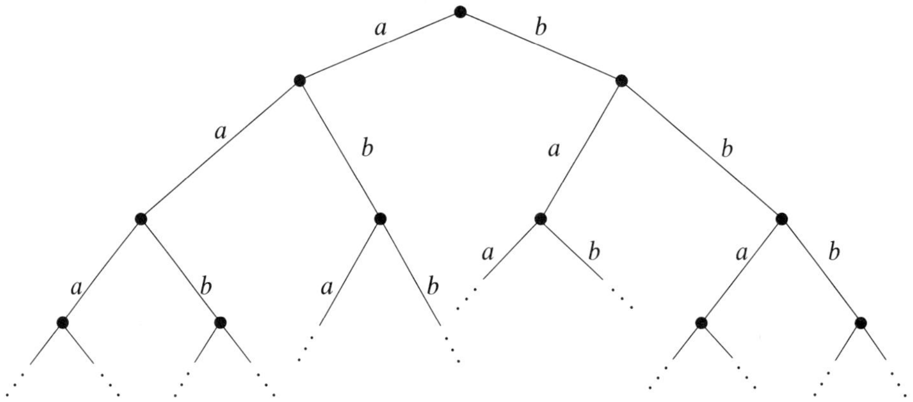

b) **AT**$[\![rec\ x.ac \oplus (bNIL + cx)]\!]$

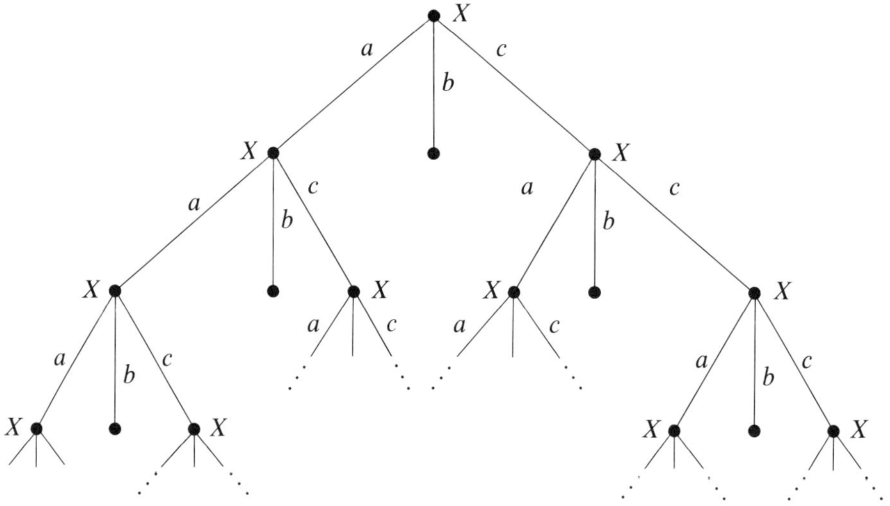

$X = \{\{a\}, \{a, b\}, \{a, c\}, \{b, c\}, \{a, b, c\}\}$

Figure 4.1
Examples of **AT**$[\![\ ]\!]$.

c) $\mathbf{AT}[\![rec\,x.aNIL + x]\!]$

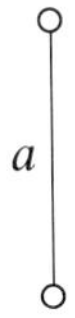

Figure 4.1 (continued)

However it is not immediately clear why it is the least fixpoint. To see this we use theorem 4.2.10. $A[\![r]\!]$ is the limit of the sequence $A[\![r^n]\!]$, $n \geq 0$. These are the sequence of trees

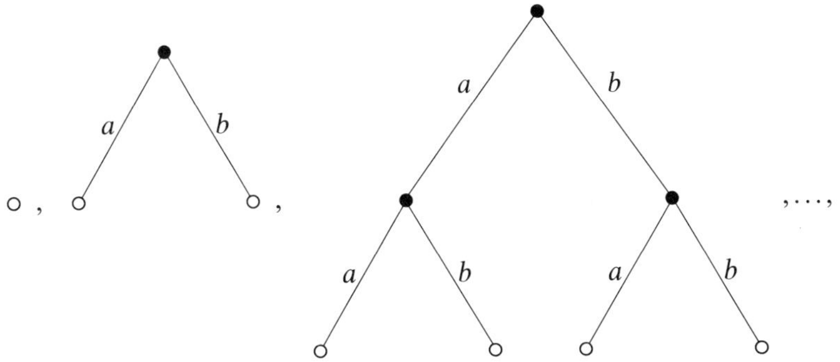

the limit of which is obviously the infinite tree we have described. Similarly reasoning will show that $\mathbf{AT}[\![rec\,x.ax \oplus (bNIL + cx)]\!]$ is the tree described in figure 4.1 b). The interpretation is the same in $\mathbf{AT_S}$.

Now consider $\mathbf{AT}[\![r]\!]$ with $r = rec\,x.aNIL + x$. This tree is the limit of the sequence $\mathbf{AT}[\![r^n]\!]$, $n \geq 0$, i.e. the sequence

$\mathbf{AT}[\![\Omega]\!]$, $\mathbf{AT}[\![aNIL + \Omega]\!]$, $\mathbf{AT}[\![aNIL + (aNIL + \Omega)]\!]$, ...

Because $\mathbf{AT}$ satisfies the equations +1, +2, +3, every element in this sequence, apart from the first, collapses to $\mathbf{AT}[\![aNIL + \Omega]\!]$. Therefore the limit is simply $\mathbf{AT}[\![aNIL + \Omega]\!]$, i.e.

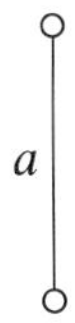

On the other hand, $\mathbf{AT_S}[\![aNIL + \Omega]\!]$ is the trivial tree $\circ$ so that $\mathbf{AT_S}[\![rec\, x.aNIL + x]\!]$ is simply $\circ$.

4.3 Proof Systems

In part I the proof systems used in the construction of the initial algebra I_E apply directly to the syntax of the appropriate language $T_\Sigma(X)$. In part II this is no longer the case: the syntax has been extended to $REC_\Sigma(X)$ and the proof system $\mathbf{\Omega DED}(E)$ used to construct CI_E, applies only to a subset of it, the finite terms. In this section we extend this proof system so as to apply to the entire language $REC_\Sigma(X)$. We add various forms of induction and extend the form of substitution allowed; these additions enable us to prove nontrivial properties of recursive processes and if we add a sufficiently powerful form of induction we obtain a completeness theorem with respect to the interpretation CI_E. This proof system will completely characterize CI_E in the same way as $\mathbf{DED}(E)$ characterizes I_E and $\mathbf{\Omega DED}(E)$ characterizes PI_E.

Consider the rules given in figure 4.2. Rules 1, 2, 3 a), 4, and 5 are from $\mathbf{DED}(E)$ as in rule 6, if we use the conventions introduced at the end of chapter 3. The five rules constitute a proof system for deriving statements of the form $t \leq u$, where t, u, $\in \mathbf{REC}_\Sigma(X)$. It is reasonable to refer to this system also as $\mathbf{DED}(E)$ although it applies to the extended language. This system is not very useful for deriving statements about recursive terms. The situation is improved if we add the two rules 3 b) and 7. The first is simply substitution into recursive terms and the second states that $rec\, x.t$ is a fixpoint of the equation $x = t$. We use $\mathbf{rDED}(E)$ to denote this proof system. Finally, if we add rule 8, ω-Induction, we obtain the largest system which we call $\boldsymbol{\omega}\mathbf{DED}(E)$. Strictly speaking, the substitution rule 3 b) is not required; it is a derived rule in $\boldsymbol{\omega}\mathbf{DED}(E)$—see Q16 of this chapter. However it is convenient to include it in the subsystem $\mathbf{rDED}(E)$. Rule 8 is not actually a rule in the normal sense of the word. For most terms t it has an infinite number of premises! A proof as we have defined it in chapter 1, and as it is usually defined, is a finite sequence of statements each of which can be derived from statements earlier in the sequence using one of the rules. So rule 8 can never be applied. In other words, it is not effective. We can, however extend the notion of a proof to allow infinite proofs. Alternatively, we say that the set of theorems of $\boldsymbol{\omega}\mathbf{DED}(E)$ is the least set of X which satisfies, for each rule 1 to 8,

if every premise is in X, then the conclusion is in X.

As before, we write $\vdash_{\mathbf{DED}(E)} t \leq t'$, $\vdash_{\mathbf{rDED}(E)} t \leq t'$, $\vdash_{\boldsymbol{\omega}\mathbf{DED}(E)} t \leq t'$ to mean that $t \leq t'$ is a theorem in the systems $\mathbf{DED}(E)$, $\mathbf{rDED}(E)$, $\boldsymbol{\omega}\mathbf{DED}(E)$ respectively. Each of these

1. *Reflexivity* $$\frac{}{t \le t}$$

2. *Transitivity* $$\frac{t \le t',\ t' \le t''}{t \le t''}$$

3. *Substitution*

   a) $$\frac{\underline{t} \le \underline{t}'}{f(\underline{t}) \le f(\underline{t}')} \quad \text{for every } f \text{ in } \Sigma$$

   b) $$\frac{t \le t'}{rec\, x.t \le rec\, x.t'}$$

4. *Instantiation* $$\frac{t \le t'}{t\rho \le t'\rho} \quad \text{for every substitution } \rho$$

5. *Inequations* $$\frac{}{t \le t'} \quad \text{for every inequation } t \le t' \text{ in } E$$

6. *Ω-rule* $$\frac{}{\Omega \le x}$$

7. *REC* $$\frac{}{rec\, x.t = t[rec\, x.t/x]}$$

8. *ω-Induction* $$\frac{\text{for every } d \in \mathrm{App}(t),\ d \le t'}{t \le t'}$$

Figure 4.2
The Proof System **ωDED**(E).

will be abbreviated to $\vdash t \leq t'$ when it is clear from the context which proof system is being referred to. Sometimes it will also be convenient to use $t \leq_E t'$, $t \leq_{Er} t'$, $t \leq_{E\omega} t'$, respectively, to mean that $t \leq t'$ is derivable in the appropriate proof system. Because of the first three rules each of these relations is a preorders. Finally, we will find some use for the particular instance of **rDED**(E) when E is empty. We will use $\vdash_r$ and $\leq_r$ as abbreviations for $\vdash_{\mathbf{rDED}(\emptyset)}$ and $\leq_{\mathbf{rDED}(\emptyset)}$ respectively. Note that $\leq_r$ can be viewed as an extension to recursive terms of the partial order $\preceq$ defined in §3.3. We will use the notation developed for **DED**(E) in part I for the new proof systems. For example $t =_{E\omega} t'$ means that we have a proof of both $t \leq t'$ and $t' \leq t$ in **ωDED**(E), and we will use the usual derived rules for this equality symbol. We will also continue the rather relaxed approach to proof systems taken in part I. We rarely give actual proofs in **ωDED**(E) as this tends to be rather tedious. Instead we use informal arguments to convince the reader that a given statement can be proved without actually exhibiting the proof; this often takes the form of showing how the proof can be constructed, knowing that related expressions have already been derived in the proof system.

We first examine some theoretical questions concerning **ωDED**(E), in particular soundness and completeness with respect to the interpretation CI_E. We will then consider more reasonable proof systems obtained by replacing ω-Induction with an effective form of induction.

Every rule in figure 4.2 is of the form

$$\frac{t_i \leq t_i',\ i \in I}{t \leq t'}.$$

The set $\{t_i \leq t_i',\ i \in I\}$ is the set of premises and $t \leq t'$ is the conclusion. We say such a rule is *sound* with respect to a relation R over $REC_\Sigma(X)$ if

$\langle t_i, t_i' \rangle \in R$ for every $i \in I$ implies $\langle t, t' \rangle \in R$.

We say it is sound with respect to the interpretation A if it is sound with respect to the relation $\leq_A$. The rules are chosen so as to make the following lemma trivial to prove.

LEMMA 4.3.1 If $A \in \mathscr{CC}(E)$ then every rule in **ωDED**(E) is sound with respect to A.

Proof Rules 1, 2, 3, 5, 6 are immediate. The Substitution lemma implies the soundness of rule 4; corollary 4.2.9 implies that of rule 7. Only rule 8 remains.

Suppose $A[\![d]\!] \leq A[\![t']\!]$ for every d in App(t). Then $\bigvee[\![\mathrm{App}(t)]\!] \leq A[\![t']\!]$. Now applying theorem 4.2.11 we obtain $A[\![t]\!] \leq A[\![t']\!]$. □

As an immediate corollary we have the soundness of the proof system with respect to any interpretation which satisfies the appropriate equations.

COROLLARY 4.3.2 For every t, t' in $REC_\Sigma(X)$, if $\vdash_{\boldsymbol{\omega}\mathbf{DED}(E)} t \leq t'$ then $A[\![t]\!] \leq A[\![t']\!]$ for every interpretation in A in $\mathscr{C}\mathscr{C}(E)$.

Proof By induction on the length of the proof of the theorem $t \leq t'$, using the previous lemma. □

We now show that the proof system $\boldsymbol{\omega}\mathbf{DED}(E)$ completely characterizes the interpretation of REC_Σ in the model CI_E in exactly the same way as $\mathbf{DED}(E)$ characterizes the interpretation of T_Σ in the model I_E: the relations $\leq_{CI_E}$ and $\leq_{E\omega}$ coincide over closed terms, REC_Σ. Along the way we will also prove a partial completeness theorem for the restricted system $\mathbf{rDED}(E)$.

LEMMA 4.3.3 For every $n \geq 0$, $\vdash_r t^n \leq t$.

Proof We first remark that if u is a finite term, $u \in FREC_\Sigma$, then in any of the three proof systems $\vdash t \leq t'$ implies $\vdash u[t/x] \leq u[t'/x]$. This can easily be proved by structural induction on u. (see Q18 of chapter 1). The proof of the main result is by induction on n. For the case $n = 0$ rule 6 is sufficient. So we prove $\vdash_r t^{k+1} \leq t$ on the assumption that $\vdash_r u^k \leq u$ for every term u. Structural induction is used and we examine two cases:

i) t is $f(\underline{u})$. Then t^{k+1} is $f(\underline{u}^{k+1})$. By structural induction $\vdash_r \underline{u}^{k+1} \leq \underline{u}$. Applying rule 3 to f we obtain $\vdash_r f(\underline{u}^{k+1}) \leq f(\underline{u})$.

ii) t is $rec\, x.u$. Here t^{k+1} is $u^{k+1}[t^k/x]$ and by induction we have both $\vdash_r u^{k+1} \leq u$ and $\vdash_r t^k \leq t$. Then

$u^{k+1}[t^k/x] \leq u^{k+1}[t/x]$	by the remark at the beginning of the proof
$\leq u[t/x]$	by rule 4
$= t$	by rule 7. □

This is just a technical result which facilitates the proof of a partial completeness result for $\mathbf{rDED}(E)$. It is the syntactic analogue of lemma 4.2.10.

THEOREM 4.3.4 For $d \in FREC_\Sigma$ and $q \in REC_\Sigma$ if $CI_E[\![d]\!] \leq CI_E[\![q]\!]$, then $\vdash_{\mathbf{rDED}(E)} d \leq q$.

Proof Suppose $CI_E[\![d]\!] \leq CI_E[\![q]\!]$. Then by the Finite Approximation theorem, 4.2.11, there is a finite approximation q^n such that $CI_E[\![d]\!] \leq CI_E[\![q^n]\!]$. Because of the construction of CI_E this means $d \leq_E q^n$. From the previous lemma $q^n \leq_{Er} q$ and since $\mathbf{DED}(E)$ is included in $\mathbf{rDED}(E)$ this means that $d \leq_{Er} q$. □

Thus the system **rDED**(E) is sound and complete when restricted to $FREC_\Sigma \times REC_\Sigma$. The more general completeness result follows easily from it. This forms part of the following general characterization result.

THEOREM 4.3.5 (ω-Equational-Logic Theorem) a) **ωDED**(E) is sound with respect to $\leq_{CI_E}$ over $REC_\Sigma(X)$.

b) **ωDED**(E) is complete with respect to $\leq_{CI_E}$ over REC_Σ.

Proof Part a) follows from corollary 4.3.2. We prove the completeness, part b). Suppose $CI_E[\![p]\!] \leq CI_E[\![q]\!]$ where both p and q are in REC_Σ. Let $d \in \mathrm{App}(p)$. Then from lemma 4.2.10 $CI_E[\![d]\!] \leq CI_E[\![p]\!]$ and therefore $CI_E[\![d]\!] \leq CI_E[\![q]\!]$. Applying the previous completeness theorem we have $d \leq_{Er} q$, i.e. $d \leq_{E\omega} q$. This is true for every $d \in \mathrm{App}(p)$ and therefore we may apply ω-Induction to conclude that $p \leq_{E\omega} q$. □

This result gives a certain amount of interest to the proof system **ωDED**(E). Informally, it states that the equations together with a very powerful form of induction are sufficient to prove all true statements about closed terms in the initial interpretation CI_E; however from a practical point of view it is more realistic to have an effective induction rule. We now investigate such rules. In general when we substitute these rules for ω-Induction we lose completeness. This of course depends on the signature and the equations but this topic will not be pursued further.

Without any form of induction, i.e. in the proof system **rDED**(E), we can only manage to prove rather uninteresting theorems about recursive terms, such as (using the equations **SA2**)

$\vdash rec\, x.ax \oplus bx \leq b(rec\, x.ax + bx).$

Using the axiom $+\oplus 1$ we obtain $ax \oplus bx \leq ax + bx$, and applying the new rule 3 b) we obtain

$rec\, x.ax \oplus bx \leq rec\, x.ax + bx.$

Applying REC and the strong axiom **S** we obtain $rec\, x.ax \oplus bx \leq b\, rec\, x.ax \oplus bx$, from which the result follows. In a similar vein we can derive

$\vdash rec\, x.ax \oplus bx \leq a\, rec\, x.ax \oplus (ax + bx)$

$\vdash b\, rec\, x.ax \oplus bx \leq b\, rec\, x.ax.$

We could also augment this simple proof system with further rules based on the following lemma.

LEMMA 4.3.6 a) $t =_\alpha u$ implies $t = u$ is a theorem in **ωDED**(E).

b) If $\rho(x) \leq \rho'(x)$ is a theorem in **rDED**(E) for every x, $t\rho \leq t\rho'$ is also a theorem.

Proof See Q16 in the exercises at the end of this chapter. □

This lemma can be restated to say that the two proof rules

$$\frac{\rho \leq \rho'}{t\rho \leq t\rho'} \quad \text{and} \quad \frac{}{t = u} \quad \text{whenever } t =_\alpha u$$

are derived rules in **rDED**(E): every theorem derivable in the proof system augmented by these rules can also be derived without using the rules. We will refer to the first rule as *generalized substitution*. The premiss $\rho \leq \rho'$ is merely shorthand for the collection of premises $\{\rho(x) \leq \rho'(x), x \in X\}$.

These rules would not significantly increase the power of the proof system. For example

$$rec\, x.ax \leq rec\, x.aax \qquad (**)$$

cannot be derived even though it is obviously true in our models. It cannot be derived because $rec\, x.aax$ cannot be transformed into $rec\, x.ax$ using substitution or equations nor can their respective bodies, aax and ax, be so transformed. To prove such statements a form of induction is required.

One simple form of induction, which we will call *Recursion Induction* or RI, is based on the idea that $rec\, x.t$ is interpreted as the least fixpoint of the equation $x = t$. This least fixpoint is in fact the least pre-fixpoint, i.e. the least p which satisfies $t[p/x] \leq p$.

$$\text{RI} \quad \frac{t[u/x] \leq u}{rec\, x.t \leq u}$$

We now give some simple examples of its use.

EXAMPLE 4.3.7 Let A, AA denote $rec\, x.ax$, $rec\, x.aax$ respectively. We prove $\vdash AA \leq A$ using RI.

$aaA = aA$ using REC

$\quad = A$ using REC again.

In particular $\vdash aaA \leq A$, i.e. $\vdash aax[A/x] \leq A$. Therefore $rec\, x.aax \leq A$ by RI.

In the foregoing example we used considerable shorthand in order to avoid exhibiting tedious proofs in detail. For example, the proof of $aaA \leq A$ would be more formally exhibited by the sequence

1.	$aA \leq A$	rule 8 (REC)
2.	$aaA \leq aA$	rule 3 applied to a
3.	$aaA \leq A$	rule 2 lines 1, 2.

However, we feel that the shorthand employed is self-explanatory and relieves the boredom somewhat.

EXAMPLE 4.3.8 Let P, Q denote $rec\, x.ax \oplus bx$, $(rec\, x.ax) \oplus (rec\, x.bx)$ respectively. We prove $\vdash P \leq Q$, using the equations **SA2**. To do so we must derive $aQ \oplus bQ \leq Q$:

$$\begin{aligned}
aQ \oplus bQ &= a(rec\, x.ax) \oplus a(rec\, x.bx) \oplus bQ && \text{from } \oplus 4\\
&\leq a(rec\, x.ax) \oplus bQ && \mathbf{S}\\
&= (rec\, x.ax) \oplus bQ && REC\\
&= (rec\, x.ax) \oplus b(rec\, x.ax) \oplus b\, rec\, x.bx && \oplus 4\\
&\leq rec\, x.ax \oplus b\, rec\, x.bx && \mathbf{S}\\
&= Q && REC.
\end{aligned}$$

We have derived $aQ \oplus bQ \leq Q$ and therefore an application of RI gives $\vdash P \leq Q$.

EXAMPLE 4.3.9 The converse of the theorem proved in example 4.3.7 is not so immediate, $\vdash A \leq AA$. First consider the term aAA. Ae show $\vdash AA \leq aAA$ using RI.

$$aAA = aaaAA \quad \text{from} \quad REC$$

In particular	$aaaAA \leq aAA$	
i.e.	$aax[aAA/x] \leq aAA$	
and so	$AA \leq aAA$	RI
Therefore	$aAA \leq aaAA$	rule 3
	$\leq AA$	REC
This has the form	$ax[AA/x] \leq AA$	
and therefore	$A \leq AA$	RI.

EXAMPLE 4.3.10 The technique used in the previous example can be used to prove a general theorem about the proof system based on RI. Let t be any term in $REC_{\Sigma}(X)$. We prove, using RI,

$\vdash rec\,x.t = rec\,x.(t[t/x])$.

We must prove two statements:

i. $\vdash rec\,x.(t[t/x]) \leq rec\,x.t$

ii. $\vdash rec\,x.t \leq rec\,x.(t[t/x])$.

We leave i. in the hands of the reader. The proof follows the lines of that in example 4.3.7. The proof of ii. follows that of the previous example and uses the fact that for any terms u, u', u'',

$$u[u'/x][u''/x] = u[u'[u''/x]/x]. \qquad (\dagger)$$

The proof is in two parts:

a) $\vdash u \leq t[u/x]$ where u denotes $rec\,x.(t[t/x])$

b) use a) to prove $\vdash rec\,x.t \leq u$.

a) We use RI. For convenience let r denote $t[u/x]$. Then

$$\begin{aligned} t[t/x][r/x] &= t[t[r/x]/x] && \text{from } (\dagger) \\ &= t[t[t/x][u/x]/x] && \text{using } (\dagger) \text{ again} \\ &= t[u/x] && REC \end{aligned}$$

i.e. $t[t/x][r/x] = r$.

In particular $t[t/x][r/x] \leq r$. Applying RI we obtain $u \leq r$.

b) Applying generalized substitution we obtain

$$t[u/x] \leq t[r/x]$$

i.e. $$\begin{aligned} t[u/x] &\leq t[t[u/x]/x] \\ &= t[t/x][u/x] && \text{from } (\dagger) \\ &= u && REC. \end{aligned}$$

Applying RI we obtain

$rec\,x.t \leq rec\,x.t[t/x]$.

EXAMPLE 4.3.11 The general result derived in the previous example can be quite useful: it allows us to unwind recursive definitions. We give an example of its application to derive, using **SA2**,

$\vdash rec\, x.ax \oplus bx \leq rec\, x.abx.$

First notice that by applying $\oplus 4$ and **S** we can derive

$a(ax \oplus bx) \oplus b(ax \oplus bx) \leq abx.$

Applying rule 3 b) we obtain

$rec\, x.a(ax \oplus bx) \oplus b(ax \oplus bx) \leq rec\, x.abx.$

However applying example 4.3.10, with t equal to $ax \oplus bx$, we have

$rec\, x.ax \oplus bx \leq rec\, x.a(ax \oplus bx) \oplus b(ax \oplus bx)$

from which it follows by transitivity that $rec\, x.ax \oplus bx \leq rec\, x.abx.$

We consider one more example of RI.

EXAMPLE 4.3.12 We prove $\vdash rec\, x.ax \oplus b \leq (rec\, x.ax) \oplus (rec\, x.ax \oplus b)$ using RI and the laws **A2**. For convenience let p denote $rec\, x.ax \oplus b$. Then

$$\begin{aligned} a((rec\, x.ax) \oplus p) \oplus b &= (a\, rec\, x.ax) \oplus ap \oplus b && \text{from} \quad \oplus 4 \\ &= rec\, x.ax \oplus ap \oplus b && REC \\ &= rec\, x.ax \oplus p && REC. \end{aligned}$$

In particular $a((rec\, x.ax) \oplus p) \oplus b \leq rec\, x.ax \oplus p$ and we may apply RI to obtain $\vdash p \leq rec\, x.ax \oplus p.$

We now turn our attention to a different induction principle which is stronger than RI, called *Scott Induction* (SI). Recursion Induction is restricted in its scope of application; it can only be applied to derive theorems of the form $rec\, x.t \leq u$. Although one can often transform a general statement $t \leq t'$ using the equations and *REC* so that it is of this form, it may not always be possible. For example it is not clear if $(rec\, x.ax) \oplus (rec\, x.ax + bx) \leq rec\, x.ax + bx$ can be derived from the equations **A2** in the system based on RI. Recursion Induction is also weak in that its soundness demands very little of the interpretation; it requires that the function symbols be interpreted as monotonic functions, but not necessarily continuous. Scott Induction will, however, demand that the interpretation is a Σ-*cpo*. To explain this new rule we need some new notation. It is different in kind than any of the proof rules we have previously encountered in that its premises are stated in terms of the ability to prove theorems from certain assumptions. If L is a list of statements such as

$u_1 \leq u_1', u_2 \leq u_2', \ldots, u_k \leq u_k'$,

we write $L \vdash t \leq t'$ to mean that we can derive $t \leq t'$ as a theorem using the rules of the proof system together with the statements in L; in short $t \leq t'$ can be proven from the assumption in L. For example

$x \leq x \oplus y \vdash ax \leq ax + ay$

is true in the proof system **DED(A2)**. We now define Scott Induction, which uses premises of this form. Let $\underline{z} = \langle z_1, \ldots, z_k \rangle$ be k distinct variables and $u = \langle u_1, \ldots, u_k \rangle$ be k terms in $REC_\Sigma(X)$ with the property that if $z_i \in FV(u_j)$ then $i = j$, i.e. the only variable from $\underline{z}$ which can appear free in u_j is z_j. Such a vector of terms $\underline{u}$ is called *noninterfering* with respect to $\underline{z}$. For convenience we use $rec\, \underline{z}.\underline{u}$ to denote the sequence $\langle rec\, z_1.u_1, \ldots, rec\, z_k.u_k \rangle$ which is non-interfering with respect to $\underline{z}$ provided $\underline{u}$ is.

SI: Suppose $\underline{u}$ is noninterfering with respect to $\underline{z}$, a sequence of distinct variables. If we can prove

i) $\vdash t[\Omega/\underline{z}] \leq t'[\Omega/\underline{z}]$

ii) $t \leq t' \vdash t[\underline{u}/\underline{z}] \leq t'[\underline{u}/\underline{z}]$

then we can conclude

$\vdash t[rec\, \underline{z}.\underline{u}/\underline{z}] \leq t'[rec\, \underline{z}.\underline{u}/\underline{z}]$.

This looks more complicated than it actually is. Some examples will clarify the notation.

EXAMPLE 4.3.13 We prove $\vdash (rec\, x.ax) \oplus (rec\, y.ay + by) \leq rec\, y.ay \oplus by$ using SI and the equations **A2**. The particular application of SI we need is with

$\underline{z} = \langle x, y \rangle$

$\underline{u} = \langle ax, ay \oplus by \rangle$

$t = x \oplus y$

$t' = y$.

To apply SI we must prove

i) $\vdash t[\underline{\Omega}/\underline{z}] \leq t'[\underline{\Omega}/\underline{z}]$, i.e. $\vdash \Omega \oplus \Omega \leq \Omega$ which follows from $\oplus 3$

and

ii) $t \leq t' \vdash t[\underline{u}/\underline{z}] \leq t'[\underline{u}/\underline{z}]$

i.e.

$x \oplus y \leq y \vdash ax \oplus (ay \oplus by) \leq ay \oplus by.$

This follows because

$ax \oplus ay \oplus by = a(x \oplus y) \oplus by$ by $\oplus 4$

$\leq ay \oplus by$ using the assumption $x \oplus y \leq y$.

Because of i. and ii., SI allows us to conclude

$\vdash t[rec\ \underline{z}.\underline{u}/\underline{z}] \leq t'[rec\ \underline{z}.\underline{u}/\underline{z}],$

i.e.

$\vdash (rec\ x.ax) \oplus rec\ y.ay \oplus by \leq rec\ y.ay \oplus by$

To sum up this application of SI, the premises are

i. $\vdash \Omega \oplus \Omega \leq \Omega$
ii. $x \oplus y \leq y \vdash ax \oplus (ay \oplus by) \leq ay \oplus by$

and the conclusion is

$\vdash (rec\ x.ax) \oplus (rec\ y.ay \oplus by) \leq rec\ y.ay \oplus by.$

EXAMPLE 4.3.14 We use SI and the equations **A2** to prove

$\vdash (rec\ x.ax + bx) \oplus (rec\ y.ay \oplus by) \leq rec\ y.(ay \oplus by).$

By SI this will follow if we can prove

i. $\vdash \Omega \oplus \Omega \leq \Omega$
ii. $x \oplus y \leq y \vdash (ax + bx) \oplus (ay \oplus by) \leq ay \oplus by.$

The premise i. is obvious so we concentrate on ii. By applying Der3 of chapter 2 we have

$$(ax + bx) \oplus (ay + by)$$

$$= (a(x \oplus y) + b(x \oplus y)) \oplus (a(x \oplus y) + b(x \oplus y))$$

$$= a(x \oplus y) + b(x \oplus y) \quad \text{by } \oplus 3. \tag{1}$$

Also Der1 gives

$$ay \oplus by = ay \oplus by \oplus (ay + by). \tag{2}$$

Therefore

$$(ax + bx) \oplus ay \oplus by = (ax + bx) \oplus (ay + by) \oplus ay \oplus by \quad \text{from (2)}$$

$$= (a(x \oplus y) + b(x \oplus y)) \oplus ay \oplus by \quad \text{from (1)}$$

$$\leq (ay + by) \oplus ay \oplus by \quad \text{from the assumption } x \oplus y \leq y$$

$$= ay \oplus by \quad \text{from (2).}$$

So the premises i. and ii. above are true; SI allows us to conclude

$$\vdash (rec\, x.ax + bx) \oplus (rec\, y.ay \oplus by) \leq rec\, y.ay \oplus by.$$

We now show that each of the proof rules we have introduced are sound. We do this indirectly by comparing the relative power of the various induction rules. Let **PS**, **PS′** be two proof systems. We say **PS** is *at least as powerful as* **PS′**, written **PS′** < **PS**, if for every t, t' in $REC_\Sigma(X)$,

$$\vdash_{\mathbf{PS'}} t \leq t' \quad \text{implies} \quad \vdash_{\mathbf{PS}} t \leq t',$$

i.e. every theorem in **PS′** is also a theorem in **PS**. It should be obvious that if **PS** is as least as powerful as **PS′** and **PS** is sound then **PS′** is also sound. We compare four different proof systems:

rDED(E) (the basic proof system obtained from the rules 1–7 in figure 4.2)

rDED(E) + RI (adding Recursion Induction)

rDED(E) + SI (adding Scott Induction)

rDED(E) + ω-Induction (this is the sound and complete proof system **ωDED**(E)).

Each of these proof systems except **rDED**(E) + SI is obtained by augmenting the set of applicable rules. We take the theorems of **rDED**(E) + SI to be those statements which can be derived by an application of SI whose premises have proofs in the basic system **rDED**(E). This rules out nested applications of SI. We do this for convenience only, as a proper formulation of the more general system would require some additional development and we do not wish to pursue the formalization of these proof systems any further. Note however that the theorems of **rDED**(E) are contained in those of **rDED**(E) + SI; they can be derived by a trivial application of SI, where the vector of variables $\underline{z}$ is empty.

These systems are related as follows:

rDED$(E) \leq$ **rDED**(E) + RI $\leq$ **rDED**(E) + SI $\leq$ **rDED**(E) + ω-Induction.

PROPOSITION 4.3.15 **rDED**(E) + RI $\leq$ **rDED**(E) + SI.

Proof In general to show **PS**$'$ < **PS** it is sufficient to prove that every application of a proof rule in **PS**$'$ can be duplicated in **PS**. Here we must show RI can be duplicated by SI. An instance of RI allows us to conclude

$rec\, x.t < u$

from the fact that $t[u/x] < u$ is a theorem. We show that the same conclusion can be arrived at using an instance of SI. Because of lemma 4.3.6 a) (or more accurately Q16) we may assume that x does not occur in u.

We apply SI to the statement $x \leq u$. The two premises are

i) $\vdash \Omega \leq u$

ii) $x < u \vdash t \leq u$.

The first is trivial. The second is derived as follows:

$x \leq u$ by assumption

$t[x/x] \leq t[u/x]$ generalized substitution

i.e. $t \leq t[u/x]$

$\leq u$ since $\vdash t[u/x] \leq u$.

SI therefore allows us to conclude

$\vdash rec\, x.t \leq u[rec\, x.t/x]$.

Now x does not occur in u and so $u[rec\, x.t/x] =_\alpha u$ and the result follows by lemma 4.3.6 a). □

PROPOSITION 4.3.16 **rDED**(E) + SI $\leq$ **ωDED**(E).

Proof We use the notation of the statement of SI. We know

i) $\vdash t[\underline{\Omega}/\underline{z}] \leq t'[\underline{\Omega}/\underline{z}]$

ii) $t \leq t' \vdash t[\underline{u}/\underline{z}] \leq t'[\underline{u}/\underline{z}]$.

We must use ω-Induction to conclude

$$\vdash t[rec\ \underline{z}.\underline{u}/\underline{z}] \leq t'[rec\ \underline{z}.\underline{u}/\underline{z}]. \qquad (\dagger)$$

For convenience we let r_i denote $rec\, z_i.u_i$ and therefore $\underline{r}$ denotes $rec\, \underline{z}.\underline{u}$. We also use

$$\vdash (t[\underline{u}/\underline{z}])^n \leq t^n[\underline{u}^n/\underline{z}],$$

which we leave the reader to verify. Unfortunately we cannot work directly with principal approximations as previously defined. We need to define a modified form:

a) $w^{(0)} = \Omega$ for every term w

b) i) $t^{(n+1)} = t$
   ii) $f(\underline{u})^{(n+1)} = f(\underline{u}^{(n+1)})$
   iii) $(rec\, x.u)^{(n+1)} = u[(rec\, x.u)^n/x]$.

Note that $t^{(n)}$ need not necessarily be a finite term. But although they may contain occurrences of $rec\, x.$ these new kind of approximations are easier to manipulate using ω-Induction than the finite approximations t^n. However we can also prove by induction on n that

$$\vdash_r t^n \leq t^{(n)} \quad \text{and} \quad \vdash_r t^{(n)} \leq t.$$

We also leave these to the reader. At this stage their proofs should be routine and are similar in spirit to that of lemmas 4.1.6, 4.2.10, and 4.3.3.

Now suppose we have already shown

$$\text{for every } n \geq 0, \vdash t[\underline{r}^{(n)}/\underline{z}] \leq t'[\underline{r}^{(n)}/z]. \qquad (*)$$

In fact the reason the new approximations are introduced is because the corresponding statement with $r_i^{(n)}$ replaced by the principal approximations is difficult to derive. We show that the desired conclusion (†) can be derived from (∗) using ω-Induction. Consider an arbitrary principal approximation to $t[\underline{r}/\underline{z}]$, $(t[\underline{r}/\underline{z}])^n$:

$$\begin{aligned}
\vdash t[\underline{r}/\underline{z}]^n &\leq t^n[\underline{r}^n/\underline{z}] && \text{from above}\\
&\leq t^n[\underline{r}^{(n)}/\underline{z}] && \text{generalized substitution}\\
&\leq t[\underline{r}^{(n)}/\underline{z}] && \text{rule 4}\\
&\leq t'[\underline{r}^{(n)}/\underline{z}] && \text{by } (*)\\
&\leq t'[\underline{r}/\underline{z}] && \text{generalized substitution again.}
\end{aligned}$$

So $\vdash d \leq t'[rec\ \underline{z}.\underline{u}/\underline{z}]$ for every $d \in \mathrm{App}(t[rec\ \underline{z}.\underline{u}/\underline{z}]$ and therefore ω-Induction gives

$\vdash t[rec\ \underline{z}.\underline{u}/\underline{z}] \leq t'[rec\ \underline{z}.\underline{u}/\underline{z}]$.

It remains to show $(*)$ and it is here that we use the assumptions i) and ii). Because of the way we defined the proof system **rDED**(E) + SI we know that these derivations are entirely within the proof system **rDED**(E). It is easy to show that for this proof system entire proofs can be instantiated, i.e.

$t_1 \leq t_2 \vdash_{\mathbf{rDED}(E)} t_1' \leq t_2'$ implies $t_1[u/z] < t_2[u/z] \vdash_{\mathbf{rDED}(E)} t_1'[u/\underline{z}] \leq t_2'[u/\underline{z}]$.

Applying this remark to ii) we obtain

$t[\underline{r}^{(k)}/\underline{z}] < t'[\underline{r}^{(k)}/\underline{z}] \vdash t[\underline{u}/\underline{z}][\underline{r}^{(k)}/\underline{z}] < t'[\underline{u}/\underline{z}][\underline{r}^{(k)}/\underline{z}]$.

However

$t[\underline{u}/\underline{z}][\underline{r}^{(k)}/\underline{z}] = t[\underline{u}[\underline{r}^{(k)}/\underline{z}]/\underline{z}]$ by the Substitution lemma.

Because $\underline{u}$ is non-interfering with respect to z the sequence $\underline{u}[\underline{r}^{(k)}/\underline{z}]/\underline{z}]$ simplifies to $u_1[\underline{r}_1^k/z_1] \ldots u_k[\underline{r}_n/z_n]$, i.e. $\underline{r}^{(k+1)}$. It follows that

$t[\underline{u}/\underline{z}][\underline{r}^{(k)}/\underline{z}] = t[\underline{r}^{(k+1)}/\underline{z}]$.

Applying the same reasoning to t' we obtain

iii) $t[\underline{r}^{(k)}/\underline{z}] < t'[\underline{r}^{(k)}/\underline{z}] \vdash t[\underline{r}^{(k+1)}/\underline{z}] < t'[\underline{r}^{(k+1)}/\underline{z}]$.

The required $(*)$ now follows. For $n = 0$ it is exactly assumption i). Assuming it is true for $n = k$ the case $n = k + 1$ follows from iii). □

Whether or not the full power of the complete system, **rDED**(E) + ω-Induction, can be obtained by the effective systems **rDED**(E) + RI, **rDED**(E) + SI will in general depend on the underlying set of equations. For the sets of equations we are interested in, **A2**, **SA2**, etc., these questions are unsolved.

We end this section with a brief discussion of a powerful induction rule which is unfortunately unsound. We call it *Unique Fixpoint Induction* (UFI).

$$\text{UFI} \quad \frac{t[u/x] = u}{rec\, x.t = u}$$

It is unsound basically because it assumes every equation $x = t$ always has a unique fixpoint. As we have seen, this is not necessarily so.

EXAMPLE 4.3.18 With the equations **A2** it is easy to establish

$$bNIL + aNIL + aNIL = bNIL + aNIL$$

i.e. $(x + aNIL)[bNIL + aNIL/x] = bNIL + aNIL.$

Then UFI allows us to conclude

$\vdash rec\, x.x + aNIL = bNIL + aNIL.$

However this is not true, for example in the model **AT**; so applying UFI allows us to prove theorems which are false. Nevertheless UFI is of interest because we can put restrictions on its application, at least for our systems of equations, in such a way that it is always sound. We restrict its use to cases where every occurrence of x in t is guarded, i.e. occurs within a prefix $a.$. We will see in the exercises that with these restrictions UFI is sound. Used within these limitations UFI is still a very powerful rule; all the examples in this section which use SI can also be derived using the restricted form of UFI.

4.4 Testing Recursive Processes

In this section we apply the Testing preorders, explained in part I, to the augmented language $REC_{\Sigma^2}(X)$. In this context we will often refer to the set of closed terms REC_{Σ^2} as $\mathbf{RM}_2(Act)$ or $\mathbf{RM}_2$ or often simply **RM**, the set of recursive machines over Σ^2. The necessary technical machinery has already been developed in §§2.1, 2.2 and 2.5. To apply it we need only formalize the operational behavior of $\mathbf{RM}_2$ as a labeled transition system. This is quite straightforward apart from the fact that we must decide on an appropriate notion of operational behavior for partial terms such as $aNIL + \Omega$. Perhaps surprisingly, the alternative characterizations of part I apply almost un-

i) $ap \xrightarrow{a} p$

ii) $p \xrightarrow{a} p'$ implies $p + q \xrightarrow{a} p'$

$q \xrightarrow{a} q'$ implies $p + q \xrightarrow{a} q'$

iii) $p \oplus q \succ\!\!\longrightarrow p$

$p \oplus q \succ\!\!\longrightarrow q$

iv) $p \succ\!\!\longrightarrow p'$ implies $p + q \succ\!\!\longrightarrow p' + q$

$q \succ\!\!\longrightarrow q'$ implies $p + q \succ\!\!\longrightarrow p + q'$

v) $rec\,x.t \succ\!\!\longrightarrow t[rec\,x.t/x]$

vi) $\Omega \succ\!\!\longrightarrow \Omega$

Figure 4.3
Operational semantics of **RM**.

changed to the resulting Testing preorders. In particular, we need only finite experiments to distinguish processes which are differentiated via any of the Testing preorders. The remainder of this chapter is devoted to showing that the various models in chapter 3 are fully abstract with respect to the appropriate Testing preorder.

The operational semantics of a language is given by viewing it as a labeled transition system or, more precisely, an extended labeled transition system: intuitively $p \xrightarrow{a} p'$ means the process p may evolve to the process p' by performing the action a, whereas $p \succ\!\!\longrightarrow p'$ means that p may evolve autonomously to p' without the intervention of any external process. With this in mind the operational semantics given for $\mathbf{M_2}$ in figure 2.5 are extended to **RM** in figure 4.3. It is a very mild extension. No new clauses are introduced for $\xrightarrow{a}$. The only new rules are for the internal relation $\succ\!\!\longrightarrow$:

$$rec\,x.t \succ\!\!\longrightarrow t[rec\,x.t/x]$$

$$\Omega \succ\!\!\longrightarrow \Omega.$$

The first allows recursive processes to unwind their definitions silently, at least at the topmost level. For example,

$$rec\,x.(ax + bx) \succ\!\!\longrightarrow a(rec\,x.ax + bx) + b(rec\,x.ax + bx).$$

So that although $rec\,x.ax + bx$ cannot immediately perform any external actions it can evolve silently to a state where it can perform either an a action or a b action.

Specifically,

$$rec\, x.ax + bx >\!\!\longrightarrow \xrightarrow{a} rec\, x.ax + bx$$

$$rec\, x.ax + bx >\!\!\longrightarrow \xrightarrow{b} rec\, x.ax + bx.$$

It follows that by ignoring silent or internal actions $rec\, x.ax + bx$ can perform any sequence of actions from $\{a, b\}^*$.

The second clause needs some explanation. Ω is interpreted in every interpretation as the least element; it is always semantically equivalent to $rec\, x.x$. If we examine the behavior of $rec\, x.x$ under the other rules we see that the only possible action, internal or external, is

$$rec\, x.x >\!\!\longrightarrow rec\, x.x.$$

Since Ω is nothing more than a syntactically finite representation of $rec\, x.x$, it is perfectly reasonable to include the rule $\Omega >\!\!\longrightarrow \Omega$.

With this view of **RM** as an *lts* we can automatically generate an Experimental System for testing processes.

Let $\mathscr{E}\mathscr{S}(\mathbf{RM})$ be the Experimental System

$$\mathscr{E}\mathscr{S}(\langle \mathbf{RM}, Act, \longrightarrow, >\!\!\longrightarrow \rangle, \langle \mathbf{RM}, Act \cup \{1, w\}, \longrightarrow, >\!\!\longrightarrow \rangle),$$

where $\longrightarrow$ and $>\!\!\longrightarrow$ are given in figure 4.3 and 1 and w are as usual. We borrow the notation developed in part I for these systems, and $\underset{\sim}{\sqsubset}_{\mathrm{MAY}}$, $\underset{\sim}{\sqsubset}_{\mathrm{MUST}}$, $\underset{\sim}{\sqsubset}$ are the induced Testing preorders over **RM**.

EXAMPLE 4.4.3

1. $a + \Omega$ *may* aw but $a + \Omega$ $\mathit{m\!\!/ust}$ aw because of the unsuccessful application

$$aw \parallel a + \Omega \rightarrow aw \parallel a + \Omega \rightarrow \cdots.$$

This computation is possible because $a + \Omega$ has an infinite internal computation $a + \Omega >\!\!\longrightarrow a + \Omega >\!\!\longrightarrow \cdots$.

2. Let p denote $rec\, x.a(x + \Omega)$. Then

$$p\ \mathit{may}\ aaw,\ p\ \mathit{must}\ aw,\ p\ \mathit{mu\!\!/st}\ aaw.$$

The latter follows because $p + \Omega$ has an infinite internal computation and so

$$aaw \parallel p \rightarrow aaw \parallel a(p + \Omega) \rightarrow aw \parallel p + \Omega \rightarrow aw \parallel p + \Omega \rightarrow \cdots$$

is an unsuccessful computation.

3. Let p denote $rec\,x.a(rec\,y.by + y + x) + ax$. Then p *may* aaw but p $\mathit{m\not{u}st}$ aaw because $rec\,y.by + y + p$ has an infinite internal computation, and can unwind itself continuously:

$$\begin{aligned} rec\,y.by + y + p &\rightarrowtail b(rec\,y.by + y + p) + (rec\,y.by + y + p) + p \\ &\rightarrowtail b(\ldots) + b(\ldots) + (rec\,y.by + y + p) + p + p \\ &\rightarrowtail \cdots . \end{aligned}$$

Therefore

$$\begin{aligned} aaw \parallel p \rightarrow aaw \parallel a(rec\,y.by + y + p) + ap &\rightarrow aw \parallel rec\,y.by + y + p \\ &\rightarrow aw \parallel \ldots \\ &\rightarrow \cdots \end{aligned}$$

is an unsuccessful computation.

4. Let p denote $rec\,x.a((x + bNIL) \oplus \Omega)$. Then p $\mathit{m\not{u}st}$ abw because of the unsuccessful computation

$$\begin{aligned} abw \parallel p \rightarrow abw \parallel a((p + bNIL) \oplus \Omega) &\rightarrow bw \parallel (p + bNIL) \oplus \Omega \\ \rightarrow bw \parallel \Omega \rightarrow bw \parallel \Omega &\rightarrow \cdots . \end{aligned}$$

These examples indicate that the presence of Ω or partial terms does not affect the weak theories based on "*may*." However, they are catastrophic from the point of view of "*must*." Their presence, at least at the topmost level, prevents the process from passing a test in the "must" sense. In this series of examples we have also used only a simple variety of experimenters, all taken from $\mathbf{M}_1$. In fact we shall see that a slight extension of the set of tests E used in part I is sufficient to distinguish all processes from **RM** which are distinguishable. In particular, all the experimenters used are finite; so that if two processes are distinguishable they are distinguishable using a finite experimenter. The proof of these facts relies on an alternative characterization of the Testing preorders, very similar to that in part I. First some examples.

EAMPLES 4.4.4

1. Ω is a least element with respect to each of the preorders because no matter what e is, $e \parallel \Omega$ has the infinite computation

$$e \parallel \Omega \rightarrow e \parallel \Omega \rightarrow \cdots .$$

Indeed this is the only application of e to Ω apart from those variations obtained by e evolving with respect to 1 actions or internal decisions. Ω satisfies, for example in the *must* case, such an experiment only if e is already in a successful state; in which case every process satisfies it.

2. $aNIL + \Omega \eqsim_{MAY} aNIL$. This is easily checked by analysis on the possible reaction to tests. A similar analysis will show that $p + \Omega \eqsim_{MAY} p$ for any p whatsoever. On the other hand, $aNIL + \Omega \eqsim_{MUST} \Omega$. From 1. we know that $\Omega \lesssim_{MUST} (aNIL + \Omega)$. Conversely, it is easy to check that $aNIL + \Omega$ *must* e only for those trivial e which are already in a successful state; in which case Ω *must* e also.

The same reasoning can be applied to show $p + \Omega \eqsim_{MUST} \Omega$.

3. $rec\, x.ax \not\lesssim_{MUST} rec\, x.ax + x$. For example $rec\, x.ax$ *must* aw but $rec\, x.ax + x$ *mu$\not{s}$t* aw because $rec\, x.ax + x$ diverges internally:

$$\begin{aligned} rec\, x.ax + x &\rightarrowtail a(rec\, x.ax + x) + rec\, x.ax + x \\ &\rightarrowtail a(rec\, x.ax + x) + a(rec\, x.ax + x) + rec\, x.ax + x \\ &\rightarrowtail \cdots . \end{aligned}$$

So

$$aw \parallel rec\, x.ax + x \rightarrow aw \parallel a(\ldots) + rec\, x.ax + x \rightarrow \cdots$$

is an unsuccessful application.

On the other hand, $rec\, x.ax \eqsim_{MAY} rec\, x.ax + x$ and $rec\, x.ax + x \lesssim_{MUST} rec\, x.ax$. These are not immediate to see but will follow readily from the alternative characterizations which follow.

4. $rec\, x.(ax \oplus bNIL) + rec\, x.ax$ *must* aw while $rec\, x.ax \oplus bNIL$ *mu$\not{s}$t* aw because of the computation

$$aw \parallel rec\, x.ax \oplus bNIL \rightarrow aw \parallel a(rec\, x.ax \oplus bNIL) \oplus bNIL \rightarrow aw \parallel bNIL.$$

Note that the last internal move is not available to the other term $rec\, x.ax \oplus bNIL + rec\, x.ax$. Essentially this can evolve via $\rightarrowtail$ to $bNIL + rec\, x.ax$ or $a(rec\, x.ax \oplus bNIL) + rec\, x.ax$, both of which can perform the desired action a.

It follows that $rec\, x.(ax \oplus bNIL) + rec\, x.ax \not\lesssim_{MUST} rec\, x.ax \oplus bNIL$. On the other hand, we will see that the converse is true.

Most of these examples have concentrated on the behavior of pathological processes. This is not unreasonable. Very often a theory can best be understood by analyzing its treatment of pathological cases. Besides, much of the phenomena exhibited by processes under testing have already been examined in chapter 2. However we have yet to see many examples of infinite processes identified by these various equivalences. This awaits their characterization in a more amenable form. In anticipation the reader might try to distinguish via $\eqsim_{\mathrm{MUST}}$ the following pairs:

i) $rec\, x.ax \oplus bx, (rec\, x.ax) \oplus (rec\, x.bx)$

ii) $(rec\, x.ax \oplus b) \oplus rec\, x.ax, rec\, x.ax \oplus b$

iii) $rec\, x.ax \oplus b, (rec\, x.ax + ab) \oplus b.$

Let us now turn our attention to this alternative characterization, generalizing the corresponding result in chapter 2. We use the notation of §2.5, in particular the double arrows $\stackrel{s}{\Longrightarrow}$ and the acceptance sets $\mathscr{A}(p, s)$. In addition we need information on the ability of processes to diverge internally. Internal divergence is the key to the generalization of the alternative characterization from the finite case to the new language **RM** and it is worthwhile to introduce a specific notation for it: we write

$p\uparrow$ if p has an infinite internal computation,

$$p \rightarrowtail p_0 \rightarrowtail p_1 \cdots \rightarrowtail p_k \rightarrowtail \cdots.$$

$p\downarrow$ if p has no infinite internal computation.

If $p\downarrow$ we say that p is *convergent* and if $p\uparrow$ we say it is *divergent*.

These can be relativized to sequences of actions by defining

$p\downarrow\varepsilon$ if $p\downarrow$

$p\downarrow as$ if $p\downarrow$ and $p \stackrel{a}{\Longrightarrow} p'$ implies $p'\downarrow s$

and $p\uparrow s$ if $p\downarrow s$ is not true.

So $p\downarrow s$ means that by performing subsequences of s on p we will always evolve to convergent terms while $p\uparrow s$ means that p can evolve to a term which diverges by performing some subsequence of actions from s. For example

$a(b\Omega + cNIL)\downarrow a$ but $a(b\Omega + cNIL)\uparrow ab$

if p is $a(rec\, x.bx + rec\, y.y)$ then $p\downarrow$ but $p\uparrow a$

if p is $rec\, x.a(bx + rec\, y.y + ax)$ then $p\downarrow$ but $p\uparrow a$.

It should be emphasized that $p\downarrow s$ does not necessarily imply that p can actually perform the sequence of actions s. For example $NIL\downarrow s$ for every s and the same is true for $rec\,x.ax + bx$.

We use this predicate $\downarrow s$ to generalize the definition of $<<$.

DEFINITION 4.4.5 For $p, q \in \mathbf{RM}$ let

i) $p <<_{\mathrm{MAY}} q$ if $L(p) \subseteq L(q)$

ii) $p <<_{\mathrm{MUST}} q$ if $p\downarrow s$ implies
   (a) $q\downarrow s$
   and (b) $\mathscr{A}(q, s) \subset\subset \mathscr{A}(p, s)$

iii) $p << q$ if both $p <<_{\mathrm{MAY}} q$ and $q <<_{\mathrm{MUST}} q$.

Note that in the case where $p\downarrow s$ for every s, this definition coincides with that of chapter 2. The expected theorem is true:

THEOREM 4.4.6 (Alternative Characterization for **RM**) For $p, q \in \mathbf{RM}$,

a) $p \sqsubseteq_{\mathrm{MAY}} q$ if and only if $p <<_{\mathrm{MAY}} q$
b) $p \sqsubseteq_{\mathrm{MUST}}$ if and only if $p <<_{\mathrm{MUST}} q$
c) $p \sqsubseteq q$ if and only if $p << q$. □

As in the finite case this gives a relatively simple method of checking if two processes are behaviorally related. For example, let p, q denote $rec\,x.ax \oplus bx$, $(rec\,x.ax + bx) \oplus (rec\,x.ax \oplus bx)$ respectively. Then $p\downarrow s$, $q\downarrow s$ for every s and both have the same language $\{a, b\}^*$. Also for every $s \in \{a, b\}^*$, $\mathscr{A}(p, s) = \{\{a\}, \{b\}\}$ and $\mathscr{A}(q, s) = \{\{a\}, \{b\}, \{a, b\}\}$. It follows from this characterization theorem 4.4.6 that $p \eqsim_{\mathrm{MUST}} q$. Similarly if we examine $p = rec\,x.ax \oplus bx$ and $q = (rec\,x.ax) \oplus rec\,y.ay$ it is easy to see their relationship. For every s, $p\downarrow s$ and $q\downarrow s$. Their languages are slightly different: $L(p) = \{a, b\}^*$ whereas $L(q) = \{a\}^* \cup \{b\}^*$.

Also $\mathscr{A}(q, \varepsilon) = \mathscr{A}(p, \varepsilon) = \{\{a\}, \{b\}\}$ and $\mathscr{A}(p, s) = \{\{a\}, \{b\}\}$ for any s, whereas $\mathscr{A}(q, a^n) = \{\{a\}\}$ and $\mathscr{A}(q, b^n) = \{\{b\}\}$. From these facts it emerges that $p \sqsubseteq_{\mathrm{MUST}} q$ but not the converse.

We have the following corollary.

COROLLARY 4.4.7 $\sqsubseteq_{\mathrm{MUST}}$ is a Σ^2-preorder.

Proof A simple examination of each operator using $<<_{\mathrm{MUST}}$. For example suppose $p <<_{\mathrm{MUST}} q$. We show $p + r <<_{\mathrm{MUST}} q + r$.

i) If $(p + r)\downarrow s$ then $p\downarrow s$ and $r\downarrow s$ and therefore $q\downarrow s$. It follows that $(q + r)\downarrow s$.

ii) Suppose $A \in \mathscr{A}(q + r, s)$. If $s \neq \varepsilon$ then either $A \in \mathscr{A}(q, s)$ or $A \in \mathscr{A}(r, s)$. In either case it is trivial to find a $B \in \mathscr{A}(p + r, s)$ such that $B \subseteq A$. But if $s = \varepsilon$ then A has the form $S(q') \cup S(r')$ where $q \overset{\varepsilon}{\Longrightarrow} q'$ and $r \overset{\varepsilon}{\Longrightarrow} r'$. Since $p <<_{\text{MUST}} q$ we can find a p' such that $p \overset{\varepsilon}{\Longrightarrow} p'$ and $S(p') \subseteq S(q')$. Then $S(p') \cup S(r') \subseteq S(q') \cup S(r')$ and $S(p') \cup S(r') \in \mathscr{A}(p + r, \varepsilon)$. □

It also follows that our notation is consistent: the preorders $\sqsubseteq$, $\sqsubseteq_{\text{MUST}}$, and $\sqsubseteq_{\text{MAY}}$ defined in part I and II agree when applied to finite processes in $\mathbf{M_2}$. This is because for processes in $\mathbf{M_2}$ the relations in definition 4.4.5 agree with those given in definition 2.2.7. Perhaps the most important consequence is that to distinguish processes finite experiments, not involving recursive terms, are sufficient. This will become apparent when we prove the alternative characterization theorem.

The proof of theorem 4.4.6 follows exactly the lines of the corresponding theorem in part I, the only new ingredient being how to deal with $\downarrow s$. Notice that if p *must* $1w$ then $p\downarrow$. More generally p *must* $c(s)$ implies $p\downarrow s$ where $c(s)$ is the experiment defined by

$$c(\varepsilon) = 1w$$

$$c(as) = ac(s) + 1w.$$

The converse is also true.

Lemma 4.4.8 $p\downarrow s$ if and only if p *must* $c(s)$.

Proof Suppose $p\downarrow s$. Then in every computation from $c(s) \parallel p$ whenever a state $e_k \parallel p_k$ is reached $p_k\downarrow$, i.e. each p_k encountered cannot diverge internally. So each computation must reach a state of the form $1w \parallel p'$ or $(c(s') + 1w) \parallel p'$ where the only next move is to the successful state $w \parallel p$. Therefore every computation from $c(s) \parallel p$ must be successful.

Conversely, suppose $p\uparrow s$, i.e. $p \overset{s'}{\Longrightarrow} p'$ where $p'\uparrow$ and s' is a prefix of s, $s = s's''$. Then,

$$c(s) \parallel p \to \cdots c(s') \parallel p' \to c(s'') \parallel p' \to \cdots$$

is an unsuccessful computation, i.e. p ~~*must*~~ $c(s)$. □

Let F be the set of all experiments of the form $c(s)$ together with those in E, the set used in part I. So F consists of all the experiments $c(s)$, $e(s, a)$ and $e(s, A)$ where s ranges over sequences in Act^*, a over Act, and A over finite subsets of Act. We prove $p \sqsubseteq_q^F q$ if and only if $p << q$.

One direction is straightforward.

LEMMA 4.4.9 For $p, q \in \mathbf{RM_2}$, $p \sqsubseteq^F_{\text{MUST}} q$ implies $p \ll_{\text{MUST}} q$.

Proof Suppose $p \downarrow s$. By the previous lemma $q \downarrow s$ also. Let $A \in \mathscr{A}(q, s)$. Then $s \in L(q)$. The tests of the form $e(s', a)$ may now be used as in proposition 2.2.11 to show $s \in L(p)$. However the proof now relies on the fact that $p \downarrow s$. We may continue, once more as in that proposition, to use the tests $e(s, L)$ to establish that $B \subseteq A$ for some $B \in \mathscr{A}(s, p)$. □

The proof of the converse is also similar to that of part I except that one must take into consideration the possibility of processes diverging. We first outline a very useful induction principle which applies to convergent terms.

A property P of terms is called *i-inductive* (or internal-inductive) if it satisfies: for every p such that $p\downarrow$, ($P(q)$ for every q in $\text{Im}(p)$) implies $P(p)$. Recall that $\text{Im}(p) = \{q, p \rightarrowtail q\}$.

LEMMA 4.4.10 If P is *i*-inductive then P is true of every convergent term.

Proof The proof is by contradiction. We suppose $p\downarrow$ and $P(p)$ is false. The contradiction is established by showing that p can diverge internally. Because P is *i*-inductive there must be some p_1 such that $p \rightarrowtail p_1$ and $P(p_1)$ is false. Because $p\downarrow$ we can assume $p_1\downarrow$ also and we may then apply the same reasoning to p_1 to obtain p_2 such that $p_1 \rightarrowtail p_2$, $p_2\downarrow$ with $P(p_2)$ false. Continuing in this manner we can construct an infinite sequence

$$p \rightarrowtail p_1 \rightarrowtail p_2 \quad \cdots \quad \rightarrowtail p_n \rightarrowtail \quad \cdots$$

with the property that $P(p_n)$ is false for every $n \geq 0$. This means that $p\uparrow$. □

As an example of the use of this form of induction prove the following lemma.

LEMMA 4.4.11

a) If $p\downarrow$ then $S(p)$ is finite

b) If $p \downarrow s$ then $S(p, s)$ is finite

c) If $p \downarrow s$ then $D(p, s)$ is finite.

Proof a) Let P be the property of terms defined by $P(p)$ if $S(p)$ is finite. Because of the previous lemma it now suffices to show that P is *i*-inductive. We show by structural induction on p that for every convergent p

($S(q)$ is finite for every q in $\text{Im}(p)$) implies $S(p)$ is finite.

i) p is NIL or ar. In either of these cases $S(p)$ is finite by definition.

ii) p is $p_1 \oplus p_2$. In this case $S(p) = S(p_1) \cup S(p_2)$ and Im(p) is $\{p_1, p_2\}$. So the assumption says that both $S(p_1)$ and $S(p_2)$ are finite, from which it follows that $S(p)$ is also finite.

iii) p is $p_1 + p_2$. Similar.

iv) p is *rec x.t*. This case is trivial because Im(p) is $\{t[rec\, x.t/x]\}$ and $S(p)$ is $S(t[rec\, x.t/x])$.

b) We use induction on s and the definition of the predicate $\downarrow s$. The details are left to the reader.

c) The proof is similar to that of parts (a) and (b). Use i-induction to prove $D(p, \varepsilon)$ is finite and then proceed by induction on s. □

This lemma has a useful corollary. It is easy to see from the definition of $<<_{\text{MUST}}$ that if $p <<_{\text{MUST}} q$ and $p\downarrow s$ then $s \in L(q)$ implies $s \in L(p)$. We show that this is true even for infinite sequences. For such a sequence u let $u(k)$ denote the finite subsequence consisting of the first k elements of u. Then $p\downarrow u$ is taken to mean $p\downarrow u(k)$ for every $k \geq 0$. We also use $p \overset{u}{\Longrightarrow}$ to mean that there is an infinite u-derivation from p.

$$p \rightarrowtail^* \xrightarrow{a_1} \rightarrowtail^* p_1 \xrightarrow{a_2} \rightarrowtail^* p_2 \ldots,$$

where u is $a_1 a_2 \ldots$.

LEMMA 4.4.12 If $p <<_{\text{MUST}} q$ and $p\downarrow u$ where u is an infinite sequence of actions then

$$q \overset{u}{\Longrightarrow} \quad \text{implies} \quad p \overset{u}{\Longrightarrow}.$$

Proof Consider the computation tree from p confined to actions from the sequence u. This records all possible computations from p using subsequences of u. The nodes are labeled by processes and the branches by actions. The root is decorated by p and in general if a node is labeled by the process r and the path from the root is labeled by the sequence of actions s, then s is an initial subsequence of u and $r \in D(p, s)$. By the previous lemma this tree is finite-branching. However, it is infinite because $p <<_{\text{MUST}} q$ and $q \overset{u}{\Longrightarrow}$ implies $u(k) \in L(p)$ for every $k \geq 0$. Therefore, by König's lemma (Knuth 1975) there is an infinite path through the tree, i.e. $p \overset{u}{\Longrightarrow}$. □

LEMMA 4.4.13 For $p, q \in \mathbf{RM}$, $p <<_{\text{MUST}} q$ implies $p \sqsubseteq_{\text{MUST}} q$.

Proof The proof is similar to that in theorem 2.2.12. Suppose $p <<_{\text{MUST}} q$ and p *must* e. We show q *must* e. Let

$$e \parallel q = e_0 \parallel q_0 \rightarrow e_1 \parallel q_1 \rightarrow \cdots \qquad (*)$$

by an arbitrary computation, which may be infinite. We must show that $e_k \in Success$ for some $k \geq 0$. The computation (*) can be decomposed into

$$e_0 \stackrel{a_1}{\Longrightarrow} e_1 \stackrel{a_2}{\Longrightarrow} e_2 \cdots \qquad (*1)$$

$$q_0 \stackrel{a_1}{\Longrightarrow} q_1 \stackrel{a_2}{\Longrightarrow} q_2 \cdots \qquad (*2)$$

where for convenience $\stackrel{1}{\longrightarrow}$ moves are absorbed in (*1). Let u denote the possibily infinite sequence $a_1 a_2 a_3 \ldots$. We may assume that $p \downarrow u$ for otherwise we could construct an unsuccessful computation from $e \parallel p$ which contradicts p *must* e.

a) The sequence u is infinite. From the previous lemma it follows that $p \stackrel{u}{\Longrightarrow}$. Any infinite u-derivation from p can be combined with (*1) to obtain a computation

$$e_0 \parallel p_0 \rightarrow e_1 \parallel p_1 \rightarrow \cdots .$$

Since p *must* e it now follows that $e_k \in Success$ for some $k \geq 0$.

b) the sequence s is finite. In this case the proof proceeds as outlined in the corresponding parts of theorem 2.2.12 and theorem 2.5.5. □

Combining this lemma with its converse, lemma 4.4.9, we have the alternative characterization theorem in the *must* case:

$$p \sqsubseteq_{\text{MUST}} q \quad \text{if and only if} \quad p <<_{\text{MUST}} q.$$

We leave the reader to establish the *may* case and combining both we obtain the general case.

We end this section by emphasizing three points.

1. Only the set of tests F are necessary to characterize the Testing preorders.

2. The alternative characterization theorem does not depend on the fact that the language used is **RM**. A closer examination of the proof will reveal the fact that it remains true in the general setting of an Experimental System generated by compatible *lts*'s, as in definition 2.5.1, provided the set of experiments contain at least experiments which correspond to the experiments in F. The proof of lemma 4.4.9 is actually language-independent. It can be duplicated when both processes and experiments are described by *lts*'s. The same is true of the converse, lemma 4.4.13, except that we need *lts*'s which correspond to each of the experiments in F. In an arbitrary setting the *lts*'s must be finite branching. This will be true when we extend **RM** in part III.

3. The theorem holds provided the set of experiments contain the minimum distinguishing power embodied in F. In particular it follows that adding extra experiments does not increase the power of discrimination between processes.

4.5 Full Abstraction

We now wish to connect the behavioral relations $\sqsubseteq$, $\sqsubseteq_{\mathrm{MUST}}$, $\sqsubseteq_{\mathrm{MAY}}$ developed in the previous section, with the semantic interpretations of chapter 3. The connection is, of course, via full abstraction. We will prove for every $p, q \in \mathbf{RM}$, that

$$p \sqsubseteq_{\mathrm{MUST}} q \quad \text{if and only if} \quad \mathbf{AT_S}[\![p]\!] < \mathbf{AT_S}[\![q]\!],$$

i.e. processes which are identified in the semantic domain cannot be distinguished by testing. Similar results will link the domains $\mathbf{AT}$, $\mathbf{AT_W}$ with $\sqsubseteq$, $\sqsubseteq_{\mathrm{MAY}}$ respectively. Each of these domains is initial in an equational class of continuous algebras and, because of the intimate connection between initial algebras and equational proof systems, these full abstractness results lead automatically to sound and complete proof systems. Indeed, in the proof of full abstractness we make full use of this connection. In §3.5 we saw that the interpretation $\mathbf{AT_S}$ is initial in the class of interpretations which satisfy the equations **SA2s**. This means that the relations $\leq_{\mathbf{AT_S}}$ and $\leq_{\mathbf{SA2s\omega}}$ coincide over **RM**. To prove full abstractness we show that $\sqsubseteq_{\mathrm{MUST}}$ also coincides with $\leq_{\mathbf{SA2s\omega}}$.

This approach to relating the behavioral, proof-theoretic, and denotational theories is a little different than that of part I. There we relate

i) the denotational with the proof-theoretic, theorem 2.4.7 and theorem 1.4.10

and

ii) the behavioral with the denotational, theorem 2.3.7 and theorem 2.5.7.

In part II, for recursive processes, we relate

iii) the denotational with the proof-theoretic also, theorem 3.4.5 and theorem 4.3.5

but

ii) the behavioral with the proof-theoretic.

There is no reason for the slight change other than variety. However, in comparing the proof-theoretic relations with the behavioral relations it will help us technically if the latter were extended to open terms. They really only make sense for closed terms but a natural extension is:

for $t, u \in REC_{\Sigma^2}(X)$, $t \sqsubseteq u$ if for every closed substitution ρ, $t\rho \sqsubseteq u\rho$.

Similarly for $\sqsubseteq_{\mathrm{MUST}}$ and $\sqsubseteq_{\mathrm{MAY}}$.

The essential notion behind the proof is that of an algebraic relation: a relation R over REC_{Σ} is *algebraic* if

$$\langle p, q\rangle \in R \Longleftrightarrow \text{for every } d \in \text{App}(p) \text{ there exists an } e \in \text{App}(q) \text{ such that } \langle d, e\rangle \in R.$$

This means that R is completely determined by its restriction to the finite elements of REC_{Σ}. More precisely, if two algebraic relations are identical on $FREC_{\Sigma}$ they are also identical.

The proof of full abstractness is carried out as follows:

1) show $\leq_{\mathbf{SA2s\omega}}$ is algebraic
2) show $\lesssim_{\text{MUST}}$ is algebraic
3) prove $\leq_{\mathbf{SA2s\omega}}$ and $\lesssim_{\text{MUST}}$ coincide on finite terms, for $d, e \in FREC_{\Sigma^2}$ $d \leq_{\mathbf{SA2s\omega}} e$ if and only if $d \lesssim_{\text{MUST}} e$.

Condition 1) is established very simply from general principles and 3) is a very mild generalization of the soundness and completeness of **DED(SA2)** for $\lesssim_{\text{MUST}}$ over T_{Σ^2}. The only difference between T_{Σ^2} and $FREC_{\Sigma^2}$ is the presence of the undefined operator Ω, and these extra partial terms are catered for by the extension of **DED(SA2)** to **ΩDED(SA2s)**. But we shall continue to abbreviate **ΩDED**(*E*) by **DED**(*E*). The second condition is the most difficult and requires a careful analysis of *must*-testing. The central technical device is *head normal forms*, a generalization of the normal forms used in the completeness proof of part I. Here we cannot use normal forms directly as the process of normalizing may not terminate for recursive terms in general. Instead we use head normal forms: these are terms which look like normal forms at the topmost level. In what follows we use the notation developed in part I for normal forms.

DEFINITION 4.5.1

i) *NIL* is a *hnf* (*head normal form*)
ii) If $\mathscr{A}$ is saturated, any term of the form $\mathop{\textstyle\sum\!\!\!\!\bigcirc}\{\sum ap(a), a \in A, A \in \mathscr{A}\}$ is a *hnf*.

The crucial difference between these and normal forms is that the terms $p(a)$ need not be *hnf*'s.

PROPOSITION 4.5.3 (Head Normal Form Theorem) Every convergent term has a *hnf*, i.e.

$p\!\downarrow$ implies $\vdash_{\mathbf{rDED(A2)}} t = h$, where h is a *hnf*.

Proof We use the form of induction developed in the previous section in lemma 4.4.10 for convergent terms: we show that having a *hnf* is *i*-inductive. So we assume that if

$p \succ\!\!\longrightarrow q$, then q has a *hnf*. From this assumption we prove p has a *hnf*. The proof is by structural induction on p and follows the proof of the normal form theorem, theorem 2.4.6.

i) p is NIL or aq. In this case p already is in *hnf*.

ii) p is $q + r$. Now $q\downarrow$ and $r\downarrow$ and by structural induction we may assume that these have *hnf*'s h, h' respectively. We can now apply the syntactical manipulations used in case c) of theorem 2.4.6 to transform $h + h'$ into a *hnf*.

iii) p is $q \oplus r$. Similar to ii) but using case d).

iv) p is $rec\, x.t$. Now $rec\, x.t \succ\!\!\longrightarrow t[rec\, x.t/x]$ and so by the hypothesis we may assume $t[rec\, x.t/x]$ has a *hnf* h. Using rule 7 we obtain $\vdash_{\mathbf{rDED(A2)}} rec\, x.t = h$. □

We have an immediate corollary if we use a slightly extended type of normal form for terms in $FREC_{\Sigma 2}$:

i) Ω and NIL are Ω-*nf's* (Ω-*normal forms*)

ii) if $\mathscr{A}$ is saturated, and $d(a)$ is an Ω-*nf* for every $a \in S(\mathscr{A})$ then $\textstyle\sum\{d(A), A \in \mathscr{A}\}$ is also an Ω-*nf*.

COROLLARY 4.5.4

a) if $d \in FREC_{\Sigma 2}$ and $d\uparrow$ then $d =_{\mathbf{SA2s}} \Omega$

b) every term in $FREC_{\Sigma 2}$ has an Ω-*nf*, i.e. for every $d \in FREC_{\Sigma 2}$ there exists an Ω-*nf* n such that $d =_{\mathbf{SA2s}} n$.

Proof a) A simple proof by structural induction on d. Although the predicate $\uparrow$ is defined behaviorally for finite terms it is a structural property: $d\uparrow$ if it contains an occurrence of Ω not contained in an occurrence of the form ad'.

b) If $d\uparrow$, then use part a). Otherwise by the previous proposition d has a head normal form $\textstyle\sum\{h(A), A \in \mathscr{A}\}$. By induction each $h(a)$ can be transformed into an Ω-*nf* $n(a)$, and therefore $\textstyle\sum\{n(A), A \in \mathscr{A}\}$ is the required Ω-*nf* for d. □

If we replace the normal forms of part I with these Ω-normal forms the completeness proofs can be borrowed more or less directly.

PROPOSITION 4.5.5 **DED(SA2s)** is sound and complete with respect to $\lessapprox_{\mathrm{MUST}}$ over $FREC_{\Sigma 2}$.

Proof a) *Soundness*: This is more naturally checked over $FREC_{\Sigma 2}(X)$ and it is simply a matter of examining each individual rule of **DED(SA2s)** in turn. (It is for this reason that we extended the behavioral relations to act over open terms.) Referring to figure 4.2, the rules 1, 2 and 3 a) preserve $\lesssim_{\mathrm{MUST}}$ because it is a Σ^2-preorder; and rule 4 is justified by the Substitution lemma. Ω is the least element with respect to $\lesssim_{\mathrm{MUST}}$ and so rule 6 is correct. Finally, to check rule 5 each axiom of **SA2s** must be considered and it is best to use $<<_{\mathrm{MUST}}$ rather than $\lesssim_{\mathrm{MUST}}$ for this purpose.

b) *Completeness*: A minor variation on the completeness proof in §2.4. Suppose $d \lesssim_{\mathrm{MUST}} e$. Then if n, m are their Ω-normal forms it follows from the soundness of **DED(SA2s)** that $n \lesssim_{\mathrm{MUST}} m$, or more helpfully $n <<_{\mathrm{MUST}} m$. Minor adjustments to lemma 2.4.4 will give $n \leq_{\mathbf{SA2s}} m$ from which it follows that $d \leq_{\mathbf{SA2s}} e$. □

With a small amount of extra effort a stronger result can be derived:

for $d \in FREC_{\Sigma^2}$, $q \in \mathbf{RM}$,

$d \lesssim_{\mathrm{MUST}} q$ if and only if $d \leq_{\mathbf{SA2sr}} q$.

However, the above proposition establishes condition 3 with which we are content.

Head-normal forms have already been used to some extent but their significant application is in the next lemma.

LEMMA 4.5.6 If e is a finite experiment and p *must* e then d *must* e for some $d \in \mathrm{App}(p)$.

Proof By induction on the size of e. Let us assume that the statement is true of every experiment whose size is less than e. There are two cases.

i) p is in *hnf*: $\oplus\{\Sigma\{ap(a), a \in A\}, A \in \mathscr{A}\}$. The idea is to replace each $p(a)$ uniformly with a finite approximation $p(a)^n$ so that the new term still satisfies e. This constructed term will then have the form p^{n+1}. Suppose $e \stackrel{a}{\Longrightarrow} e'$ for no a in $A(\mathscr{A})$. Then each $p(a)$ may be replaced by Ω and the resulting term p^1 *must* satisfy e. Otherwise if $e \stackrel{a}{\Longrightarrow} e'$, where $a \in A(\mathscr{A})$, then $p(a)$ *must* e'. By induction there exists a k such that $p(a)^k$ *must* e'. For each such a and e' such a k exists. Let n be the maximum of these k. Then $p(a)^n$ *must* e' is also true since $p^k \preceq p^{k+1}$. It follows that p^{n+1} *must* e.

ii) p is not in *hnf*. If p *must* e then p converges and therefore it has a *hnf* h. Also h *must* e and by part i) d *must* e for some $d \in \mathrm{App}(h)$. Consider the denotations of h and p in the model $\mathbf{AT_S}$. $\mathbf{AT_S}[\![p]\!] = \mathbf{AT_S}[\![h]\!]$ and therefore $\mathbf{AT_S}[\![d]\!] \leq \mathbf{AT_S}[\![p]\!]$. Because $\mathbf{AT_S}$ is initial, $\mathbf{AT_S}[\![d]\!]$ is finite and since $\mathbf{AT_S}[\![p]\!] = \bigvee \mathbf{AT_S}(\mathrm{App}(p))$, it follows that $\mathbf{AT_S}[\![d]\!] \leq \mathbf{AT_S}[\![d']\!]$ for some $d' \in \mathrm{App}(p)$. Because $\mathbf{AT_S}$ is initial in the class $\mathscr{CC}(\mathbf{SA2s})$ we have that $d \leq_{\mathbf{SA2s}} d'$. Each of the equations in **SA2s** is sound with respect to $\lesssim_{\mathrm{MUST}}$ and therefore d' *must* e. □

LEMMA 4.5.7 For $d \in FREC_{\Sigma^2}$ and $p \in \mathbf{RM}$, if $d \lesssim_{\mathrm{MUST}} p$ then $d \lesssim_{\mathrm{MUST}} d'$ for some $d' \in \mathrm{App}(p)$.

Proof Recall the set of experiments F which are used in the proof of the alternative characterization theorem. All these are finite and we know that to establish $p \precsim_{\mathbf{MUST}} q$ for arbitrary p, q it is sufficient to restrict attention to tests in F. For $d \in FREC_{\Sigma^2}$ let $F(d)$ denote the finite of set of the experiments in F which only use actions from d and whose depth (i.e. the length of its longest path) is at most one more than that of d. Because of the structure of the experiments in F in order to show $d \precsim_{\mathbf{MUST}} q$ it is sufficient to show q *must* e for those e in $F(d)$ such that d *must* e.

Let F' denote this subset of $F(d)$. From the hypothesis we know that $d \precsim_{\mathbf{MUST}} p$ and so p *must* e for every e in F'. From the previous lemma we can find for each such e a finite $d(e) \in \mathrm{App}(p)$ such that $d(e)$ *must* e. Let d' be an upper bound in $\mathrm{App}(p)$ of the set $\{d(e), e \in F'\}$. Then d' *must* e for every e in F' and therefore $d \precsim_{\mathbf{MUST}} d'$. □

These two results are sufficient to show that $\precsim_{\mathbf{MUST}}$ is algebraic.

PROPOSITION 4.5.8 The relation $\precsim_{\mathbf{MUST}}$ is algebraic over **RM**.

Proof a) Suppose $p \precsim_{\mathbf{MUST}} q$ and $d \in \mathrm{App}(p)$. We must find an $e \in \mathrm{App}(q)$ such that $d \precsim_{\mathbf{MUST}} e$. If $d \in \mathrm{App}(p)$, then by lemma 4.3.3 $\vdash_r d \leq p$. Each of the proof rules in this proof system trivially preserve $\precsim_{\mathbf{MUST}}$ and therefore it follows that $d \precsim_{\mathbf{MUST}} p$. Using the hypothesis $d \precsim_{\mathbf{MUST}} q$ and the required e is obtained from the previous lemma.

b) Conversely suppose $d \precsim_{\mathbf{MUST}} q$ for every d in $\mathrm{App}(p)$. We show $p \precsim_{\mathbf{MUST}} q$. Suppose p *must* e. From proposition 4.5.6 there exists some $d \in \mathrm{App}(p)$ such that d must e. From the hypothesis $d \precsim_{\mathbf{MUST}} q$ and therefore q *must* e. □

We leave the remaining ingredient of the proof of full abstractions as exercise Q20:

PROPOSITION 4.5.9 The relation $\leq_{\mathbf{SA2s\omega}}$ is algebraic.

Combining these two propositions with proposition 4.5.5, as in the introduction to this section, we obtain:

THEOREM 4.5.10 (*Full Abstraction*) The interpretation $\mathbf{AT_S}$ is fully abstract with respect to $\precsim_{\mathbf{MUST}}$ over **RM**.

We know that $\mathbf{AT_S}$ is isomorphic to the initial algebra $CI_{\mathbf{SA2s}}$ and by applying the Recursive Equational Logic theorem we obtain:

THEOREM 4.5.11 (*Soundness and Completeness*) The proof system **ωDED(SA2s)** is sound and complete with respect to $\precsim_{\mathbf{MUST}}$ over REC_{Σ^2}.

Throughout this section we have centred on the strong model $\mathbf{AT_S}$ and the strong axioms **SA2s**. However exactly the same techniques apply to the other two cases: **AT**,

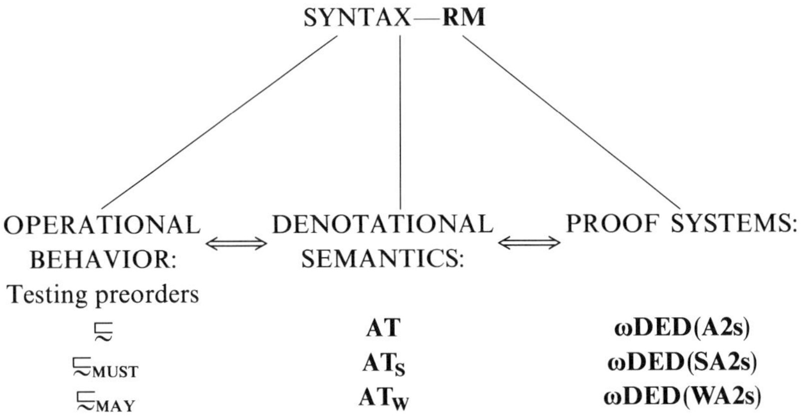

Figure 4.4
Schema of results.

$\mathbf{AT_W}$ are fully abstract with respect to $\underset{\sim}{\sqsubset}$, $\underset{\sim}{\sqsubset}_{MAY}$ respectively and consequently **ωDED(A2s)** and **ωDED(WA2s)** are sound and complete proof systems. All of these results are schematized in figure 4.4.

Exercises

Q1 a) Prove that $FV(t)$ is finite for every term t.
b) Prove that $x \in FV(t\rho)$ if and only if $x \in FV(\rho(y))$ for some $y \in FV(t)$.
c) Show $t\rho = t\rho'$ whenever $\rho(x) = \rho'(x)$ for every $x \in FV(t)$.
d) Show $(rec\, x.t)\rho = (rec\, y.tI[x \to y])\rho$ if $y = new\, xt\rho$.

Q2 a) Prove $t =_\alpha u$ implies $FV(t) = FV(u)$.
b) For any set of variables Y let $\rho =_\alpha^Y \rho'$ if for every $y \in Y$, $\rho(y) =_\alpha \rho'(y)$. Prove

$$\rho =_\alpha^{FV(t)} \rho' \quad \text{implies} \quad t\rho =_\alpha t\rho'.$$

c) Prove $t[u/x] =_\alpha t$ whenever $x \notin FV(t)$.
d) Prove $t =_\alpha u$ implies $t^n = u^n$ for every $n \geq 0$.

Q3 (Substitution Lemma) a) Let $y_1 = new\, xt\rho$, $y = new\, y_1(t\rho[x \to y])\rho'$ and $z = new\, xt(\rho' \circ \rho)$. Prove $y = z$.
b) Prove

$$\rho'[y_1 \to y] \circ \rho[x \to y_1] =^{FV(t)} \rho' \circ \rho[x \to y].$$

c) Assuming $(t\sigma)\sigma' = t\sigma' \circ \sigma$ for all substitutions σ, σ' prove

$((rec\, x.t)\rho)\rho' = (rec\, x.t)\rho' \circ \rho.$

d) Prove $(t\rho)\rho' = t(\rho' \circ \rho)$ for all terms t.

Q4 Using the Substitution lemma prove:

a) $t[u/x][v/x] = t[u[v/x]/x]$.

b) $t[y/x][u/y] = t[u/x]$ if $y \notin FV(t)$.

c) More generally $t[y/x]\rho[y \to u] = t\rho[x \to u]$ if $y \notin FV(t)$.

d) $(t\rho[x \to y])I[y \to u] = t\rho[x \to u]$ if $y \notin FV(\rho(z))$ for every z in $FV(t)$ different from x.

Q5 Use the Substitution lemma to show $(\rho_1 \circ \rho_2) \circ \rho_3 = \rho_1 \circ (\rho_2 \circ \rho_3)$.

Q6 Prove $t =_\alpha t'$ if and only if $tI = t'I$.

Q7 Let ρ^n be the substitution defined by $\rho^n(x) = \rho(x)^n$. Show that $(t\rho)^n \preceq (t^n)\rho^n$.

Q8 Let $\prec\!\!\prec$ be the least Σ-congruence over $REC_\Sigma(X)$ which satisfies

i) $\Omega \prec\!\!\prec t$

ii) $t[rec\, x.t/x] \prec\!\!\prec rec\, x.t.$

a) Show that $t \prec\!\!\prec u$ implies $t\rho \prec\!\!\prec u\rho$.

b) Prove that for every n, $t^n \prec\!\!\prec t$.

c) Prove that if $t \prec\!\!\prec u$ then for every $n \geq 0$ there exists some $m \geq 0$ such that $t^n \prec\!\!\prec u^m$.

Q9 Let $\text{Fin}(t) = \{d \in FREC_\Sigma(X), d \prec\!\!\prec t\}$.

a) Use Q8 c) to show that for every $d \in \text{Fin}(t)$ there exists some n such that $d \preceq t^n$.

b) Prove that $\text{Fin}(t)$ is directed with respect to $\prec$.

c) Prove $A[\![t]\!] = \bigvee_A A[\![\text{Fin}(t)]\!]$.

Q10 Describe the meanings of the following terms in the interpretations **AT** and $\mathbf{AT_S}$.

a) $(rec\, x.ax) \oplus (rec\, x.bx)$

b) $rec\, x.(ax \oplus bx)$

c) $rec\, x.(ax + x)$

d) $rec\, x.ax + b(rec\, y.cy + y + x)$.

Q11 (Simultaneous Recursion) For $f, g : D \times D \longrightarrow D$ let $\langle f, g\rangle : D \times D \longrightarrow D \times D$ be the function defined by $\langle f, g\rangle(d, e) = (f(d, e), g(d, e))$.

a) Show that $\langle f, g\rangle$ is continuous whenever both f and g are continuous.

b) Let $d_0 \in D$ denote $Y\lambda d.f(d, Y\lambda e.g(d, e))$ and $e_0 \in D$ denote $Y\lambda e, g(d_0, e)$. Prove that the pair (d_0, e_0) is the minimal fixpoint of the function $\langle f, g\rangle$.

c) Extend the syntax of $REC_\Sigma(X)$ by allowing terms of the form $rec_1(x, y).(t, u)$ and $rec_2(x, y).(t, u)$. Intuitively these represent the simultaneous definitions

$x \Longleftarrow t$

$y \Longleftarrow u.$

Define $A[\![rec_i(x, y).(t, u)]\!]\rho$, for $i = 1, 2$ to be the ith component of

$Y\lambda(d, e).(A[\![t]\!]\rho[d/x, e/y], A[\![u]\!]\rho[d/x, e/y]).$

The problem is to show that there are already terms in $REC_\Sigma(X)$ which have the same meaning.

Let t_0 denote $rec\, x.(t[rec\, y.u/y])$ and u_0 denote $rec\, y.(u[t_0/x])$. Prove

$$A[\![t_0]\!] = A[\![rec_1(x, y).(t, u)]\!] \quad \text{and} \quad A[\![u_0]\!] = A[\![rec_2(x, y).(t, u)]\!].$$

d) Prove that $A[\![rec\, x.t[rec\, y.u/y]]\!] = A[\![rec\, x.(t[rec\, y.(u[rec\, x.t/x])/y])]\!]$.

Q12 Show that $\downarrow$ is the least predicate on **RM** which satisfies

i) $NIL\downarrow, ap\downarrow$

ii) $p\downarrow, q\downarrow$ implies $(p + q)\downarrow, (p \oplus q)\downarrow$

iii) $t[rec\, x.t/x]\downarrow$ implies $rec\, x.t\downarrow$

Q13 There are two ways of generalizing $\downarrow$ to open terms.

—Let $\Downarrow x$ be the least relation over $REC_{\Sigma^2}(X)$ which satisfies, in addition to i)–iii) of Q12, iv) $y \Downarrow x$ if $y \neq x$.

—Let $\downarrow x$ be the least relation over $REC_{\Sigma^2}(X)$ which satisfies i), ii), iii)′, and iv) where iii)′ is

$$t\downarrow x, t\downarrow y \quad \text{implies} \quad rec\, y.t\downarrow x.$$

a) Prove $t\rho\downarrow x$ implies that for every y either $t\downarrow y$ or $\rho(y)\downarrow x$

b) Use a) to show $t \Downarrow x$ implies $t\downarrow x$

c) Prove that $t \Downarrow x_i$, $1 \leq i \leq n$, implies $t\rho \Downarrow x_i$, $l \leq i \leq n$ if $\rho(y) \Downarrow x_i$, $l \leq i \leq n$, for every y different from $x_1, \ldots, x_n$.

d) Use c) to show $t\downarrow x$ implies $t \Downarrow x$

e) Deduce that for $p \in$ **RM**

i) $p\downarrow$ implies $p\downarrow x$ for every x

ii) $p\downarrow x$ for any x implies $p\downarrow$

iii) if p has the form $rec\,x.t$ then $p\downarrow$ if and only if $t\downarrow x$.

Q14 For $t \in \mathbf{AT}$ or $\mathbf{AT_S}$ let t_n be the tree obtained by removing nodes at depth greater than or equal to n. (In general t_n will not be elements of $\mathbf{AT}$, $\mathbf{AT_S}$ respectively.) We say $f: \mathbf{AT} \longrightarrow \mathbf{AT}$ (or $\mathbf{AT_S} \longrightarrow \mathbf{AT_S}$) is contractive if for $n \geq 0$ $t_n = t'_n$ implies $f(t)_{n+1} = f(t')_{n+1}$.

a) Show that every contractive function has a unique fixpoint.

b) Show that the limit of contractive maps in $[\mathbf{AT} \longrightarrow \mathbf{AT}]$ or $[\mathbf{AT_S} \longrightarrow \mathbf{AT_S}]$ is also contractive.

c) Prove that if $t \in FREC_{\Sigma^2}(X)$ and $t\downarrow x$ then for any ρ the function $\lambda a.\mathbf{AT}[\![t]\!]\rho[a/x]$ has a unique fixpoint.

d) More generally prove that for any $t \in REC_{\Sigma^2}(X)$ if $t\downarrow x$ then for any ρ the function $\lambda a.\mathbf{AT}[\![t]\!]\rho[a/x]$ has a unique fixpoint.

e) Show that the proof rule UFI is sound when restricted to terms $rec\,x.t$ where $t\downarrow x$.

Q15 a) For $t \in REC_{\Sigma^2}(X)$ prove

i) $t\downarrow x$ and $y \notin FV(t)$ implies $t[y/x]\downarrow y$

ii) $t\downarrow x$ and $y \neq x$ implies $t[y/z]\downarrow x$

b) Show that if $t =_\alpha u$ and $t\downarrow x$ then $u\downarrow x$.

Q16 a) Prove that if $t \leq_r u$ and $t\downarrow x$ then $u\downarrow x$.

b) Show that if $t =_\alpha u$ then $t = u$ in $\mathbf{rDED}(\emptyset)$ + RI and $\mathbf{rDED}(\emptyset)$ + SI.

c) Consider the proof rule

$$\frac{\rho \leq \rho'}{t\rho \leq t\rho'}$$

where $\rho \leq \rho'$ means $\rho(x) \leq \rho'(x)$ for every variable x. Show that this is a derived rule in $\mathbf{rDED}(E)$.

d) Show that rule 3 b) is a derived rule in the proof system $\boldsymbol{\omega}\mathbf{DED}(E)$.

Q17 Extend the syntax of $REC_\Sigma(X)$ as in Q11, allowing terms of the form $rec_1(x, y).(t, u)$ and $rec_2(x, y).(t, u)$. For convenience let these be denoted by r_1, r_2 respectively.

a) Using the semantics outlined in Q11 show that

i) $r_1 = t[r_1/x, r_2/y]$, $r_2 = u[r_1/x, r_2/y]$ are sound as proof rules.

ii) $\dfrac{t[r/x, r'/y] \leq r,\ u[r/x, r'/y] \leq r'}{r_1 \leq r,\ r_2 \leq r'}$ is a sound proof rule.

b) Using these rules derive the theorems

$r_1 < t_0, \quad r_2 < u_0,$

where

$t_0 = rec\,x.t[rec\,y.u/y], \quad u_0 = rec\,y.u[t_0/x].$

Q18 Prove that the following are theorems in the proof system: **rDED**($\varnothing$):

i) $t^n \leq t^{(m)}$

ii) $t^{(n)} \leq t$

iii) $(t\rho)^n \leq t^n \rho^n$.

Q19 For each of the following pairs of processes decide whether or not they are related via $\sqsubseteq$ or $\sqsubseteq_{\mathrm{MUST}}$. If they are, give a proof using RI or SI and the appropriate equations. If not give experiments which differentiate between them.

i) $rec\,x.ax \oplus bx$ $\qquad (rec\,x.ax) \oplus (rec\,x.bx)$

ii) $(rec\,x.ax \oplus b) \oplus (rec\,x.ax)$ $\qquad rec\,x.ax \oplus b$

iii) $rec\,x.ax \oplus b$ $\qquad (rec\,x.ax + b) \oplus b$

iv) $(rec\,x.a(x + b)) + (rec\,x.a(x + c))$ $\qquad (rec\,x.a(ax + b) + c) + (rec\,x.a(ax + c) + b)$

v) $rec\,x.a(rec\,y.bx \oplus cy) + rec\,x.aby$ $\qquad rec\,x.a(rec\,y.bx \oplus cy)$

Q20 a) Let A be an algebraic *cpo*. Under what circumstances is $\leq_A$ algebraic?

b) Show that algebraic relations are determined by their restrictions to finite elements: that is, if R_1 and R_2 coincide over $FREC_\Sigma$ they also coincide over REC_Σ.

c) Prove that the interpretation CI_E is fully abstract with respect to the algebraic relation R if and only if **ΩDED**(E) is sound and complete with respect to R.

d) Without using the completeness theorem show that $\sqsubseteq$, $\sqsubseteq_{\mathrm{MUST}}$, $\sqsubseteq_{\mathrm{MAY}}$ are all algebraic.

Q21 Prove that the interpretations **AT** and $\mathbf{AT_W}$ are fully abstract with respect to $\sqsubseteq$, $\sqsubseteq_{\mathrm{MAY}}$ (respectively) over **RM**.

III COMMUNICATING PROCESSES

5 Communicating Processes

In this final chapter we return to the example language **EPL** for describing processes. Here we view it as an extension of the nondeterministic language **RM**, obtained by adding two new combinators. In addition the set of actions *Act* is interpreted in a particular fashion. We assume some predefined set of channel names *Chan*. There are then two different kinds of actions, one representing input from a channel, the other output to a channel. Communication is viewed as the simultaneous occurrence of two actions: one input and the other output. (This approach has already been thoroughly discussed in the introduction.) One new operator is the parallel combinator $|$: $t_1 \mid t_2$ is a new term which presents a process consisting of two independent subprocesses running in parallel. The final combinator, or, more formally, a parametrized combinator, is a mechanism for hiding channels. For each $c \in Chan$ there is a restriction combinator $\backslash c : p \backslash c$ acts like the process p except that the channel c cannot be used for input or output with an external process.

The result is a powerful language for describing processes which is essentially **EPL**, as described in the introduction, except that there is an extra nondeterministic operator $\oplus$, and restriction is with respect to single channel names rather than sets of channel names. The informal description of the meaning of processes given in the introduction is formalized by viewing **EPL** as a labeled transition system. This is the subject of §5.1, where we also extend the Testing preorders to the new language. In §5.2 we see that the proof system **ωDED(SA2s)** can be extended so as to be sound and complete for **EPL** with respect to $\sqsubseteq_{\mathrm{MUST}}$ over **EPL**. We do so by adding new sets of equations for $|$ and $\backslash c$. Similar results hold for the other Testing preorders $\sqsubseteq$ and $\sqsubseteq_{\mathrm{MAY}}$. In §5.3 we show that the model $\mathbf{AT_S}$ is also fully abstract for **EPL**; we merely extend this model by interpreting the new combinators as functions over the domain. In §5.4, we indicate that **EPL** is only one of a large collection of algebraic process languages which can be explained using the mathematical foundations we have developed. The Testing preorders are very general and may be defined on any language which can be given an operational semantics as a labeled transition system. The relationship between these behavioral preorders and the equational proof systems remains stable even if we vary the kind of parallel, communication, or hiding combinators in the language.

5.1 Operational Semantics

The processes discussed informally in the introduction can perform two different distinct kinds of actions:

—output a signal to a channel c
—input a signal from a channel c.

Furthermore, we introduced a convention that input actions are identified by "barred" labels, such as $\bar{a}$, $\bar{b}$, $\bar{c}$, etc., whereas output actions are labeled without bars. To incorporate this within the framework we have developed let *Chan* be an arbitrary set of channel names, *In* be the set $\{\bar{c}, c \in Chan\}$, and *Out* be simply *Chan*. That is, we use c to represent the action of outputting a signal to channel c and $\bar{c}$ to represent input from this channel. We now assume *Act*, the set of actions, has the form $Out \cup In$, i.e. every action is either an input action or an output action. We use *name*(a) to refer to the name of the channel used in the action a, i.e. $name(c) = name(\bar{c}) = c$. The actions, $c, \bar{c}$ are said to be complementary and communication is modeled as the simultaneous occurrence of those complementary actions. It always involves one process outputting a signal and another simultaneously inputting this signal.

We now describe the language formally. Let *Act* be as described above. We will find it convenient to extend the complementation notation defined above by allowing $\bar{\bar{a}}$ to denote a for every $\bar{a} \in Out$ and $\bar{B}$ to denote $\{\bar{b}, b \in B\}$ for every subset B of *Act*. The allowed operators Σ^3 are

arity 0 : NIL, Ω
arity 1 : prefixing, $a\_$ for every $a \in Act$
: restriction, $\backslash c$ for every $c \in Chan$
arity 2 : external nondeterminism, $+$
internal nondeterminism, $\oplus$
parallelism, $|$

Let **EPL** denote REC_{Σ^3}, the set of closed recursive terms over Σ^3. Essentially it is obtained from **RM** by adding the combinators $|$ and $\backslash c$ for every channel c. This represents a slight extension of the language in the introduction since we have added the very useful operator $\oplus$. We continue to use the notation developed in chapter 3 for recursive terms in general. To preserve compatibility with the introduction we use $\backslash c$ in a postfix manner, writing $p\backslash c$ instead of $\backslash c(p)$. Finally the order of precedence is:

$$a\_, \quad \backslash c, \quad |, \quad \oplus, \quad +.$$

The intuitive semantics discussed in the introduction can be formalized by embedding **EPL** in a labeled transition system. This amounts to extending the corresponding definition for **RM**, by giving clauses for the new operators. These are given in figure 5.1. The rules for restriction are unremarkable. They merely state that in the process $p\backslash c$ the channel c cannot be used for external actions. The first four rules for $|$ state that the subprocesses may act independently; in $p \mid q$ either p or q can perform internal or external moves without hindrance from its partner. The fifth rule says they may

$p \succ\!\!\longrightarrow q$ implies $p \backslash c \succ\!\!\longrightarrow q \backslash c$

$p \xrightarrow{a} q$ implies $p \backslash c \xrightarrow{a} q \backslash c$ if $a \neq c, \bar{c}$

$p \succ\!\!\longrightarrow r$ implies $p \mid q \succ\!\!\longrightarrow r \mid q$

$q \succ\!\!\longrightarrow r$ implies $p \mid q \succ\!\!\longrightarrow p \mid r$

$p \xrightarrow{a} r$ implies $p \mid q \xrightarrow{a} r \mid r$

$q \xrightarrow{a} r$ implies $p \mid q \xrightarrow{a} p \mid r$

$p \xrightarrow{a} p'$, $q \xrightarrow{\bar{a}} q'$ implies $p \mid q \succ\!\!\longrightarrow p' \mid q'$

Figure 5.1
New transition rules for $\backslash c$ and $|$.

cooperate if they wish, by communicating. This happens by the two components simultaneously performing complementary actions.

To see the effect of these rules, consider the example NAB from the introduction. Let

$X1$ denote $rec\, x.cadx$
$X2$ denote $rec\, x.\bar{c}\bar{d}x$
$X3$ denote $rec\, x.cbdx$
NAB denote $((X1 \mid X2) \mid X3) \backslash c \backslash d$.

We will see later that $|$ is associative with respect to our semantics so that NAB can be written simply as $(X1 \mid X2 \mid X3) \backslash c \backslash d$. Its first possible external moves are derived in figure 5.2 from those of its components. It is seen that although $X1$, $X2$ and $X3$ can make external moves the only initial moves that the overall process can make are internal. There are essentially two derivations although minor variations exist:

$NAB \succ\!\!\longrightarrow^* (adX1 \mid \bar{d}X2 \mid X3) \backslash c \backslash d$

$NAB \succ\!\!\longrightarrow^* (X1 \mid \bar{d}X2 \mid bdX3) \backslash c \backslash d.$

These represent the possibility of the subprocess $X2$ synchronizing along the internal channel c with either $X1$ or $X3$.

Further analysis of these two derivatives shows that each can only perform one possible external action, once more because of $\backslash c \backslash d$:

$$X1 \rightarrowtail cadX1 \xrightarrow{c} adX1$$

$$X2 \rightarrowtail \bar{c}\bar{d}X2 \xrightarrow{\bar{c}} \bar{d}X2$$

$$X3 \rightarrowtail cbdX3 \xrightarrow{c} bdX3$$

$$\begin{aligned}
X1 \mid X2 &\rightarrowtail \xrightarrow{c} adX1 \mid X2 \\
\text{or} &\rightarrowtail \xrightarrow{\bar{c}} X1 \mid \bar{d}X2 \\
\text{or} &\rightarrowtail \rightarrowtail \rightarrowtail adX1 \mid \bar{d}X2
\end{aligned}$$

$$\begin{aligned}
(X1 \mid X2) \mid X3 &\rightarrowtail \xrightarrow{c} (adX1 \mid X2) \mid X3 \\
\text{or} &\rightarrowtail \xrightarrow{\bar{c}} (X1 \mid \bar{d}X2) \mid X3 \\
\text{or} &\rightarrowtail \xrightarrow{c} (X1 \mid X2) \mid bdX3 \\
\text{or} &\rightarrowtail \rightarrowtail \rightarrowtail (adX1 \mid \bar{d}X2) \mid X3 \\
\text{or} &\rightarrowtail \rightarrowtail \rightarrowtail (X1 \mid \bar{d}X2) \mid bdX3
\end{aligned}$$

$$\begin{aligned}
((X1 \mid X2) \mid X3)\backslash c\backslash d &\rightarrowtail \rightarrowtail \rightarrowtail ((adX1 \mid \bar{d}X2) \mid X3)\backslash c\backslash d \\
\text{or} &\rightarrowtail \rightarrowtail \rightarrowtail ((X1 \mid \bar{d}X2) \mid bdX3)\backslash c\backslash d
\end{aligned}$$

Figure 5.2
First moves of *NAB*.

$$(adX1 \mid \bar{d}X2 \mid X3)\backslash c\backslash d \xrightarrow{a} (dX1 \mid \bar{d}X2 \mid X3)\backslash c\backslash d$$

$$(X1 \mid \bar{d}X2 \mid bdX3)\backslash c\backslash d \xrightarrow{b} (X1 \mid \bar{d}X2 \mid dX3)\backslash c\backslash d.$$

In these states the only possible move is another internal synchronization along the channel d; in the first case between the first and second components and in the second case between the second and third components:

$$(dX1 \mid \bar{d}X2 \mid X3)\backslash c\backslash d \rightarrowtail NAB$$

$$(X1 \mid \bar{d}X2 \mid dX3)\backslash c\backslash d \rightarrowtail NAB.$$

In short, NAB acts very much like the term $rec\, x.ax \oplus bx$. It exhibits a high degree of internal nondeterminism. This accords with the intuitive semantics of the introduction. The reader should examine other examples in order to be convinced that these formal rules do indeed capture the intuitive explanations given in the introduction. One interesting example to consider is $(rec\, x.cax \mid rec\, x.\bar{c}x \mid rec\, x.cbx)\backslash c$.

The addition of these new operators adds considerable power to the language. For example, one can show that any term p in **RM** can only lead to a finite number of derivations. Specifically $\{q, p \overset{s}{\Longrightarrow}{}^* q \text{ for some } s\}$ is finite. This is not so for **EPL**.

EXAMPLE 5.1.1 Let *up.*, *down.* be two distinct actions and C denote the term $rec\, x.up.(x \mid down.NIL)$.

Then C can perform

$$C \overset{up.}{\Longrightarrow} C \mid down.NIL \overset{up.}{\Longrightarrow} C \mid down.NIL \mid down.NIL \overset{up.}{\Longrightarrow} \cdots.$$

In short, C acts like a counter and has an infinite number of derivatives. If we use C_n, $n > 0$, to denote $C \mid down.NIL \mid \ldots \mid down.NIL$, where $\mid down.NIL$ is repeated n times, then if we identify p with $p \mid NIL$ these terms satisfy

$$C \overset{up.}{\Longrightarrow} C_1$$

$$C_n \overset{up.}{\Longrightarrow} C_{n+1}, \quad n > 0,$$

$$C_n \overset{down.}{\Longrightarrow} C_{n-1}, \quad n > 0.$$

Having defined **EPL** as a labeled transition system we immediately obtain Testing preorders. As usual we use the Experimental System obtained from the two labeled transition systems:

processes : $\langle \mathbf{EPL}, Act, \longrightarrow, \rightarrowtail \rangle$

experimenters : $\langle \mathbf{EPL}, Act \cup \{1, w\}, \longrightarrow, \rightarrowtail \rangle$.

We continue to use $\sqsubseteq$, $\sqsubseteq_{\text{MUST}}$, and $\sqsubseteq_{\text{MAY}}$ to refer to these preorders. This is a little presumptuous as it may be that these relations are different for **RM**, considered as a subset of **EPL**, than those in chapter 4: the set of experimenters has been enlarged.

However, we refer the reader to the discussion at the end of §4.4, from which it follows that the Alternative Characterization theorem also holds in this extended setting. This in turn justifies the continued use of $\sqsubseteq$ etc.

THEOREM 5.1.2 (Alternative Characterization) For $p, q \in \mathbf{EPL}$,

a) $p \sqsubseteq_{\text{MAY}} q$ if and only if $p \ll_{\text{MAY}} q$
b) $p \sqsubseteq_{\text{MUST}} q$ if and only if $p \ll_{\text{MUST}} q$
c) $p \sqsubseteq q$ if and only if $p \ll q$.

Because of the complexity of the new operator | this alternative characterization is not as easily applicable as before. For example, the following corollary is not completely trivial.

COROLLARY 5.1.3 The Testing preorders $\sqsubseteq_{\text{MAY}}$, $\sqsubseteq_{\text{MUST}}$, and $\sqsubseteq$ are preserved by all the operators of **EPL**.

Proof We need only consider the new operators | and $\backslash c$. Hiding is straightforward and we merely show that $<<_{\text{MUST}}$ is preserved by |. Suppose $p <<_{\text{MUST}} q$. We show $p \mid r <<_{\text{MUST}} q \mid r$.

a) Suppose $p \mid r \downarrow s$. We must prove $q \mid r \downarrow s$. We do so by contradiction: we assume $q \mid r \uparrow s'$ for some prefix s' of s and derive a contradiction. If $q \mid r \uparrow s'$ then there is an infinite derivation of the form

$$q \mid r \stackrel{s'}{\Longrightarrow} q_0 \mid r_0 \rightarrowtail q_1 \mid r_1 \rightarrowtail q_2 \mid r_2 \rightarrowtail \cdots, \qquad (*)$$

where s' is a prefix of s. We first show that $q_n \downarrow$ and $r_n \downarrow$ for every $n \geq 0$. Consider the derivation

$$q \mid r \stackrel{s'}{\Longrightarrow} q_n \mid r_n. \qquad (\dagger)$$

This can be "unzipped" to give two derivations

$$q \stackrel{s_1}{\Longrightarrow} q_n \qquad (*1)$$

$$r \stackrel{s_2}{\Longrightarrow} r_n. \qquad (*\text{r})$$

We will not define the unzipping of derivations precisely although in §5.3 we will need to define the exact relationship between the strings s', s_1, and s_2; it suffices to say that (∗1) and (∗r) contain all the history of the contributions of q, r, respectively, to (†). In general both s_1 and s_2 will be longer than s' because internal moves in (†), which are communications between q and r, will be unzipped to give external actions in (∗1) and (∗r).

Now $p \downarrow s_1$ for otherwise a computation

$$p \stackrel{s_1'}{\Longrightarrow} p',$$

where $p' \uparrow$, and s_1' is a prefix of s_1, could be zipped to (∗r) to obtain

$$p \mid r \stackrel{s''}{\Longrightarrow} p' \mid r',$$

where $p' \mid r' \uparrow$. This contradicts $p \mid r \downarrow s$. It follows from this that $q_n \downarrow$.

Also $r_n\downarrow$; otherwise we could proceed as above to get the same contradiction. However in this case the p' is obtained from the fact that $p <<_{\text{MUST}} q$ and $p\downarrow s_1$.

We have shown that in (∗) every q_n and r_n is convergent. Therefore (∗) can only occur by q and r interacting infinitely often. In this case (∗) itself can be unzipped to obtain two infinite computations, which we denote by

$$q \stackrel{u}{\Longrightarrow}$$

$$r \stackrel{\bar{u}}{\Longrightarrow}. \qquad (**)$$

Here u is an infinite sequence from Act. Because $p <<_{\text{MUST}} q$ and $p\downarrow u$ we know from lemma 4.4.12 that $p \stackrel{u}{\Longrightarrow}$ also. This infinite derivation can be zipped with (∗∗) to obtain an infinite derivation from $p \mid r$. This contradicts the fact that $p \mid r\downarrow s$.

b) Let $A \in \mathscr{A}(q \mid r, s)$ where $p \mid r\downarrow s$. We must find some $B \in \mathscr{A}(p \mid r, s)$ such that $B \subseteq A$. We know

$$q \mid r \stackrel{s}{\Longrightarrow} q' \mid r'$$

where $A = S(q' \mid r')$. Now let $q'' \mid r''$ be such that

$$q' \mid r' \stackrel{\varepsilon}{\Longrightarrow} q'' \mid r'' \not\rightarrowtail.$$

Such terms are called *stable* in Q6 at the end of chapter 2 and we know at least one exists because $q \mid r\downarrow s$. Moreover $A \supseteq S(q'' \mid r'') = S(q'') \cup S(r')$. We can unzip the combined derivation

$$q \mid r \stackrel{s}{\Longrightarrow} q'' \mid r''$$

to obtain

$$q \stackrel{s_1}{\Longrightarrow} q''$$

$$r \stackrel{s_2}{\Longrightarrow} r''.$$

Using the fact that $p <<_{\text{MUST}} q$ we can obtain a derivation

$$p \stackrel{s_1}{\Longrightarrow} p'$$

where $S(p') \subseteq S(q'')$. Once more extend this to a stable state p'' to obtain the derivation

$$p \stackrel{s_1}{\Longrightarrow} p' \rightarrowtail^* p'' \not\rightarrowtail.$$

This derivation can now be zipped to that from r to obtain

$$p \mid r \stackrel{s}{\Longrightarrow} p'' \mid r''.$$

Now $S(p'') \subseteq S(p') \subseteq S(q'')$ and therefore $p'' \mid r'' \not\rightarrowtail$. The required B is $S(p'' \mid r'')$ since this is $S(p'') \cup S(r'')$. □

We end this section with some examples. They are mostly examples of equivalences between terms rather than inequivalences.

EXAMPLE 5.1.4 (Infinite Chatter) Consider the term IC defined in the introduction. Here it is represented by $((rec\, x.bx) \mid (rec\, x.\bar{b}x))\backslash b$. The only possible derivation from it is

$$IC \rightarrowtail IC \rightarrowtail IC \rightarrowtail \cdots.$$

Consequently IC *must* e only if $e \xrightarrow{w}$, so that $IC \sqsubseteq p$ for every p. In fact this is sufficient to show that $IC \approx \Omega$; infinite chatter corresponds to the totally undefined process. It is also distinguished from NIL because NIL *must* $1w$ whereas we know Ω *mu$\not{s}$t* $1w$.

EXAMPLE 5.1.5 Consider ABB, again from the introduction, defined as $((rec\, x.ax + bx) \mid (rec\, x.\bar{b}x))\backslash b$. ABB also has an infinite internal derivation and therefore ABB *mu$\not{s}$t aw*. On the other hand, $rec\, x.ax$ *must aw*, so that $rec\, x.ax \not\sqsubseteq_{\text{MUST}} ABB$. The converse is trivially true (using $<<_{\text{MUST}}$) because $ABB\downarrow s$ for no s in Act^*. Note also that $ABB \approx_{\text{MAY}} rec\, x.ax$; once more with *may* testing possible divergences are ignored. Using the Alternative Characterization it is not very difficult to see that $ABB \approx rec\, x.ax + \Omega$ and $ABB \approx_{\text{MUST}} \Omega$.

EXAMPLE 5.1.6 We have already examined to some extent NAB, $(rec\, x.cadx \mid rec\, x.\bar{c}\bar{d}x \mid rec\, x.cbdx)\backslash c\backslash d$. By inspection we can see that $NAB\downarrow s$ for every s and $L(NAB)$ is $\{a, b\}^*$. Also for every s, $\mathscr{A}(NAB, s)$ is the same set $\{\{a\}, \{b\}, \{a, b\}\}$. These are exactly the characteristics of $rec\, x.ax \oplus bx$. Therefore $NAB \approx rec\, x.ax \oplus bx$.

EXAMPLE 5.1.7 Consider the process $p = rec\, x.(ax + bx) \mid rec\, x.cdx$ where c is not the complement of a or b. Let q denote the process $ap + bp + cp'$ where p' denotes $rec\, x.(ax + bx) \mid d\, rec\, x.cdx$. Note that a, b and c are all the possible moves of p and the corresponding derivatives are p, p, p' respectively. This explains the composition of q; it is obtained by unfolding the possible moves of p. Once more we can argue that p and q are identified by the alternative equivalences. They have the same languages and Acceptance sets. So $p \approx q$.

EXAMPLE 5.1.8 As a modification of the last example consider $p = rec\, x.(ax + bx) \mid rec\, x.\bar{b}dx$. Here we can also unwind the possible moves of p to obtain $q = ap + bp + \bar{b}p'$, where p' is $rec\, x.(ax + bx) \mid d\, rec\, x.\bar{b}dx$. However q does not tabulate the possible internal move or communication along channel b which also results in p'. Let r denote $(q + p') \oplus p'$. Now we can argue that $p \approx r$. Both converge on all sequences and generate the same language. Moreover, if the length of s is at least one then $\mathscr{A}(p, s)$ and $\mathscr{A}(r, s)$ obviously coincide. Finally one can calculate both $\mathscr{A}(p, \varepsilon)$ and $\mathscr{A}(r, \varepsilon)$ to be $\{\{a, b, \bar{b}, d\}, \{a, b, d\}\}$. Note that p is not equivalent to $q \oplus p'$ or $q + p'$ because in each case the Acceptance sets after ε are not correctly related.

The two foregoing examples show an essential feature of our semantic theory of processes; every parallel process can be "unwound" to an equivalent nondeterministic but purely sequential process. Each of the examples shows one step of this "unwinding", which may recursively be applied to the derivatives. If the original terms are finite this recursive application will eventually terminate in terms which contain no occurrences of the parallel combinator $|$. Otherwise we can only determine the corresponding purely nondeterministic process in the limit. The exact method of unwinding will be examined in more detail when we consider the equational theory of **EPL**.

We consider one more example of this unwinding, in the presence of restriction.

EXAMPLE 5.1.9 Now consider $p = (rec\,x.(ax + bx) \,|\, rec\,x.\bar{b}dx)\backslash b$, the same process as in the last example except that the channel b is internalized. Once more we can unwind this process but b or $\bar{b}$ actions are not allowed. In this case one can show that $p \approx (q + p') \oplus p'$ where q denotes ap and p' denotes the result of the internal communication, $rec\,x.(ax + bx) \,|\, d\,rec\,x.\bar{b}dx)\backslash b$. As in the previous example q represents the possible external actions of p and p' the possible internal actions.

EXAMPLE 5.1.10 Let p, q, r be any triple of processes. Then the two terms $(p \,|\, q) \,|\, r$ and $p \,|\, (q \,|\, r)$ generate exactly the same language; any derivation from $(p \,|\, q) \,|\, r$ can be unzipped to give three individual derivations from p, q, r respectively. These can be recomposed in a different order to give a derivation from $p \,|\, (q \,|\, r)$. A similar argument will show that $(p \,|\, q) \,|\, r \downarrow s$ if and only if $p \,|\, (q \,|\, r) \downarrow s$. Finally, suppose $(p \,|\, q) \,|\, r \overset{s}{\Longrightarrow} (p' \,|\, q') \,|\, r'$. The unzipping/zipping argument once more shows that $p \,|\, (q \,|\, r) \overset{s}{\Longrightarrow} (p' \,|\, q') \,|\, r'$. This shows that for every s, $\mathscr{A}((p \,|\, q) \,|\, r, s) = \mathscr{A}(p \,|\, (q \,|\, r), s)$. It follows that $(p \,|\, q) \,|\, r \approx p \,|\, (q \,|\, r)$.

Similar arguments show that

$$p \,|\, q \approx q \,|\, p$$

and $p \,|\, NIL \approx p$,

which are natural properties one would expect of a parallel combinator.

5.2 Proof Systems

In this section we extend the results of part II to the language **EPL**; we give sound and complete proof systems for the Testing preorders. In the next section we show that **AT**, $\mathbf{AT_S}$ and $\mathbf{AT_W}$ may also act as fully abstract models for this extended language. The general method is to reduce the communicating processes in **EPL** to equivalent nondeterministic processes in **RM**. We have already had some indication in examples 5.1.7–9 of how this can be done. Intuitively each process in **EPL** can be systematically unwound by enumerating at each stage all its possible next actions and their respective derivatives. This set of next actions will always be finite, provided the process does not diverge internally (see lemma 4.4.11) and so the result of each stage can be described syntactically in **RM**. However care must be taken to ensure that the acceptance sets are preserved as the process is unwound.

Before defining exactly how this winding may be performed, let us examine why it leads more or less automatically to a complete proof system. Although there is a general explanation, we will remain specific, giving definitions relative to Σ^2 and **RM**. The only generalization which we consider is to examine arbitrary languages which extend **RM**; the interesting example of such a language is, of course, **EPL**.

Let Σ be an arbitrary signature containing Σ^2. Then the language REC_Σ contains **RM**. We assume that REC_Σ has been given an operational semantics, as a labeled transition system, by extending that of **RM**. For each new combinator rules are introduced for $\xrightarrow{a}$ and $\succ\!\!\longrightarrow$ in some reasonable manner. In particular the rules for **RM** (figure 4.3) remain untouched. In this way we obtain Testing preorders $\lesssim_{\text{MUST}}$, $\lesssim_{\text{MAY}}$, and $\lesssim$ for REC_Σ. The Alternative Characterization will hold for the extended language provided the new labeled transition system is finite branching. This in turn implies that the Testing preorders are determined by the set of finite experiments F; this fact plays an important role in the proof of the main result below. We concentrate on $\lesssim_{\text{MUST}}$ but the others can be treated analogously. In this extended setting most of the proof rules of the system **ωDED** (figure 4.2) automatically remain sound with respect to $\lesssim_{\text{MUST}}$. These are rules 1., 2., 4., 6., and 7..

We now show how to augment **SA2s** to a new set of equations E in such a way that **ωDED**(E) is both sound and complete.

Syntactically the unwinding of processes to their equivalent purely nondeterministic versions corresponds to eliminating all occurrences of function symbols which are not in Σ^2. This procedure of elimination proceeds by applying a set of equations E in some systematic manner. One property which reflects the ability to eliminate the new function symbols is

A) for every $d \in FREC_{\Sigma}$ there exists a $d' \in FREC_{\Sigma^2}$ such that $d =_E d'$.

For infinite terms we will not demand that all new function symbols are eliminated; only that progress should be always possible in the unwinding of a term into a term from **RM**. Specifically we require that the equations E be powerful enough to eliminate the top-level occurrences of the new function symbols. By repeated application these function symbols are gradually pushed further down into the terms. A term with no top-level occurrence of a new function symbol will be called a *Σ^2-head term* or simply a *head term*; it looks like a term from T_{Σ^2} at the topmost level. Formally

i) NIL is a head-term
ii) ap is a head-term for any p in REC_{Σ}
iii) if $f \in \Sigma^2$ and p are head terms then $f(p)$ is also a head term.

In the third clause of this definition the only possible function symbols are $+$ and $\oplus$.

Let E be a set of equations over Σ, which we assume contains **SA2s**. We say E is *reductive* if it satisfies A) above and

B) for every $p \in REC_{\Sigma}$, $p\downarrow$ implies $p =_{Er} q$ for some head term q.

Unfortunately A) does not follow from B) because we require A) to be true for divergent terms. On the other hand, it is too strong to demand B) of divergent terms also.

The usefulness of a reductive set of equations stems primarily from the fact that they can be used to transform every convergent term in REC_{Σ} into what are essentially a head-normal form. All the results in §4.5 for the language **RM** flow naturally from the existence of head-normal forms. These results can now be transferred virtually for free to the extended language REC_{Σ}. Combining these we will obtain a proof of the main result:

THEOREM 5.2.1 (Reduction Theorem) If E is reductive and **DED**(E) is sound with respect to $\sqsubseteq_{\text{MUST}}$ over REC_{Σ} then **ωDED**(E) is both sound and complete with respect to $\sqsubseteq_{\text{MUST}}$ over REC_{Σ}.

We first show how to prove this theorem and then apply it to the particular language **EPL**.

The proof proceeds along the lines of that in §4.5 and so many of the details will be omitted here. We know that the relation $\leq_{E\omega}$ is algebraic over REC_{Σ} (see Q20 at the end of chapter 4). We prove

1) $\sqsubseteq_{\text{MUST}}$ is algebraic over REC_{Σ}
2) **DED**(E) is sound and complete with respect to $\sqsubseteq_{\text{MUST}}$ over $FREC_{\Sigma}$.

The theorem will then follow because $\leq_{E\omega}$ coincides with $\leq_E$ on $FREC_\Sigma$ and algebraic relations are determined by how they behave on finite elements, as explained in §4.5. The proof of 1) and 2) both depend on E satisfying the conditions of the Reduction theorem, and it is convenient to call such equations *normal*; E is *normal* if it is reductive and **DED**(E) is sound with respect to $\underset{\sim}{\sqsubset}_{\mathbf{MUST}}$ over REC_Σ.

Lemma 5.2.2 If E is normal then **DED**(E) is sound and complete with respect to $\underset{\sim}{\sqsubset}_{\mathbf{MUST}}$ over $FREC_\Sigma$.

Proof Soundness is implied by the definition of normal. We show completeness. Suppose $d \underset{\sim}{\sqsubset}_{\mathbf{MUST}} e$, with $d, e \in FREC_\Sigma$. Because E is reductive there exist d', e' in $FREC_{\Sigma^2}$ such that $d =_E d'$ and $e =_E e'$. Because E is normal $d' \underset{\sim}{\sqsubset}_{\mathbf{MUST}} e'$ and we can apply the completeness theorem, proposition 4.5.5, to deduce $d' \leq_{\mathbf{SA2s}} e'$ and therefore $d \leq_E e$. □

To prove the second condition above a slight generalization of head-normal forms are needed. A *Σ-hnf* is any term of the form

i) NIL
or ii) $\sum \{p(A), A \in \mathscr{A}\}$ where $\mathscr{A}$ is saturated.

These terms look just like head-normal forms except that the derivations $p(a)$ may contain combinators not in the original signature Σ^2. In what follows they also play the same role as that played by head-normal forms played in §4.5.

Lemma 5.2.3 Every head term has a *Σ-hnf*, i.e. if p is a head term then there exists a *Σ-hnf* q such that $p =_{\mathbf{SA2s}} q$.

Proof We use structural induction on p. There are four different forms p may have. The first two, NIL and ar, are trivial since they are already in *Σ-hnf*. Otherwise it may have the form $p_1 + p_2$ or $p_1 \oplus p_2$. By induction we may assume that both p_1 and p_2 have already been transformed into *Σ-hnf*. We may now proceed as in the Head-Normal Form theorem 4.5.3, which in turn uses the syntactic manipulations from the Normal Form theorem 2.4.6. □

As an immediate corollary we have the following proposition.

Proposition 5.2.4 If E is reductive then every convergent term has a *Σ-hnf*, i.e. for every $p \in REC_\Sigma$,

$p\downarrow$ implies $p =_{Er} q$ for some *Σ-hnf* q.

Lemma 4.5.6 can now be repeated with Σ-head-normal forms used in place of head-normal forms to derive:

if e is a finite experiment and $p \in REC_\Sigma$ then p *must* e implies d *must* e for some d in App(p).

Also the proof of lemma 4.5.7 does not depend on the fact that the process language is **RM**; it only uses the result that the set of experiments F determines $\lesssim_{\mathrm{MUST}}$, which is also true for the extended language, once more provided the labeled transition system is finite-branching. Therefore it can be adapted to show

if $d \in FREC_\Sigma$ and $p \in REC_\Sigma$ then $d \lesssim_{\mathrm{MUST}} p$ implies $d \lesssim_{\mathrm{MUST}} d'$ for some $d' \in$ App(p).

These two results are combined as in proposition 4.5.8 to obtain:

PROPOSITION 5.2.5 The relation $\lesssim_{\mathrm{MUST}}$ is algebraic over REC_Σ.

We have now derived both conditions 1) and 2) above and the Reduction theorem follows.

Before proceeding to the main application of this result let us divert our attention to an interesting corollary. It states that if E is normal then **ωDED**(E) is *conservative* over **ωDED**(**SA2s**).

COROLLARY 5.2.6 For every p, q in **RM**, $p \leq_{E\omega} q$ implies $p \leq_{\mathbf{SA2s\omega}} q$, whenever E is normal.

Proof If $p \leq_{E\omega} q$, then by the Reduction theorem $p \lesssim_{\mathrm{MUST}} q$. Because **ωDED(SA2s)** is complete with respect to $\lesssim_{\mathrm{MUST}}$ over **RM** it follows that $p \leq_{\mathbf{SA2s\omega}} q$.

Let us now turn our attention to the application of this theorem to **EPL**. The new equations for the extra operators are given in figure 5.3. Most are self-explanatory. For example $\backslash c$ commutes with both nondeterministic operators $+$ and $\oplus$, and it can be pushed inside a guard provided the name of the guard is different from c. The parallel operator also commutes with internal nondeterminism but its relation with $+$ is more complicated. This equation $+\,|$, often referred to as an *interleaving* law, allows us to unwind parallel processes into purely nondeterministic ones. It has already been motivated in the examples of the previous section. To express it succinctly we use a predicate *comm*: *comm*(p, q) is true of two processes if there is an immediate possibility of communication between them. Due to the form of x and y in $+\,|$, *comm*(x, y) is true whenever there exists at least one pair of indices $i \in I$ and $j \in J$ such that $a_i = \overline{b}_j$. The law states that if there is no possibility of communication then $x \,|\, y$ is equal to the

$$(x + y)\backslash c = x\backslash c + y\backslash c \qquad +\backslash c$$

$$(x \oplus y)\backslash c = x\backslash c \oplus y\backslash c \qquad \oplus\backslash c$$

$$ax\backslash c = \begin{cases} a(x\backslash c) & \text{if name } (a) \neq c \\ NIL & \text{otherwise} \end{cases} \qquad a\backslash c$$

$$NIL\backslash c = NIL \qquad NIL\backslash c$$

$$\Omega\backslash c = \Omega \qquad \Omega\backslash c$$

$$(x \oplus y) \mid z = x \mid z \oplus y \mid z \qquad \oplus\mid$$

$$z \mid (x \oplus y) = z \mid x \oplus z \mid y$$

$$x \mid (y + \Omega) = (x + \Omega) \mid y = (x \mid y) + \Omega \qquad \Omega\mid$$

If x, y represent $\sum\{a_i x_i, i \in I\}, \sum\{b_j y_j, j \in J\}$ then

$$x \mid y = \begin{cases} ext(x, y) & \text{if } comm(x, y) \text{ is false} \\ (ext(x, y) + int(x, y)) \oplus int(x, y) & \text{otherwise} \end{cases} \qquad +\mid$$

where $ext(x, y) = \sum\{a_i(x_i \mid y), i \in I\} + \sum\{b_j(x \mid y_j), j \in J\}$

and $int(x, y) = \bigoplus\{x_i \mid y_j, a_i = \overline{b_j}\}$

Figure 5.3
Equations for parallelism and restriction.

interleaving of the possible external actions of x and y represented by $ext(x, y)$. If communication is possible, then some consideration must be taken of the result of these moves. This is the role of $int(x, y)$ which is the nondeterministic sum of all the states to which $x \mid y$ can evolve immediately by internal communication between x and y. The precise form of the term in this case has been motivated in example 5.1.8. It is worth pointing out that this law holds force even when one or both of x and y are empty sums. In these cases it reduces to

$$NIL \mid NIL = NIL, \quad x \mid NIL = x, \quad \text{and} \quad NIL \mid y = y.$$

There is also a much stronger version of $\Omega\mid$ which is valid for $\eqsim_{\mathrm{MUST}}$: namely, $x \mid \Omega = \Omega \mid x = \Omega$. However, this is not true of $\eqsim$ or $\eqsim_{\mathrm{MAY}}$ and for this reason we use the weaker equation. The stronger version can easily be derived from it using the strong equation S.

Our aim is to apply the Reduction theorem 5.2.1 to show that the addition of these new axioms results in a sound and complete proof system for **EPL**. Let **SA3s** denote the set of axioms obtained by this addition to **SA2s**. Note that $+|$ is not strictly speaking an axiom but is an axiom schema, or an infinite set of axioms, one for each choice of the finite sets $\{a_i, i \in I\}$, $\{b_j, j \in J\}$. To apply the theorem we must show that **SA3s** is normal, i.e. it is reductive and that **DED(SA3s)** is sound with respect to $\lesssim_{\mathbf{MUST}}$ over REC_Σ. To show the latter it is sufficient to show

—$\lesssim_{\mathbf{MUST}}$ is preserved by the new operators

—the extra equations in **SA3s** are sound with respect to $\lesssim_{\mathbf{MUST}}$.

The definition of $\lesssim_{\mathbf{MUST}}$ is such that it is automatically a preorder and therefore these two conditions ensure that **DED(SA3s)** is sound.

LEMMA 5.2.7 **DED(SA3s)** is sound with respect to $\lesssim_{\mathbf{MUST}}$ over REC_Σ.

Proof We have already shown that $\lesssim_{\mathbf{MUST}}$ is preserved by the operators in **EPL** (corollary 5.1.3). So it is sufficient to prove that the new axioms are sound. As usual $<<_{\mathbf{MUST}}$ is used in place of $\lesssim_{\mathbf{MUST}}$. The only nontrivial axiom is $+|$ and here the only difficulty is to show that $\mathscr{A}(p, \varepsilon)$ and $\mathscr{A}(q, \varepsilon)$ are related in the appropriate manner whenever $p = q$ is an instance of this axiom; this is simply a matter of calculation. □

We now show that **SA3s** is reductive. We first show that every finite term in the extended language can be reduced to terms in the basic language $\mathbf{M_2}$. This essentially amounts to showing that when d, d' are terms in $\mathbf{M_2}$ the extended terms $d\backslash c$, $d \,|\, d'$ can be reduced to terms in $\mathbf{M_2}$ using the equations **SA3s**.

LEMMA 5.2.8 For every $d \in FREC_{\Sigma^3}$ there exist an $e \in FREC_{\Sigma^2}$ such that $d =_{\mathbf{SA3s}} e$.

Proof We look at each of the possible syntactic forms of d.

a) d is $d_1\backslash c$.

By induction we may assume that d_1 has already been transformed into a term in $FREC_{\Sigma^2}$ and by corollary 4.5.4 we may further assume that it is in Ω-normal form. Here we apply a further induction. We use induction on the size of d_1 to show $d_1\backslash c$ can be transformed into a term $FREC_{\Sigma^2}$. If it is NIL or Ω the axioms $NIL\backslash c$ or $\Omega\backslash c$ may be applied to $d_1\backslash c$ to obtain the required e. Otherwise d has the form $\bigoplus\{d(A), A \in \mathscr{A}\}$, where as usual $d(A)$ stands for $\sum\{ad(a), a \in A\}$. Applying $\oplus\backslash c$, $d\backslash c$ may be transformed into $\oplus\{d(A)\backslash c, A \in \mathscr{A}\}$. Each individual sum $d(A)\backslash c$ may be in turn transformed into $\sum\{ad(a)\backslash c, a \in A\}$, using $+\backslash c$. Each individual term $(ad(a))\backslash c$ may be further treated by the axioms $a\backslash c$ so that each $d(A)\backslash c$ is transformed in turn into a term of the form $\sum\{a(d(a)\backslash c), a \in A'\}$. By applying the inner inductive hypothesis we may assume each $d(a)\backslash c$ can be transformed into a term in $FREC_{\Sigma^2}$, say $e(a)$. The required e is then

$$\oplus\{\sum\{ae(a), a \in A'\}, A \in \mathscr{A}\}.$$

b) d is $d_1 \mid d_2$.

The structure of the proof in this case is similar to that used in a). We assume both d_1 and d_2 are in Ω-normal form and use induction on their combined size to prove that d can be reduced to the required form. If either d_1 or d_2 are Ω the axiom $\Omega\mid$ may be applied. Otherwise we can assume that they are of the form $\oplus\{d_1(A), A \in \mathscr{A}\}$, $\oplus\{d_2(B), B \in \mathscr{B}\}$ respectively. By applying $\oplus\mid$ (possibly zero times) d can be transformed to

$$\oplus\{d_1(A) \mid d_2(B), A \in \mathscr{A}, B \in \mathscr{B}\} \qquad (*)$$

It is now sufficient to show how to reduce each $d_1(A) \mid d_2(B)$ to a term not involving $\mid$, and of course $+\mid$ is employed. Using it in a left-to-right fashion we reduce each such $d_1(A) \mid d_2(B)$ to a term where all the occurrences of $\mid$ are of the form $d_1' \mid d_2'$ where the combined size of d_1', d_2' is less than that of $d_1(A)$, $d_2(A)$ respectively. Using the inner inductive hypothesis these can all be reduced to a term in $FREC_{\Sigma^2}$. Substituting all of these results into $(*)$ we obtain the required e.

c) Otherwise d has the form $f(\underline{d})$ where $f \in \Sigma^2$. By induction $\underline{d}$ may be transformed into terms in $FREC_{\Sigma^2}$ and therefore the same holds for d. □

The proof of the same result for arbitrary convergent terms is very similar except that we use the form of induction associated with convergent terms, developed in lemma 4.4.10.

LEMMA 5.2.9 For every convergent term p in **EPL** there exists a Σ-head term q such that $p =_{\mathbf{SA3r}} q$.

Proof We abbreviate $=_{\mathbf{SA3r}}$ to $\equiv$ for convenience. We show that the property of being transformable into a head-term is i-inductive: let us assume that $p \rightarrowtail p'$ implies $p' \equiv h'$ for some head term h. From this assumption we deduce $p \equiv h$ for some head term h.

To make this deduction we need a further level of induction, this time structural induction on p. Here the inductive hypothesis is that this statement is also true for all terms structurally less than p.

i) p is NIL or aq. Here p is already a head term.

ii) p is $q + r$. By structural induction we know $q \equiv h_1, r \equiv h_2$ for head terms h_1, h_2 and so $p \equiv h_1 + h_2$, which is a head term.

iii) p *is* $q \oplus r$. This case is similar to ii).

iv) p is $q \backslash c$. Again, $p\downarrow$ implies $q\downarrow$ and by structural induction we can assume $q \equiv h$ for a head term h. Applying lemma 5.2.3 h may be assumed to be Σ-*hnf* and so we can apply the equations $\oplus\backslash c$, $+\backslash c$, and $a\backslash c$, in this order, to transform $h\backslash c$ into a head form. The details here are similar to case a) of the previous lemma.

v) p is $rec\, x.t$. In this case $p \succ\!\!\longrightarrow t[rec\, x.t/x]$ and by the outer induction we may assume $t[rec\, x.t/x] \equiv h$ for some head-term h. Then using the equation REC, p can also be transformed into h.

vi) p is $q \mid r$. Here $q\downarrow$ and $r\downarrow$ and therefore by structural induction we may assume $q \equiv h_1, r \equiv h_2$ for some Σ-*hnf*, h_1 and h_2, again invoking lemma 5.2.3.

We can assume that h_1, h_2 are $\sum\{h_1(A), A \in \mathscr{A}\}, \sum\{h_2(B), B \in \mathscr{B}\}$, respectively, and we may proceed exactly as in the case b) of the previous lemma. □

Combining these three lemmas and applying the Reduction theorem we obtain:

THEOREM 5.2.10 **ωDED(SA3s)** is both sound and complete for $\sqsubseteq_{\text{MUST}}$ over **EPL**.

This is quite a powerful result for the language **EPL** but its limitations should also be kept in mind. For example, the completeness only applies to closed terms and even the most elementary statements about arbitrary processes, such as

$$(x \mid y) \mid z = x \mid (y \mid z)$$

are not derivable. So any practical proof system based on **ωDED(SA3s)** would have to augment the set of basic axioms considerably with useful but nonessential laws. A minimal set can be found in the appendix of Milner 1980 and some are examined in the exercises at the end of this chapter.

5.3 Fully Abstract Models of EPL

Here we show that the model $\mathbf{AT_S}$ can be adopted to serve as a fully abstract denotational model for the extended language **EPL** with respect to $\sqsubseteq_{\text{MUST}}$. The method uses the fact that **SA3s** is reductive with respect to **SA2s** which, in turn, determines the model $\mathbf{AT_S}$.

As in the previous section we first develop some general results, this time about reductive sets of equations and fully abstract models, and then apply them to our particular language **EPL**.

As an immediate corollary to the Reduction theorem and the ω-Equational-Logic theorem 4.3.5 we have:

PROPOSITION 5.3.1 If E is a normal set of equations then CI_E is fully abstract with respect to $\sqsubseteq_{\mathbf{MUST}}$ over REC_Σ.

We will show that $\mathbf{AT_S}$ can be extended automatically to a representation of this fully abstract model CI_E. Let $\langle \mathbf{AT_S}, \leq_{\mathbf{AT_S}}, \Sigma_{Arb}\rangle$ be a domain, i.e. the carrier is $\langle \mathbf{AT_S}, \leq_{\mathbf{AT_S}}\rangle$ and for each $f \in \Sigma$ there is a continuous function f_{Arb} of the appropriate arity over $\mathbf{AT_S}$. Such a domain is called an *extension* (of $\mathbf{AT_S}$) if for every $f \in \Sigma^2$, f_{Arb} coincides with $f_{\mathbf{AT_S}}$, the functions defined in §3.4. Thus extensions are defined by simply associating new continuous functions over $\mathbf{AT_S}$ with each new function symbol added to the signature. The name is appropriate because if A is an extension of $\mathbf{AT_S}$ and $t \in REC_{\Sigma^2}(X)$ then $A[\![t]\!] = \mathbf{AT_S}[\![t]\!]$.

Suppose E is a normal set of equations. We show how to associate a continuous function f_{Arb} over $\mathbf{AT_S}$ for each new symbol f, in a completely automatic fashion. To define f_{Arb} it is sufficient to define it on the finite elements of $\mathbf{AT_S}$ and prove that it is monotonic. The unique extension guaranteed by theorem 3.3.10 gives a continuous function over the entire domain $\mathbf{AT_S}$. $\mathbf{AT_S}$ is isomorphic to CI_{SA2s} and it is now more convenient to use the latter representation. The finite elements are all of the form $[d]$ where d is in T_{Σ^2} and $[d] = \{e \in T_{\Sigma^2}, d =_{\mathbf{SA2s}} e\}$. This should be denoted by $[d]_{\mathbf{SA2}}$ but, for convenience, we omit the subscript. Now define

$f_{Arb}([\underline{d}]) = [e]$,

where $f(\underline{d}) =_E e$ and $e \in T_{\Sigma^2}$. Because E is reductive some such e is guaranteed to exist.

i) f_{Arb} is well-defined. For suppose $f(\underline{d}) =_E e$ and $f(\underline{d}) =_E e'$. Then $e =_E e'$ and by corollary 5.2.6, $e =_{\mathbf{SA2s}} e'$, i.e. $[e] = [e']$.

ii) f_{Arb} is monotonic. The same argument as in i) applies here.

This construction gives a Σ-domain which we call $\mathbf{eAT_S}$ to emphasize the fact that it is an extension of $\mathbf{AT_S}$. This domain satisfies

for $d \in FREC_\Sigma$, $e \in FREC_{\Sigma^2}$, $\mathbf{eAT_S}[\![d]\!] = [e]$ if and only if $d =_E e$.

This can be established by a simple inductive argument on the structure of d. It enables us to show that $\mathbf{eAT_S}$ is fully abstract over REC_Σ and, moreover, it is the unique such extension.

THEOREM 5.3.2 If E is normal then there is a *unique* extension of $\mathbf{AT_S}$, up to isomorphism, which is fully abstract with respect to $\sqsubseteq_{\mathrm{MUST}}$ over REC_Σ.

Proof We show that the extension $\mathbf{eAT_S}$ is in $\mathscr{CC}(E)$ and isomorphic to CI_E which will imply both that it is fully abstract and unique (up to isomorphism).

a) $\mathbf{eAT_S}$ is in $\mathscr{CC}(E)$. Let $d \leq e$ be an arbitrary equation in E. We must show $\mathbf{eAT_S}[\![d]\!]\rho \leq \mathbf{eAT_S}[\![e]\!]\rho$ for every $\mathbf{AT_S}$-assignment ρ. However, because of the continuity of the semantic function it is sufficient to consider the case when the assignment ρ is finite, i.e. assigns finite elements of $\mathbf{AT_S}$ to variables. But every finite element in $\mathbf{AT_S}$ is denotable by a finite term in $T_{\Sigma 2}$. Let σ be a substitution which uses these terms which denote the corresponding values of ρ, i.e. $\mathbf{AT_S}[\![\sigma(x)]\!] = \rho(x)$. Because of the Substitution lemma it is now sufficient to establish that $\mathbf{eAT_S}[\![d\sigma]\!] \leq \mathbf{eAT_S}[\![e\sigma]\!]$. We know from the soundness of $\mathbf{DED}(E)$, with respect to $\sqsubseteq_{\mathrm{MUST}}$, that $d\sigma \sqsubseteq_{\mathrm{MUST}} e\sigma$. Let d', $e' \in T_{\Sigma 2}$ be such that $d\sigma =_E d'$, $e\sigma =_E e'$ respectively. Then, again by soundness, $d' \sqsubseteq_{\mathrm{MUST}} e'$ and by the completeness of $\mathbf{DED(SA2s)}$ that $d' \leq_{\mathbf{SA2s}} e'$. In the model this translates to $[d'] \leq [e']$, i.e. $\mathbf{eAT_S}[\![d\sigma]\!] \leq \mathbf{eAT_S}[\![e\sigma]\!]$.

b) $\mathbf{eAT_S}$ is isomorphic to CI_E. From part a) we have a continuous Σ-homomorphism $h: CI_E \longrightarrow \mathbf{eAT_S}$. We need to define an inverse, $k: \mathbf{eAT_S} \longrightarrow CI_E$. As usual it is sufficient to restrict attention to finite elements when defining k. Let

$$k([d]_{\mathbf{SA2s}}) = [d]_E.$$

This is well-defined since E contains $\mathbf{SA2s}$ and for the same reason it is monotonic. Simple calculations show that it preserves all of the functions, i.e. it is a Σ-*po* homomorphism and therefore its extension to the full domain $\mathbf{eAT_S}$ is a continuous Σ-homomorphism. By definition h and k are inverses on finite elements and by continuity are therefore inverses also on the entire domains. □

This theorem gives an automatic method of extending the fully abstract model $\mathbf{AT_S}$ for **RM** to other languages. However, onc important property is lost in this extension: naturalness. Each of the combinators in **RM** is interpreted in a reasonably natural manner over $\mathbf{AT_S}$. With the new combinators this is no longer the case, at least if we follow the prescription in the definition of $\mathbf{eAT_S}$ above. Whether or not any natural definitions actually exist will in general depend on the operational behavior associated with the new symbols. We can only offer one helpful result which will sometimes ease the problem of searching for natural extensions. Essentially it says that any sound extension of $\mathbf{AT_S}$ is fully abstract. So in seeking natural models for the extended language it is sufficient to check that the definitions of the new functions generate a sound interpretation, i.e. satisfy the extended equations.

COROLLARY 5.3.3 If A is *any* extension of $\mathbf{AT_S}$ in $\mathscr{CC}(E)$, where E is a normal set of equations, then it is fully abstract with respect to $\sqsubseteq_{\mathrm{MUST}}$ over REC_{Σ}.

Proof Recall from §5.2 that $\sqsubseteq_{\mathrm{MUST}}$ is algebraic over REC_{Σ} because E is normal. Also it follows from Q20 of chapter 4 that $\leq_A$ is algebraic. Therefore, to prove the result it is sufficient to show that for d, e in $FREC_{\Sigma}$,

$$d \sqsubseteq_{\mathrm{MUST}} e \Longleftrightarrow A[\![d]\!] < A[\![e]\!].$$

Let $d', e' \in FREC_{\Sigma^2}$ be such that $d =_E d'$, $e =_E e'$. Then

$$\begin{aligned}
d \sqsubseteq_{\mathrm{MUST}} e &\Longleftrightarrow d' \sqsubseteq_{\mathrm{MUST}} e' && \text{because } E \text{ is reductive}\\
&\Longleftrightarrow \mathbf{AT_S}[\![d']\!] \leq \mathbf{AT_S}[\![e']\!] && \text{because } \mathbf{AT_S} \text{ is fully abstract}\\
&\Longleftrightarrow A[\![d']\!] \leq A[\![e']\!] && \text{because } A \text{ is an extension}\\
&\Longleftrightarrow A[\![d]\!] \leq A[\![e]\!] && \text{because } A \text{ satisfies the axioms } E.
\end{aligned}$$

□

These results have immediate application to our language **EPL**. Because **SA3s** is reductive, the construction of $\mathbf{eAT_S}$ gives a fully abstract model. However the search for a natural representation of this model is only partially successful. The problem lies with the operator | which is intrinsically complicated and any semantic counterpart to it on trees, such as those in $\mathbf{AT_S}$, is destined to be intricate. Nevertheless we give their definitions although we shy away from many of the related proofs which tend to involve long sequences of rather boring manipulations.

We start with the restriction operator $\backslash c$: for each tree t in $\mathbf{AT_S}$ we define a tree $t\backslash c$, which also should be in $\mathbf{AT_S}$. To define a tree t' in $\mathbf{AT_S}$ we must specify:

$L(t')$, a nonempty prefix-closed subset of Act^*

$CL(t')$, a prefix-closed subset of $L(t')$

for each $s \in CL(t')$ a finite saturated set $\mathscr{A}(t'(s))$.

Or course we should check that these are mutually consistent, i.e. that they determine an element of $\mathbf{AT_S}$.

To define the tree $t\backslash c$ we need some notation. For $s \in Act^*$ and $c \in Chan$ let $s\backslash c \in Act^*$ be defined by

$\varepsilon \backslash c = \varepsilon$

$as \backslash c = \varepsilon \qquad$ if $a = c$ or $\bar{a} = c$

$\qquad = a(s \backslash c) \quad$ otherwise.

In other words, $s \backslash c$ truncates s at the first occurrence of c or $\bar{c}$ if it exists. It is applied to sets in the natural way:

$L \backslash c = \{s \backslash c, s \in L\}$.

So $s \in L \backslash c$ if and only if it does not contain any occurrence of c or $\bar{c}$ and either s or scs' or $s\bar{c}s'$ is in L for some string s'. The prefix closure of $L \backslash c$ therefore follows from that of L.

If $A \subseteq Act$ let $A \backslash c = A - \{c, \bar{c}\}$, and for a collection of such sets $\mathscr{A}$, let $\mathscr{A} \backslash c = \{A \backslash c, A \in \mathscr{A}\}$. It is easy to check that if $\mathscr{A}$ is an S-set then $\mathscr{A} \backslash c$ is an $S \backslash c$-set. We can now define the tree $t \backslash c$:

$L(t \backslash c) = L(t) \backslash c$

$CL(t \backslash c) = CL(t) \backslash c$

$\mathscr{A}(t \backslash c(s)) = \mathscr{A}(t(s)) \backslash c.$

LEMMA 5.3.4 The function $\backslash c \in [\mathbf{AT_S} \longrightarrow \mathbf{AT_S}]$.

Proof a) We must first show that $t \backslash c$, defined above, is actually an element of $\mathbf{AT_S}$.

i) From the remarks above we know that $L(t \backslash c)$ and $CL(t \backslash c)$ are both prefix-closed and $CL(t \backslash c) \subseteq L(t \backslash c)$ because $CL(t) \subseteq L(t)$. Let $sa \in L(t \backslash c)$. We must show $s \in CL(t \backslash c)$. If $sa \in L(t \backslash c)$, then sa contains no occurrence of c or $\bar{c}$ and as a typical example suppose $sacs' \in L(t)$. Then because $CL(t)$ is prefix-closed $s \in CL(t)$ and therefore $s \in CL(t \backslash c)$.

ii) It is easy to see that $t \backslash c$ is finite-branching, if $s \in CL(t \backslash c)$ then s is also in $CL(t)$ and $S(t \backslash c(s)) \subseteq S(t(s))$.

iii) Let $s \in CL(t \backslash c)$. Then $S(t \backslash c(s)) = S(t(s)) \backslash c$ and using the fact that $\mathscr{A}(t(s))$ is an $S(t(s))$-set it follows trivially that $\mathscr{A}(t \backslash c(s))$ is an $S(t(s)) \backslash c$-set.

b) We leave the reader to check that the function is continuous. □

The definition of the function over $\mathbf{AT_S}$ corresponding to | is much more difficult and it is debatable if it can be called natural. It is essentially an encoding of the effect | has on the properties of processes which are used in the alternative characterization of $\sqsubseteq_{\mathrm{MUST}}$. Before presenting the definition, a considerable amount of notation is needed.

Let $zip \subseteq Act^* \times Act^* \times Act^*$ be the least relation which satisfies

i) $\varepsilon \in zip(\varepsilon, \varepsilon)$

ii) $s \in zip(s_1, s_2)$ implies $sa \in zip(s_1 a, s_2)$
$sa \in zip(s_1, s_2 a)$
$s \in zip(s_1 a, s_2 \bar{a})$

where $s \in zip(s_1, s_2)$ is used instead of $(s, s_1, s_2) \in zip$.

Intuitively $s \in zip(s_1, s_2)$ means that s can be obtained by interleaving the actions of s_1 and s_2 and cancelling complementary pairs. The notation is generalized to sets in the usual way:

$$zip(L_1, L_2) = \{zip(s_1, s_2), s_1 \in L_1, s_2 \in L_2\}.$$

Recall that the languages associated with trees in $\mathbf{AT_S}$ are finite-branching: for each $s \in L$ the set $\{a, sa \in L\}$ is finite. So the function to be associated with | must preserve this property. However, if L_1 and L_2 are finite-branching, it does not follow that $zip(L_1, L_2)$ also enjoys this property. For example if $L_1 = \{b^n, n \geq 0\} \cup \{b^n a_n, n \geq 0\}$ and $L_2 = \{\bar{b}^n, n \geq 0\}$, then $a_n \in zip(L_1, L_2)$ for every $n \geq 0$.

A more restrictive form of zipping languages together is given by

$$zip{\downarrow}(L_1, L_2) = \{s \in zip(L_1, L_2), \text{ the number of pairs } \langle s_1, s_2\rangle \in L_1 \times L_2 \text{ such that } s = zip(s_1, s_2) \text{ is finite}\}$$

For example, if $L_1 = \{b^n a, n \geq 0\}$, $L_2 = \{\bar{b}^n, n \geq 0\}$ then $a \in zip(L_1, L_2)$ but $a \notin zip{\downarrow}(L_1, L_2)$. If L_1, L_2 are as in the previous example, then ε is in $zip(L_1, L_2)$ but not in $zip{\downarrow}(L_1, L_2)$. In general, at least for finite-branching languages, if $s \in zip(L_1, L_2)$ and $s \notin zip{\downarrow}(L_1, L_2)$ then there exists some prefix s' of s and infinite sequences w, $\bar{w}$ such that $s' = zip(w(k), \bar{w}(k))$, $w(k) \in L_1$, $\bar{w}(k) \in L_2$ for every $k \geq 0$. (Here $w(k)$ denotes the finite prefix of w of length k.)

This new form of zipping does preserve finite-branching. More precisely,

P1: If L_1, L_2 are finite-branching and $s \in zip{\downarrow}(L_1, L_2)$ then $\{a, sa \in zip{\downarrow}(L_1, L_2)\}$ is finite.

We also require the function associated with | to preserve prefix closure. Unfortunately, this is not preserved by $zip{\downarrow}$. For example if $L_1 = \{a\} \cup \{b^n, n \geq 0\}$ and $L_2 = \{\bar{b}^n, n \geq 0\}$ then a is in $zip{\downarrow}(L_1, L_2)$ but ε is not. We need to modify the definition of $zip{\downarrow}$:

let $zip{\downarrow\downarrow}(L_1, L_2) = \{s, \text{every prefix of } s \text{ is in } zip{\downarrow}(L_1, L_2)\}$

The property P1 is also true for $zip{\downarrow\downarrow}$ and it is simple to check

P2: if L is prefix-closed then so is $zip{\downarrow\downarrow}(L)$.

Now let t_1, t_2 be trees in $\mathbf{AT_S}$. Define

$$CL(t_1 \mid t_2) = \{s \in zip{\downarrow\downarrow}(CL(t_1), CL(t_2)), \text{ if } s = zip(s_1, s_2) \text{ where } s_1 \in L(t_1), s_2 \in L(t_2) \text{ then } s_i \in CL(t_i), i = 1 \text{ and } 2\}$$

$$L(t_1 \mid t_2) = \{\varepsilon\} \cup \{sa \in zip(L(t_1), L(t_2)), s \in CL(t_1 \mid t_2)\}$$

So $CL(t_1 \mid t_2)$ contains all those sequences in $zip{\downarrow\downarrow}(CL(t_1), CL(t_2))$ with the property that if they can be decomposed into sequences from $L(t_1)$ and $L(t_2)$ these decompositions are actually in $CL(t_1)$ and $CL(t_2)$. The language $L(t_1 \mid t_2)$ is defined in a more natural manner and automatically satisfies the requirement on elements of $\mathbf{AT_S}$:

if $sa \in L(t_1 \mid t_2)$ then $s \in CL(t_1 \mid t_2)$.

The properties P1, P2 above imply the other requirements:

$L(t_1 \mid t_2)$ is finite-branching

$L(t_1 \mid t_2)$, $CL(t_1 \mid t_2)$ are prefix-closed

$CL(t_1 \mid t_2) \subseteq L(t_1 \mid t_2)$.

The acceptance sets are defined as

$$\mathcal{A}(t_1 \mid t_2, s) = c(\mathcal{B} \cup \{S(L(t_1 \mid t_2), s)\}),$$

where

$$\mathcal{B} = \{A_1 \cup A_2, A_1 \cap \bar{A}_2 = \varnothing \text{ and } A_i \in \mathcal{A}(t_i, s_i) \text{ for some } \langle s_1, s_2 \rangle \text{ such that } s = zip(s_1, s_2)\}$$

The addition of the successor set $S(L(t_1 \mid t_2), s)$ is required because otherwise we would not obtain an element in $\mathbf{AT_S}$. For example, with t_1, t_2 as defined in the first example of figure 5.4, the acceptance set associated with ε would be $\{\{b, c\}\}$ which is not an $\{a, \bar{a}, b, c\}$-set. We leave it to the reader to check that the requirements for acceptance sets are satisfied.

The three definitions $L(t_1 \mid t_2)$, $CL(t_1 \mid t_2)$ and $\mathcal{A}(t_1 \mid t_2, s)$ determine an element of $\mathbf{AT_S}$ and therefore we have defined a function $\mid$ over $\mathbf{AT_S}$. Two examples are given in figure 5.4. The proof of continuity is long and tedious, which is to be expected given the intricacy of the definitions, but it presents no serious difficulty. So we simply state:

LEMMA 5.3.5 The function $\mid \in [\mathbf{AT_S^2} \longrightarrow \mathbf{AT_S}]$.

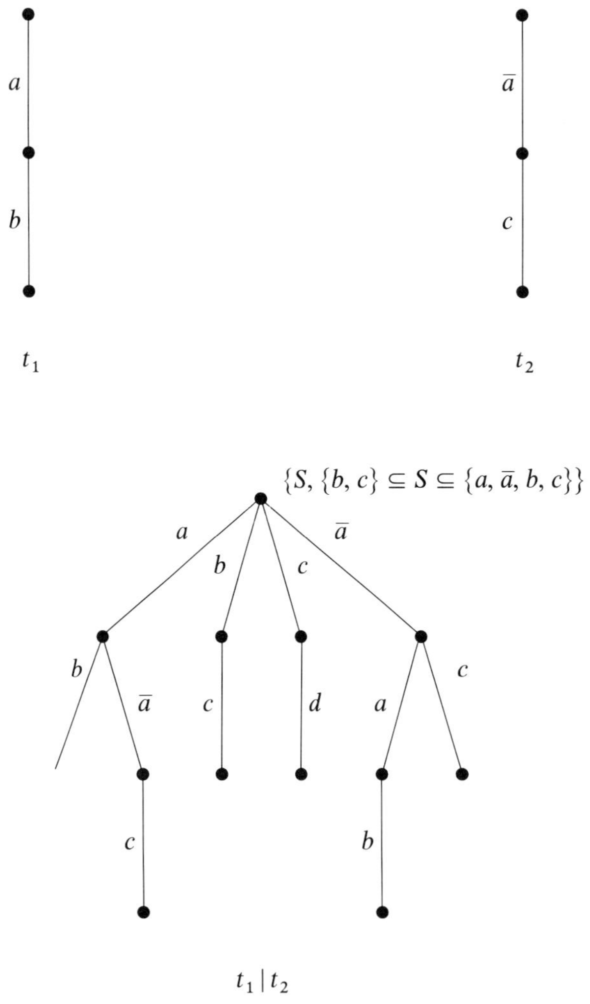

Figure 5.4
Examples of | on $\mathbf{AT}_S$.

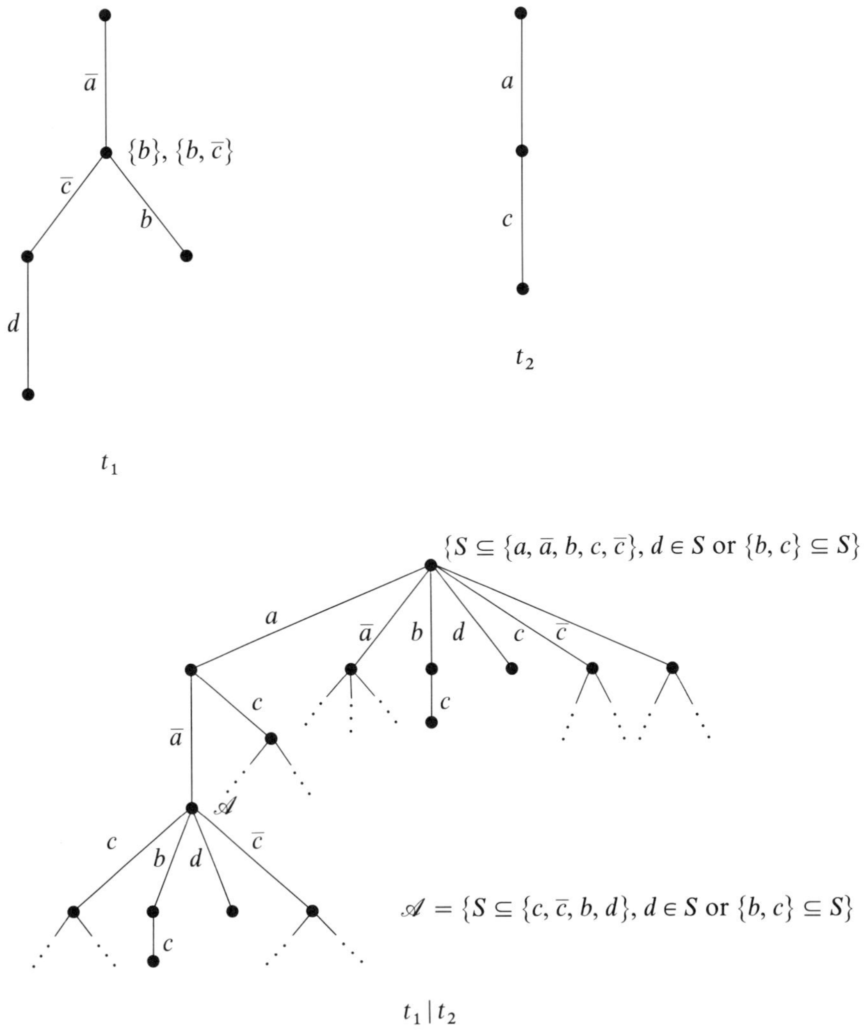

Figure 5.4 (continued)

With these definitions we have extended the model $\mathbf{AT_S}$ to a model for our language **EPL**. To show that it is fully abstract with respect to $\sqsubseteq_{\text{MUST}}$ we invoke corollary 5.3.3. To do so we must show that the new functions satisfy the axioms in figure 5.3. Once again this is straightforward but tedious: it involves grappling with many syntactic manipulations and case analyses and requires a good understanding of the functions over $\mathbf{AT_S}$.

This completes our analysis of the example language **EPL**. We have extended the schema of results in figure 4.4 to this language. Although we have concentrated on the preorder $\sqsubseteq_{\text{MUST}}$ and the related strong model $\mathbf{AT_S}$ and proof system **ωDED(SA2s)** there is no difficulty in modifying the last two sections so that similar results can be obtained for $\sqsubseteq$, **AT**, **ωDED(A2s)** and $\sqsubseteq_{\text{MAY}}$, $\mathbf{AT_W}$, and **ωDED(WA2s)**. This will be explored briefly in the exercises.

5.4 Alternative Communication Principles

Our treatment of the language **EPL** in this chapter is an example of how the basic theory of the language **RM** can be applied to develop a semantic theory and related proof system for a large class of languages. In this brief section we wish to reinforce the generality of this method by considering another language, a variant of that popularized in Hoare 1985.

In this language we also have communication between processes and a method of abstracting from their internal details. However, the mechanisms are different from those used in **EPL**, which are inherited from CCS. We represent the new form of communication by a collection of binary operators $\|_S$ where S is an arbitrary subset of the collection of actions *Act*. Unlike **EPL**, there is no need to assume that this set of actions has any structure. Instead we demand that the combined process $p \|_S q$ may only perform an action from the *synchronization set* S whenever both the subprocesses p and q simultaneously perform it; both subprocesses must synchronize on actions from S. For example, if p, q represent *rec x.acdx*, *rec x.cbdx* respectively, and S in $\{c, d\}$, then $p \|_S q$ can only perform actions in the order *acbdac*....

The fundamental difference between this form of communication or synchronization and the one employed in **EPL** is that the latter is two-way whereas the former is multiway. In any synchronization in $p_1 | p_2 | p_3$ at most two of the processes from p_1, p_2, p_3 are involved, whereas in $p_1 \|_S p_2 \|_S p_3$ all three processes may synchronize together. This occurs, for example, if we define p_i to be *rec* $x.a_i bx$ for $i = 1$, 2, and 3, and $S = \{b\}$. In this case all three actions a_1, a_2, a_3 must occur in some order; then the three-way synchronization b may occur and the behavior repeats itself.

If this primitive synchronization is used then some method of internalizing or hiding the synchronization actions is required, at least if we require the language to have abstraction features. For each action a we introduce a unary post-fix operator $- a$: intuitively the process $p - a$ acts just as p except that the a actions of p are considered as internal actions in $p - a$. So that if we consider the first example above $(p \parallel_S q) - c - d$ acts just like the process *rec x.abx*. Also, the process $(p_1 \parallel_S p_2 \parallel_S p_3) - b$, where the subprocesses are defined above, can only perform actions from $\{a_1, a_2, a_3\}$, and these are constrained so that all three actions must occur before any individual action is repeated.

We will not discuss in any more detail the motivation behind these new operators. The interested reader can consult Hoare 1985 where they, or variations of them, are considered in depth. For our purposes it suffices to say that they represent an alternative view of communicating processes and we now show how they can be accommodated in our framework.

Let Σ^4 be the signature obtained by adding the new operators to Σ^2. Then REC_{Σ^4} is a different algebraic process language and we can apply to it the results of the previous sections. The application is in four stages.

I Operational Semantics To give an operational semantics to REC_{Σ^4} we embed it in an *lts*. As with **EPL**, this is accomplished by extending the existing *lts* view of the basic language *RM*; it is sufficient to add extra rules for the new combinators. The intuition given above is captured by the rules:

i) if $a \in S, p \xrightarrow{a} p', q \xrightarrow{a} q'$ implies $p \parallel_S q \xrightarrow{a} p' \parallel_S q'$

ii) if $a \notin S$, a) $p \xrightarrow{a} p'$ implies $p \parallel_S q \xrightarrow{a} p' \parallel_S q$
b) $q \xrightarrow{a} q'$ implies $p \parallel_S q \xrightarrow{a} p \parallel_S q'$

iii) if $p \xrightarrow{a} p'$ implies $p - a \succ\!\!\longrightarrow p' - a$

iv) if $a \neq b$, then $p \xrightarrow{b} p'$ implies $p - a \xrightarrow{b} p' - a$.

II Testing Preorders Having embedded REC_{Σ^4} into an *lts* the general definition of §2.6 can be applied to obtain three semantic preorders, $\sqsubseteq$, $\sqsubseteq_{\text{MUST}}$, and $\sqsubseteq_{\text{MAY}}$. Moreover, the alternative characterization will remain valid in this extended environment because the new labeled transition system is finite-branching. Finally, we must show that the new Testing preorders are preserved by the extra operators. Although this is true for the operators $\parallel_S$ and $- a$, there is no guarantee that it will be true in

$$(x \oplus y) - a = x - a \oplus y - a$$

$$(ax + y) - a = x - a \oplus (x - a + y - a)$$

$$\text{if} \quad x = \sum\{b_i x_i, i \in I\} \quad \text{and} \quad a \neq b_i \quad \text{for} \quad i = 1, \ldots, n,$$

$$x - a = \sum\{b_i(x_i - a), i \in I\}$$

$$NIL - a = NIL$$

$$\Omega - a = \Omega$$

$$(x \oplus y) \parallel_S z = x \parallel_S z \oplus y \parallel_S z$$

$$z \parallel_S (x \oplus y) = z \parallel_S x \oplus z \parallel_S y$$

$$\text{if } x, y \quad \text{represent} \quad \sum\{a_i x_i, i \in I\}, \quad \sum\{b_j y_j, j \in J\},$$

$$x \parallel_S y = \sum\{a_i(x_i \parallel_S y), a_i \notin S\} + \sum\{b_j(x \parallel_S y_j), b_j \notin S\} + \sum\{x_i \parallel_S y_j, a_i = b_j \in S\}$$

$$\Omega \parallel_S p = p \parallel_S \Omega = \Omega$$

Figure 5.5
Equations for $-\ a$ and $\parallel_S$.

general. In the cases where it is false the theory cannot be applied directly. An example of an operator which does not preserve $\sqsubseteq_{\text{MUST}}$ is given in Q18 of this chapter.

III Equational Characterization At this stage we must find a reductive set of sound equations for the new operators. In our case they are given in figure 5.5. They are similar in design to those for | and $\backslash c$. However, the equations for $-\ a$ are somewhat more complicated than those for $\backslash c$. This is to be expected as $\backslash c$ does not seek to internalize behavior whereas that is the major concern of $-\ a$. Similarly, the equations for $\parallel_S$ are simpler than those for | because the latter introduces internal behavior.

The proof that the new set of equations **SA4s**, obtained by adding those in figure 5.5 to **SA2s**, is reductive follows the same lines as that for **SA3** in §5.3; intuitively terms in the base language **RM** are obtained by systematically applying the equations of figure 5.5 in a left-to-right fashion.

It is also easy to check that **SA4s** is normal (using the alternative characterization) and therefore by the Reduction theorem 5.2.1, we have a sound and complete proof system for $REC_{\Sigma 4}$, namely **ωDED(SA4s)**.

IV Fully Abstract Models Theorem 5.3.2 also gives an automatic extension of $\mathbf{AT_S}$ to a fully abstract model for REC_{Σ^4}. As in §5.3 for **EPL**, we can try to give a natural representation for this model. It involves associating natural operations over $\mathbf{AT_S}$ with each of the operators $\|_S$ and $-\ a$ and checking that the operations satisfy the extra equations of figure 5.5. We do not propose to do this as it is somewhat complicated. But the reader should be able to find suitable definitions by modifying those for $|$ and $\backslash c$. As might be expected, the definition for $\|_S$ is simpler than that for $|$ whereas that for $-\ a$ is more complicated than that for restriction $\backslash c$.

REC_{Σ^4} is simply one variation on the theme of **EPL**; it can be treated just as easily within our framework. Moreover we feel that we can accommodate a large class of such languages with similar ease.

Exercises

Q1 For each of the following pairs of processes decide whether or not they are related via $\sqsubseteq$, $\sqsubseteq_{\mathbf{MUST}}$, and $\sqsubseteq_{\mathbf{MAY}}$.

i) $((rec\, x.a\alpha x) \,|\, (rec\, x.b\bar{\alpha}x))\backslash\alpha \qquad rec\, x.ax + bx$

ii) $((rec\, x.\alpha ax) \,|\, (rec\, x.(\bar{\alpha}x + \bar{\beta}x)) \,|\, (rec\, x.\beta bx))\backslash\alpha\backslash\beta$
$((rec\, x.\lambda ax) \,|\, (rec\, x.\bar{\lambda}x) \,|\, (rec\, x.\lambda bx))\backslash\lambda$

iii) $rec\, x.ax \qquad (rec\, x.ax) \,|\, (rec\, x.ax)$

iv) $rec\, x.ax \,|\, (bx \oplus (bx + cx)) \qquad (rec\, x.ax) \,|\, (bx \oplus (bx + cx))) \oplus rec\, x.ax \,|\, bx.$

v) $(rec\, x.ax) \,|\, (rec\, x.bx + rec\, x.cx) \qquad (rec\, x.ax) \,|\, (rec\, x.bx) + (rec\, x.ax) \,|\, (rec\, x.cx).$

Q2 For each of the following processes give an equivalent process (with respect to $\eqsim$) in **RM**.

i) $((rec\, x.ax + \beta x) \,|\, (rec\, x.cx + \bar{\beta}x))\backslash\beta$

ii) $((rec\, x.ax \oplus \beta cx) \,|\, (rec\, x.dx + \bar{\beta}x))\backslash\beta$

iii) $(rec\, x.\alpha x \,|\, \bar{\alpha}(bx + cx))\backslash\alpha$

iv) $((rec\, x.a\alpha\lambda x) \,|\, (rec\, x.b\bar{\alpha}\beta x) \,|\, (rec\, x.c\bar{\lambda}\bar{\beta}x))\backslash\alpha\backslash\beta\backslash\lambda$

v) $((rec\, x.up.\alpha down.\beta x) \,|\, (rec\, x.in.\bar{\alpha}out.\bar{\beta}x))\backslash\alpha\backslash\beta.$

Q3 Give an example of a process in $\mathbf{RM_3}$ for which there is no equivalent process (with respect to $\eqsim$) in $\mathbf{RM_2}$.

Q4 Let *Sem* denote the term $rec\,x.\overline{p}.\overline{v}.x$, which acts like a binary semaphore, and let *Reader*, *Writer* denote $rec\,x.p.readb.reade.v.x$, $rec\,x.p.writeb.writee.v.x$ respectively.

a) Draw a "derivation diagram" which details all possible derivation sequences from the term

$(Reader \mid Sem \mid Writer)\backslash p.\backslash v.$

b) Show that

$(Reader \mid Sem \mid Writer)\backslash p.\backslash v. \eqsim rec\,x.(readb.read\,e.x \oplus writeb.write.x).$

Q5 Let *V*, *H* denote the terms $rec\,x.up.\alpha.\beta.down.x$, $rec\,x.\overline{\alpha}.in.\overline{\beta}.x$ respectively. Show that $(V \mid H)\backslash\alpha\beta \eqsim rec\,x.up.in.down.x.$

Q6 Recall from Q6 of chapter 2 that p is stable if $p \not\rightarrowtail$.

For $a \in Act$ let p *must* a if for every stable q such that $p \rightarrowtail^* q$ $q \xrightarrow{a}$. Let $p \leq_{\mathrm{m}} p'$ if for every $a \in Act$ p *must* a implies q *must* a.

a) Show that $\leq_{\mathrm{m}}$ is not a Σ^3-congruence.

b) A *context*, $C[\ ]$, is a term with a hole, $[\ ]$, in it. We use $C[t]$ to denote the term which is obtained by placing t in the hole in the context $C[\ ]$. For any relation R over $\mathbf{RM_3}$ let R^c be the relation over $\mathbf{RM_3}$ defined by

$\langle p, q\rangle \in R^c$ if for every context $C[\ ]$. $\langle C[p]C[q]\rangle \in R.$

Show that if R is reflexive and transitive then R^c is a Σ^3-preorder. Prove further that it is the largest Σ^3-preorder contained in R.

c) A process p is called *strongly convergent* if $p \stackrel{s}{\Longrightarrow} p'$ implies $p'\downarrow$ for every $s \in Act^*$. Prove that for strongly convergent p, q $p \lesssim_{\mathrm{MUST}} q$ if and only if $p \leq_{\mathrm{m}} q$.

Q7 A process p is called *stopped* if it is stable and $p \xrightarrow{a}$ for no $a \in Act$.

Let

$MTraces(p) = \{s \in Act^*, p \stackrel{s}{\Longrightarrow} p', \text{ where } p' \text{ is stopped}\}.$

For $p, q \in \mathbf{RM_3}$ let $p \leq_{\mathrm{T}} q$ if

$MTraces(q) \subseteq MTraces(p).$

Prove that for strongly convergent terms,

$p \lesssim_{\mathrm{MUST}} q$ if and only if $p \leq_{\mathrm{T}}^{c} q.$

Q8 For $p \in \mathbf{RM_3}$ we write *p may deadlock* if $p \succ\!\longrightarrow^* p'$ for some p' which is stopped. Let $p \leq_{\mathrm{d}} q$ if *q may deadlock* implies *p may deadlock.*
Prove

$p \lesssim_{\mathrm{MUST}} q$ if and only if $p \leq^{\mathrm{c}}_{\mathrm{d}} q$.

Q9 For $p, r \in \mathbf{RM_3}$ we write *p may stop r* if

$p \mid r \succ\!\longrightarrow * p' \mid r' \succ\!\not\longrightarrow$ where r' is stopped.

Let $p \leq_{\mathrm{ms}} q$ if for every $r \in \mathbf{RM_3}$ *q may stop r* implies *p may stop r*.
Prove

$p \lesssim_{\mathrm{MUST}} q$ if and only if $p \leq^{\mathrm{c}}_{\mathrm{ms}} q$.

Q10 The set of *partial sequences* over *Act* is defined to be $Act^* \cup \{s\bot, s \in Act\}$. These are ordered by $w \leq w'$ if $w = w'$ or w has the form $s\bot$ where s is a prefix of w'. This also induces a preordering on sets of partial sequences by $X < Y$ if for every $w \in Y$ there is some $w' \in X$ such that $w' \leq w$.

With each $p \in \mathbf{RM_3}$ we can associate a set of partial sequences *PM no traces Traces*(p) by:

i) for $s \in Act^*$ $s \in$ *PM no traces Traces*(p) if $p \stackrel{s}{\Longrightarrow} p'$ where p' is stopped and $p'\!\downarrow$.
ii) for $s \in Act^*$ $s\bot \in$ *PM no traces Traces*(p) if $p \stackrel{s}{\Longrightarrow} p'$ where $p'\!\uparrow$.

Let $p \leq_{\mathrm{pT}} q$ if *PM no traces Traces*$(q) \subseteq$ *PM no traces Traces*(p).
Prove $p \lesssim_{\mathrm{MUST}} q$ if and only if $p \leq^{\mathrm{c}}_{\mathrm{pT}} q$, for every p, q in $\mathbf{RM_3}$.

Q11 (Renaming) Let *Act* be of the form $In \cup Out$, as explained in §5.1. A *renaming* is a total function $R\colon Out \longrightarrow Out$ which is extended to *Act* by defining $R(\bar{c})$ to be $\overline{R(c)}$. For each R let $[R]$ be a new post-fix combinator. Intuitively $p[R]$ is a process which acts like p except that the actions are renamed by R. Let Σ^5 denote $\Sigma^3 \cup \{[R], R$ is a renaming$\}$. The operational semantics of $\mathbf{RM_3}$ is extended to REC_{Σ^5} by using the new rules:

$$p \stackrel{a}{\longrightarrow} p' \quad \text{implies} \quad p[R] \stackrel{R(a)}{\longrightarrow} p'[R]$$

$$p \succ\!\longrightarrow p' \quad \text{implies} \quad p[R] \succ\!\longrightarrow p'[R].$$

Find a set of equations E such that $\boldsymbol{\omega}\mathbf{DED}(E)$ is sound and complete with respect to $\lesssim_{\mathrm{MUST}}$ over the extended language REC_{Σ^5}.

Q12 (Arbitrary Interleaving) Extend the language REC_{Σ^4} by introducing a new parallel operator $|||$: intuitively $p \,|||\, q$ evolves by arbitrarily interleaving the possible actions of p and q. Formally the operational semantics is extended by the rules

i) $p \xrightarrow{a} p'$ implies $p \,|||\, q \xrightarrow{a} p' \,|||\, q$

$q \xrightarrow{a} q'$ implies $p \,|||\, q \xrightarrow{a} p \,|||\, q'$

ii) $p \rightarrowtail p'$ implies $p \,|||\, q \rightarrowtail p' \,|||\, q$

$q \rightarrowtail q'$ implies $p \,|||\, q \rightarrowtail p \,|||\, q'$.

Find a set of equations E such that $\boldsymbol{\omega}$**DED**(E) is sound and complete with respect to the extended language.

Q13 For any p in **EPL** let $chan(p)$ be the finite set of channel names which appear syntactically in p. Show that the following equations are valid for $\eqsim$ over $\mathbf{RM_4}$:

$$x \backslash a = x \qquad \text{if } a \notin chan(x)$$

$$x \backslash a \backslash b = x \backslash b \backslash a$$

$$(x \mid y) \backslash a = x \backslash a \mid y \backslash a \quad \text{if } a \notin chan(x) \cap chan(y).$$

Q14 Consider the extension of **EPL** outlined in Q11. Show that $\eqsim$ satisfies the following equations over this extension:

$$x[R] = x[R'] \qquad \text{if } R(a) = R'(a) \text{ for every } a \text{ in } chan(x)$$

$$(x \mid y)[R] = x[R] \mid y[R]$$

$$x[R] \backslash a = (x \backslash b)[R] \qquad \text{where } R(b) = a.$$

Q15 Let **A3s**, **WA3s** be the set of equations obtained by adding the equations in figure 5.3 to **A2s** and **WA2s** respectively.

i) Show that **A3s**, **WA3s**, respectively, are reductive with respect to **RM**, i.e.

a) for every $d \in FREC_{\Sigma 3}$, there exists a $d' \in FREC_{\Sigma 2}$ such that $d =_{\mathbf{A3s}} d'$, $d =_{\mathbf{WA3s}} d'$ respectively

b) for every $p \in REC_{\Sigma 3}$ such that $p\!\downarrow$ there exists a head term q such that $p =_{\mathbf{A3s}} q$, $p =_{\mathbf{WA3s}} q$ respectively.

ii) Modify the Reduction theorem so that part i) can be applied to deduce that $\boldsymbol{\omega}$**DED(A3s)**, $\boldsymbol{\omega}$**DED(WA3s)**, are sound and complete with respect to $\lesssim$, $\lesssim_{\mathbf{MAY}}$ respectively.

Q16 Let $\Sigma \subseteq \Sigma'$ and let E, E' be sets of Σ-equations, Σ'-equations respectively, such that $E \subseteq E'$. Let S be an algebraic relation over $REC_{\Sigma'}$ which satisfies

i) S is conservative over $<_{E\omega}$, i.e. for p, q in REC_{Σ}, $\langle p, q\rangle \in S$ implies $p <_{E\omega} q$

ii) **DED**(E') is sound with respect to S

iii) For every $d \in FREC_{\Sigma'}$, there exists some $e \in FREC$ such that $d =_{E'} e$.

Prove that **ωDED**(E') is sound and complete with respect to S.

Q17 a) Without using the Reduction theorem or completeness of **ωDED(A3s)**, **ωDED(WA3s)**, or **ωDED(SA3s)** prove that $\lesssim$, $\lesssim_{\text{MAY}}$, and $\lesssim_{\text{MUST}}$ are all algebraic over $\mathbf{RM_3}$.

b) Use part a) and Q16 to prove that **ωDED(A3s)**, **ωDED(WA3s)**, and **ωDED(SA3s)** are sound and complete with respect to $\lesssim$, $\lesssim_{\text{MAY}}$, and $\lesssim_{\text{MUST}}$ respectively over $\mathbf{RM_3}$.

Q18 We consider a modified communication connective, $\not{N}$, called 'broadcasting' in Pneuli 1985. Intuitively in the process $p \not{N} q$, p or q can perform any action at will but once such an action a is to be performed by p or q, if its partner can also perform it they both must do it together. This can be expressed by the rules

i) $p \xrightarrow{a} p', q \xrightarrow{a} q'$ implies $p \not{N} q \xrightarrow{a} p' \not{N} q'$

ii) $p \xrightarrow{a} p', q \not\xrightarrow{a}$ implies $p \not{N} q \xrightarrow{a} p' \not{N} q$.

iii) $p \not\xrightarrow{a}, q \xrightarrow{a} q'$ implies $p \not{N} q \xrightarrow{a} p \not{N} q'$.

We cannot use ii) and iii) as part of an inductive definition of $\xrightarrow{a}$ because negative premises are not allowed. However, if we confine our attention to finite terms, the definition of $\xrightarrow{a}$ may be by structural induction; in this framework we may use ii) and iii).

Show that $\lesssim_{\text{MUST}}$ is *not* preserved by $\not{N}$.

Hint: let p, q denote $a(c + be) + a(d + bf)$, $a(c + df) + a(d + be)$ respectively. Now $p \eqsim_{\text{MUST}} q$ and so it suffices to find a process r and an experiment e such that $p \not{N} r$ *must* e but $q \not{N} r$ *mu$\not{s}$t* e.

Q19 (Interleaving Theorem) Let p denote the term $(p_1 \mid p_2 \mid \ldots \mid p_m)\backslash C$ where

i) C is a finite set of channel names and $\backslash C$ is a shorthand, justified in Q13, for $\backslash c_1 \backslash c_2 \ldots \backslash c_k$, where $c_1, \ldots c_k$ is an arbitrary enumeration of C.

ii) each p_i has the form $\sum\{a_{ij}q_{ij}, j \in J_i\}$.

Let

$ext_c(p) = \sum\{a_{ij}(p_1 | \ldots | q_{ij} | \ldots | p_m)\backslash C, a_{ij} \notin C, \bar{a}_{ij} \notin C, j \in J_i, 1 \leq i \leq m\}$

$int(p) = \sum\{(p_1 | \ldots | q_{ij} | \ldots | q_{kl} | \ldots | p_m)\backslash C, a_{ij} = \bar{a}_{kl}, j \in J_i, k \in J_l, 1 \leq i \neq k \leq m.$

Assuming $int(p)$ is a nonempty summation, prove that $p \approx (ext_c(p) + int(p)) \oplus int(p)$.

Historical Note

Robin Milner was the first person to use an algebraic approach to the theory of communicating systems and much of the general framework of this book relies on his pioneering work. Various aspects of this view of processes as algebras are explored and developed in papers such as Milner 1978, 1979a, 1979b, and Milne and Milner 1979. For example, Milner (1978) advocates the development of an algebra of activities which are subject to a number of laws expressed as equations. This work led directly to **CCS** (Milner 1980), which consists of an algebraic language for processes, a behavioral equivalence, and a calculus for reasoning about processes based on a set of equations. In this work a direct link is made for the first time between the equational theory and a behavioral equivalence based on an operational semantics. This particular subject is examined in detail in Hennessy and Milner 1985. The language used is quite similar to $\mathbf{M_2}$ except that the binary internal choice operator $\oplus$ is replaced by an internal action τ, and the particular behavioral equivalence studied is called *observational equivalence*. This is much finer than any of the Testing equivalences and has been touched upon briefly in the exercises at the end of chapter 2. A more elegant version of this equivalence, *bisimulation equivalence*, is defined in Park 1981 and its application can be seen in papers such as Milner 1983. However, it is difficult to reconcile this behavioral equivalence with a model-theoretic approach to denotational semantics based on *cpos*. Some attempts are made in Hennessy and Plotkin 1980 and in Hennessy 1981 but the resulting *cpo*-based fully abstract models are all term models; unlike Acceptance Trees, they have no intuitive representation independent of the operational semantics.

Testing equivalence was developed, in DeNicola and Hennessy 1983, as an alternative behavioral equivalence for **CCS**. In subsequent papers, such as Hennessy 1983, it was realized that a cleaner algebraic theory and denotational model could be obtained by replacing the troublesome τ operator with a binary internal choice operator $\oplus$. Acceptance Trees were developed in detail in Hennessy 1985, although a slight variation already existed in DeNicola and Hennessy 1983. The argument that $\oplus$ could be used in place of τ within the context of **CCS** was first put forward in DeNicola and Hennessy 1987. Indeed the present book is essentially a detailed cohesive explanation of the material in these four papers: DeNicola and Hennessy 1983, Hennessy 1983, 1985, and DeNicola and Hennessy 1987.

Bismulation equivalence and Testing equivalence have very different formulations. The former is based on the ability of processes to simulate each other while the latter uses the idea of an experimenter testing processes. However, in Abramsky 1986 it is shown how bisimulation equivalence can be based on a notion of testing. Here one can perform much more complicated tests than we have allowed. This work actually introduces a number of different equivalences, each based on a reasonable notion of test. In contrast, Philips 1986 only slightly modifies our framework by allowing the course of an experiment to depend on the inability of the process to perform actions and obtains a new Testing equivalence, called Refusal Testing. In short, the general idea of testing processes can be formalized in many different ways and can lead to a variety of interesting equivalences.

I have concentrated on presenting one particular approach to the semantics of communicating systems to the exclusion of all others. However, I hope that large sections of the material is of general interest and will be equally applicable when the reader wishes to pursue other approaches. This at least will be true of the general algebraic framework and the approach to operational semantics using labeled transition systems. One well-developed school of thought is centered on the language **CSP**. The original version of this language appears in Hoare 1978; in subsequent work the nature of the language changed radically. Binary communication, as in **CCS**, is replaced by multiway sychronization, as developed in the language **COSY** (Shields and Lauer 1978). Communication with values is replaced by formal uninterpreted actions, as we have used, and the algebraic approach advocated by Milner is followed. The new version of **CSP**, now often called **TCSP** (theoretical **CSP**), consists of a set of combinators, together with recursive definitions of processes. This language, and a semantics, is presented in Brookes, Hoare, and Roscoe 1984, and Hoare 1986 gives a comprehensive account. The semantics is presented in terms of a denotational model called the Failures model which, in turn, is used to justify equations over the set of combinators. This also has been touched upon briefly in the exercises. An improved version of the model (see Brookes and Roscoe 1984) is actually isomorphic to one of our models, $\mathbf{AT}_S$. The reader is referred to DeNicola 1985a, 1985b, and Olderog 1986 for more details. The original presentation of the Failures model did not refer to a formal operational semantics and from this point of view this isomorphism is of interest: the Failures model can be justified operationally via the *must*-Testing equivalence. The Failures model is also analyzed operationally in Olderog and Hoare 1986 and Olderog 1986. These two references provide excellent surveys of the variety of models in existence; the authors present material very similar to ours from a different perspective.

Two other languages of note are Process Algebra (Bergstra and Klop 1984) and Meije (Austry and Boudol 1984). These both provide a set of combinators which are different yet again than those of **CCS** or **TCSP**. The semantic theories are based essentially on bisimulation equivalence; but in much research on the former, metric spaces rather than complete partial orders are used as the mathematical framework for solving fixpoint equations. A flavor of the use of metric spaces in this context may be obtained from deBakker and Zucker 1982, Golsen and Rounds 1983, and deBakker *et al.* 1986. The last report contains an excellent list of references to the extensive literature on process algebras.

There is not much point in discussing the relative merits of the various collections of combinators. I have argued in chapter 5 that most can be accommodated within our framework. Doubtless the different languages will be more or less effective depending on the area of application. Indeed, one of the major challenges is to design an effective and efficient set of combinators for particular areas of interest, such as VLSI systems, digital circuits, communication protocols, etc. It is natural to expect that these very different concerns will give rise to very different types of combinators. However, I hope that all can be encompassed within the theory of Testing which has been presented here.

The mathematics used throughout the present book is elementary. The material in chapter 1 on equational algebras may be found in standard textbooks such as Birkoff 1940, Cohn 1981, and Gratzer 1968, although references such as Messguer and Goguen 1985, or Ehrig and Mahr 1985 give more readable presentations of the mathematics we require. Ordered algebras, continuous partial orders, and the associated equational theories are not well known in the mathematical literature but have been thoroughly investigated by computer scientists. All the material we have used may be found in works such as Guessarian 1981 and Goguen *et al.* 1977, but with different notational conventions. (Nivat and Reynolds 1985 is also a useful reference.)

As this avenue of research is developed to include languages with more sophisticated features we will require a more sophisticated mathematical framework, and it is unlikely that this will be found in the existing literature. I give two examples to illustrate the point. The first concerns the infinite behavior of processes. In the present work I have been careful to demand that the infinite behavior of a process is completely determined by its finite behavior, and therefore algebraic *cpos* may be used as denotational models. However if we wish to consider descriptions of processes which have fairness constraints this demand is no longer reasonable, and a suitable mathematical framework has yet to be found. As a second example, we could take the language in Hoare 1978 or the full version of **CCS** in Milner 1980. Here values are passed along the channels so that the actions are no longer uninterpreted. Rather, the port names now act as variable binders in the same way as λ-abstraction in the λ-calculus. An adequate semantic treatment of these languages would therefore require a mathematical framework of Σ-domains which supports equational theories, variable binders, and, presumably, equational theories incorporating these binders.

Symbols and Notation

Frequently used Symbols

$\in$	is a member of
$\cup$	set union
$\cap$	set intersection
$\subseteq$	set containment
$\bigvee$	*lub*, least upper bound
$\subset\subset$	set containment for Acceptance sets
$\mathcal{u}$	pointwise union of Acceptance sets
$\circ$	functional composition
Ω	syntactic bottom
$\preceq$	syntactic partial order
$\parallel$	interaction of process and experimenter
$\rightarrow$	one step of experiment
$[\![\ ]\!]$	semantic brackets
$\vdash$	logical turnstile
ε	empty string
$\top\ \perp$	top and bottom

Algebras

$\langle A, \Sigma_A \rangle$	carrier A
	signature Σ
T_Σ	term algebra over signature Σ
I_E	initial Σ-algebra with respect to equations E
	initial Σ-*po* algebra with respect to inequations E
CI_E	initial Σ-domain with respect to inequations E
PI_E	initial Σ-predomain with respect to inequations E
$\mathbf{AT}$, $\mathbf{AT_W}$, $\mathbf{AT_S}$	Acceptance Trees, weak Acceptance Trees, and strong Acceptance Trees
$\mathbf{fAT}$, $\mathbf{fAT_W}$, $\mathbf{fAT_S}$	finite versions of Acceptance Trees

Classes of Algebras

$\mathscr{C}(E)$	the class of Σ-*po* algebras which satisfy inequations E
$\mathscr{C}\mathscr{P}(E)$	the class of Σ-predomains which satisfy inequations E
$\mathscr{C}\mathscr{C}(E)$	the class of Σ-domains which satisfy inequations E.

Languages

$T_\Sigma(X)$	finite terms over Σ and variables X
T_Σ	closed finite terms over Σ
$REC_\Sigma(X)$	recursive terms over Σ and variables X
$FREC_\Sigma(X)$	finite terms in $REC_\Sigma(X)$
REC_Σ	closed recursive terms over Σ
$FREC_\Sigma$	finite closed terms in REC_Σ
$\mathbf{M}_1$	abbreviation for T_{Σ^1}
$\mathbf{M}_2$	abbreviation for T_{Σ^2}
$\mathbf{RM}_2$, $\mathbf{RM}$	abbreviations for REC_{Σ^2}
$\mathbf{RM}_3$	abbreviation for REC_{Σ^3}

Signatures We always assume a predefined set of actions *Act* ranged over by *a*.

Σ^1	$NIL,\ a,\ +$
Σ^2	$NIL,\ a,\ +,\ \oplus$
Σ^3	$NIL,\ a,\ +,\ |,\ \backslash c$

Operational Semantics

i) $ap \xrightarrow{a} p$

ii) $p \xrightarrow{a} p'$ implies $p + q \xrightarrow{a} p'$

$q \xrightarrow{a} q'$ implies $p + q \xrightarrow{a} q'$

$p \rightarrowtail p'$ implies $p + q \rightarrowtail p' + q$

$q \rightarrowtail q'$ implies $p + q \rightarrowtail p + q'$

iii) $p \oplus q \rightarrowtail p$

$p \oplus q \rightarrowtail q$

iv) $rec\, x.t \rightarrowtail t[rec\, x.t/x]$

v) $\Omega \rightarrowtail \Omega$

vi) $p \rightarrowtail q$ implies $p\backslash c \rightarrowtail q\backslash c$

$p \xrightarrow{a} q$ implies $p\backslash c \xrightarrow{a} q\backslash c$ if $a \neq c, \overline{c}$

vii) $p \succ\!\!\longrightarrow r$ implies $p \mid q \succ\!\!\longrightarrow r \mid q$

$q \succ\!\!\longrightarrow r$ implies $p \mid q \succ\!\!\longrightarrow p \mid r$

$p \xrightarrow{a} r$ implies $p \mid q \xrightarrow{a} r \mid q$

$q \xrightarrow{a} r$ implies $p \mid q \xrightarrow{a} p \mid r$

$p \xrightarrow{a} p'$, $q \xrightarrow{\bar{a}} q'$ implies $p \mid q \succ\!\!\longrightarrow p' \mid q'$

Notation for Operational Semantics

$\rightarrow$ one step of experiment

$\xrightarrow{a}$ performing an a action

$\succ\!\!\longrightarrow$ performing an internal action

$\xrightarrow{s}$ performing a sequence of actions s

$\stackrel{s}{\Longrightarrow}$ performing a sequence of actions s, possibly interspersed by internal actions

$\downarrow\ \uparrow$ convergence and divergence predicates

$\downarrow s\ \uparrow s$ convergence and divergence after sequence s

$D(p, s)$ $\{p', p \stackrel{s}{\Longrightarrow} p'\}$, the s-derivatives of p

$D(p)$ $\{p', \text{for some } a \in A, p \stackrel{a}{\Longrightarrow} p'\}$, the derivatives of p

$\mathrm{Im}(p)$ $\{p', p \succ\!\!\longrightarrow p'\}$, the immediate internal derivatives of p

$S(p)$ $\{a, p \stackrel{a}{\Longrightarrow}\}$, the successors of p

$L(p)$ $\{s, p \stackrel{s}{\Longrightarrow}\}$, the language of p

$S(p, s)$ $\{a, p \stackrel{s}{\Longrightarrow} p' \stackrel{a}{\Longrightarrow}\}$, the successors of p after s

$\mathscr{A}(p, s)$ $\{S(p'), p \stackrel{s}{\Longrightarrow} p'\}$, the Acceptance set of p after s

Equations

$$x \oplus (y \oplus z) = (x \oplus y) \oplus z \qquad \oplus 1$$

$$x \oplus y = y \oplus x \qquad \oplus 2$$

$$x \oplus x = x \qquad \oplus 3$$

$$x + (y + z) = (x + y) + z \qquad +1$$

$$x + y = y + x \qquad +2$$

$$x + x = x \qquad +3$$

$$x + NIL = x \qquad +4$$

$$x \oplus y \leq x + y \qquad +\oplus 1$$

$$ax + ay = ax \oplus ay \qquad +\oplus 2$$

$$ax \oplus ay = a(x \oplus y) \qquad \oplus 4$$

$$x + (y \oplus z) = (x + y) \oplus (x + z) \qquad +\oplus 3$$

$$x \oplus (y + z) = (x \oplus y) + (x \oplus z) \qquad +\oplus 4$$

$$(x + y)\backslash c = x\backslash c + y\backslash c \qquad +\backslash c$$

$$(x \oplus y)\backslash c = x\backslash c \oplus y\backslash c \qquad +\backslash c$$

$$ax\backslash c = \begin{cases} a(x\backslash c) & \text{if name } (a) \neq c \\ NIL & \text{otherwise} \end{cases} \qquad a\backslash c$$

$$NIL\backslash c = NIL \qquad NIL\backslash c$$

$$\Omega\backslash c = \Omega \qquad \Omega\backslash c$$

$$(x \oplus y) \mid z = x \mid z \oplus y \mid z \qquad \oplus \mid$$

$$z \mid (x \oplus y) = z \mid x \oplus z \mid y$$

$$NIL \mid p = p \mid NIL = p \qquad NIL \mid$$

$$x \mid (y + \Omega) = (x + \Omega) \mid y = x \mid y + \Omega \qquad \Omega \mid$$

If x, y represent $\sum\{a_i x_i, i \in I\}$, $\sum\{b_j y_j, j \in J\}$ then

$$x \mid y = \begin{cases} ext(x, y) & \text{if } comm(x, y) \text{ is false} \\ (ext(x, y) + int(x, y)) \oplus int(x, y) & \text{otherwise} \end{cases} \qquad + \mid$$

where $ext(x, y) = \sum\{a_i(x_i \mid y), i \in I\} + \sum b_j(x \mid y_j), j \in J\}$

and $int(x, y) = \textstyle\sum\!\!\!\!\bigcirc \{x_i \mid y_j, a_i = \bar{b}_j\}$

$$x + \Omega \leq x \oplus \Omega \quad \textbf{(s)}$$

$$x \oplus y \leq x \quad \textbf{(S)}$$

$$x \leq x \oplus y \quad \textbf{(W)}$$

Proof Rules

1. *Reflexivity* $\dfrac{}{t \leq t}$

2. *Transitivity* $\dfrac{t \leq t', t' \leq t''}{t \leq t''}$

3. *Substitution* a) $\dfrac{\underline{t} \leq \underline{t}'}{f(\underline{t}) \leq f(\underline{t}')}$ for every f in Σ

   b) $\dfrac{t \leq t'}{rec\, x.t \leq rec\, x.t'}$

4. *Instantiation* $\dfrac{t \leq t'}{t\rho \leq t'\rho}$ for every substitution ρ

5. *Inequations* $\dfrac{}{t \leq t'}$ for every inequation $t \leq t'$ in E

6. Ω-*Rule* $\dfrac{}{\Omega \leq x}$

7. *REC* (unwinding) $\dfrac{}{rec\, x.t = t[rec\, x.t/x]}$

8. ω-Induction $\dfrac{\text{for every } d \in \text{App}(t),\ d \leq t'}{t \leq t'}$

Notation for Equations

A2 $\oplus 1 - \oplus 4$, $+1 - +4$, $+\oplus 1 - +\oplus 4$

SA2 **A2 + S**

WA2 **A2 + W**

A2s	**A2 + s**
SA2s	**SA2 + s**
WA2s	**WA2 + s**

Notation for Proof Systems

DED(E)	basic proof system with inequations E
ΩDED(E)	**DED**(E) with the extra inequation $\Omega \leq x$ Usually abbreviated to **DED**(E)
rDED(E)	**DED**(E) with unwinding of recursive definitions
ωDED(E)	**rDED**(E) with ω-Induction

Equivalences and Preorders

$\leq_A$	A a Σ-algebra: $t \leq_A t'$ means $A[\![t]\!] \leq A[\![t']\!]$
$\leq_E$	E a set of inequations: $t \leq_E t'$ means $t \leq t'$ is a theorem in **DED**(E)
$\leq_{Er}$	E a set of inequations: $t \leq_{Er} t'$ means $t \leq t'$ is a theorem in **rDED**(E)
$\leq_{E\omega}$	E a set of inequations: $t \leq_{E\omega} t'$ means $t \leq t'$ is a theorem in **ωDED**(E)
$\leq_{E\Omega}$	E a set of inequations: $t \leq_{E\Omega}$ means $t \leq t'$ is a theorem in **ΩDED**(E). (Usually abbreviated to $\leq_E$.)
$\sqsubset\!\!\!\sim_{MUST}$, $\sqsubset\!\!\!\sim_{MAY}$, $\sqsubset\!\!\!\sim$	Testing preorders over processes
$<<_{MUST}$, $<<_{MAY}$, $<<$	Alternative preorders over processes
$\eqsim_{MUST}$, $\eqsim_{MAY}$, $\eqsim$	Testing equivalences over processes
$\preceq$	syntactic partial order generated by $\Omega \leq x$
$=_\alpha$	α-equality

References

Abbreviations:

JACM Journal of the Association of Computing Machinery

LNCS Lecture Notes in Computer Science, Springer-Verlag

TCS Theoretical Computer Science

Abramsky, S., 1983 Experiments, Powerdomains and Fully-Abstract Models for Applicative Programming, *LNCS*, 158, 1–13.

—, 1986 Observational Equivalence as a Testing Equivalence, Imperial College Technical Report.

Apt, K. (ed.), 1985 Logics and Models of Concurrent Systems, Springer-Verlag.

Austry, D., and G. Boudol, 1984 Algèbre de processus et synchronisation, *TCS*, 30, 91–131.

Bergstra, J., and J. Klop, 1984 Process Algebra for Synchronous Communication, *Information and Control*, 60, 109–137.

—, 1985 Algebra of Communicating Processes with Abstraction, *TCS*, 37, 77–121.

Birkoff, G., 1940 Lattice Theory, *American Mathematical Society Colloquium Publications*, XXV.

Bloom, S., 1976 Varieties of Ordered Algebras, *Journal of Computer and System Sciences*, 13, 200–212.

Brinksma, H., and G. Karjoth, 1984 A Specification of the OSI Transport Service in LOTUS, in Yemini, Y. (ed.), *Proceedings of 4th International Workshop on Protocol Specification, Testing and Verification*, North-Holland.

Brookes, S., 1983 On the Relationship of **CCS** and **CSP**, *LNCS*, 154, 83–96.

—, 1984 A Model for Communicating Systems, Ph.D. Thesis, Oxford University.

Brookes, S., C. Hoare, and A. Roscoe, 1984 A Theory of Communicating Sequential Processes, *JACM*, 31, no. 7, 560–599.

Brookes, S., and A. Roscoe, 1984 An Improved Failures Model for Communicating Processes, *LNCS*, 197, 281–305.

Brookes, S., and W. Rounds, 1983 Behavioural Equivalence Relations induced by Programming Logics, *LNCS*, 154, 97–108.

Cohn, P., 1981 *Universal Algebra*, Reidel.

Courcelle, B., and I. Guessanan, 1978 On Some Classes of Interpretations, *Journal of Computer and System Sciences*, 17, 388–413.

Courcelle, B., and J. Raoult, 1980 Completions of Ordered Magmas, *Fundamentae Informaticae*, 3, 105–160.

deBakker, J., and J. Zucker, 1982 Processes and the Denotational Semantics of Concurrency, *Information and Control*, 54, 70–120.

deBakker, J., J. Kok, J. Meyer, E. Olderog, and J. Zucker, 1986 Contrasting Themes in the Semantics of Imperative Concurrency, Centre for Mathematics and Computer Science Technical Report CS-R8603, Amsterdam.

DeNicola, R., 1985a Fully Abstract Models and Testing Equivalences for Communicating Processes, Ph.D. Thesis, University of Edinburgh CST-36-85.

—, 1985b Two Complete Sets of Axioms for a Theory of Communicating Sequential Processes, *Information and Control*, 64, nos. 1–3, 136–176.

—, 1987 Extensional Equivalences for Transition Systems, *Acta Informatica*, 24, 211–237.

DeNicola, R., and M. Hennessy, 1983 Testing Equivalences for Processes, *TCS* 34, pp. 83–133.

—, 1987 **CCS** without τs, Proceedings of Tapsoft, '87, LNCS, 250.

Ehrig, H., and B. Mahr, 1985 *Fundamentals of Algebraic Specification 1*, Springer-Verlag.

Francez, N., C. Hoare, D. Lehman, and W. DeRoever, 1979 Semantics of Non-Determinism, Concurrency and Communication, *Journal of Computer and System Sciences*, 19, no. 3.

Francez, N., D. Lehman, and A. Pnueli, 1984 A Linear History of Semantics for Languages with Distributed Processing, *TCS*, 32, 25–46.

Goguen, J., J. Thatcher, E. Wagner, and J. Wright, 1977 Initial Algebra Semantics and Continuous Algebras, *JACM*, 24, no. 1, 68–95.

Golsen, G., 1985 A Complete Proof System for an Acceptance-Refusal Model of **CSP**, Technical Report TR85-19, Rice University.

Golsen, G., and W. Rounds, 1983 Connections between Two Theories of Concurrency: Metric Spaces and Synchronisation Trees, *Information and Control*, 57, 102–124.

Graf, S., and J. Sifakis, 1986a A Logic for the Description of Nondeterministic Programs and their Properties, *Acta Informatica*, 23, 507–525.

—, 1986b A Modal Characterisation of Observational Congruence of Finite Terms of **CCS**, *Information and Control*, 68, 254–270.

Grätzer, G., 1968 *Universal Algebra*, Springer-Verlag.

Guessanan, I., 1981 Algebraic Semantics, *LNCS*, 99.

Hennessy, M., 1981 A Term Model for Synchronous Processes, *Information and Control*, 51, no. 1, 58–75.

—, 1983 Synchronous and Asynchronous Experiments on Processes, *Information and Control*, 59, 36–83.

—, 1985 Acceptance Trees, *JACM*, 32, no. 4, 896–928.

Hennessy, M., and R. Milner, 1985 Algebraic Laws for Nondeterminism and Concurrency, *JACM*, 32, no. 1, 137–161.

Hennessy, M., and G. Plotkin, 1980 A Term Model for **CCS**, *LNCS*, 88.

Hoare, C., 1978 Communicating Sequential Processes, Communications of *ACM*, 21, no. 8.

—, 1980 A Model for Communicating Processes, in R. McKeag and A. McNaughton (eds.), *On the Construction of Programs*, Cambridge University Press, pp. 229–243.

—, 1981 A Calculus of Total Correctness, *Science of Computer Programming*, 1, 49–72.

—, 1985 *Communicating Sequential Processes*, Prentice-Hall.

Kennaway, J., 1981 Formal Semantics of Nondeterminism and Concurrency, Ph.D. Thesis, Oxford University.

Kennaway, J., and C. Hoare, 1980 A Theory of Nondeterminism, *LNCS*, 85, 338–350.

Knuth, D., 1975 *Fundamental Algorithms*, vol. 1, Addison-Wesley.

Markowsky, G., 1976 Chain-Complete Posets and Directed Sets with Applications, *Alg. Universalis*, 53–68.

Meseguer, J., and J. Goguen, 1985 Initiality, induction and computability, in M. Nivat and J. Reynolds (eds.) *Algebraic Methods in Semantics*, Cambridge University Press.

Milne, G., and R. Milner, 1979 Concurrent Processes and their Syntax, *JACM*, 26, 302–321.

Milner, R., 1973 Processes: A Mathematical Model of Computing Agents, in Rose and Shepardson (eds.), *Proceeding of Logic Colloquium*, North-Holland, pp. 157–173.

—, 1978 Synthesis of Communicating Behaviour, *LNCS*, 64, 71–83.

—, 1979a Flowgraphs and Flowalgebras, *JACM*, 26, 794–818.

—, 1979b An Algebraic Theory for Synchronisation, Proc. 4th GI Conf. on Theoretical Computer Science, LNCS, 67, pp. 27–35.

—, 1980 A Calculus of Communicating Systems, *LNCS*, 94.

—, 1983 Calculi for Synchrony and Asynchrony, *TCS*, 25, 267–310.

Nivat, M., 1980 Synchronisation of Concurrent Processes, in R. Brook (ed.), *Formal Language Theory*, Academic Press, pp. 429–451.

Nivat, M., and J. Reynolds, (eds.) 1985 *Algebraic Methods in Semantics*, Cambridge University Press.

Olderog, E., 1986 Semantics of Concurrent Processes: the search for structure and abstraction, parts I and II, Bulletin of EATCS.

Olderog, E., and C. Hoare, 1986 Specification-Oriented Semantics for Communicating Processes, *Acta Informatica*, vol. 23, pp. 9–66.

Park, D., 1981 Concurrency and Automata in Infinite Strings, *LNCS*, 104, 167–183.

Phillips, I., 1985 Refusal Testing, TCS, 50, no. 2.

Plotkin, G., 1981 A Structural Approach to Semantics, Lecture Notes, Aarhus University.

Pneuli, A., 1985 Linear and Branching Structures in the Semantics and Logics of Reactive Systems, *LNCS*, 194, 15–31.

—, 1986 Specification and Development of Reactive Systems, Proceedings of *IFIP*, 1986, Dublin.

Reisig, W., 1985 Petri Nets, Springer-Verlag.

Roges, H., 1967 Theory of Recursive Functions and Effective Computability, McGraw-Hill.

Rounds, W., 1983 On the Relationships between Scott Domains, Synchronisation Trees and Metric Spaces, University of Michigan Technical Report CRL-TR-25-83.

—, 1984 Application of Topology to Semantics of Communicating Processes, University of Michigan Technical Report.

Rounds, W., and S. Brookes, 1981 Possible Futures, Acceptances, Refusals and Communicating Processes, *Proceedings of 22nd IEEE Symposium on Foundations of Computer Science.*

Scott, D., and C. Strachey, 1971 Towards a mathematical semantics for computer languages, in J. Fox (ed.), *Proceedings on the Symposium Computers and Automata*, Polytechnic Institute of Brooklyn Press, pp. 19–46.

Shields, M., and P. Lauer, 1978 On the Abstract Specification and Formal Analysis of Synchronisation Properties of Concurrent Systems, *LNCS*, 75, 1–32.

Smyth, M., 1978 Powerdomain, *Journal of Computer and System Sciences*, 16, 23–36.

Stoughton, A., 1987 Substitution Revisited, to appear in TCS.

Index

The MIT Press, with Peter Denning, general consulting editor, and Brian Randell, European consulting editor, publishes computer science books in the following series:

ACM Doctoral Dissertation Award and Distinguished Dissertation Series

Artificial Intelligence, Patrick Winston and Michael Brady, editors

Charles Babbage Institute Reprint Series for the History of Computing, Martin Campbell-Kelly, editor

Computer Systems, Herb Schwetman, editor

Exploring with Logo, E. Paul Goldenberg, editor

Foundations of Computing, Michael Garey and Albert Meyer, editors

History of Computing, I. Bernard Cohen and William Aspray, editors

Information Systems, Michael Lesk, editor

Logic Programming, Ehud Shapiro, editor; Fernando Pereira, Koichi Furukawa, and D. H. D. Warren, associate editors

The MIT Electrical Engineering and Computer Science Series

Scientific Computation, Dennis Gannon, editor